IRISH INNOVATORS

IN

SCIENCE AND TECHNOLOGY

Edited by Charles Mollan, William Davis
and Brendan Finucane

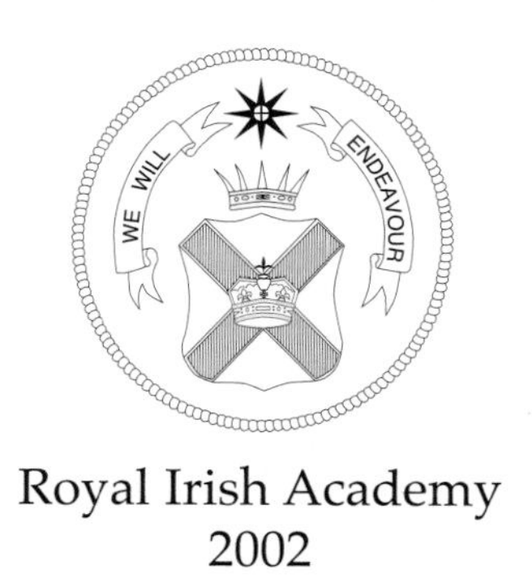

Royal Irish Academy
2002

ISBN 1 874 045 88 7

Published by the Royal Irish Academy and Enterprise Ireland
Designed and typeset by Declan Keena
Printed by W & G Baird Ltd.

PREFACE

The Royal Irish Academy, in association with the predecessor organisations of Enterprise Ireland, published *Some People and Places in Irish Science and Technology* in 1985 and, five years later, *More People and Places in Irish Science and Technology*. These books proved popular at the time and are now out of print. It is opportune to update their contents to include numerous additions that were identified since the earlier works were published.

Histories of Ireland and scientific textbooks have not emphasised sufficiently the contribution that Irish men and women have made to the advancement of science. There is a keen awareness of the Irish contribution to the world of literature, but many people may be surprised to learn that in Ireland in the seventeenth, eighteenth and nineteenth centuries there were many scholars, both amateur and professional, keenly interested in scientific and technical matters. We had renowned astronomers, biologists, chemists, mathematicians and physicists, technologists and instrument makers, many of whom had a lasting influence in Ireland and elsewhere in shaping the evolution of their particular areas of expertise.

This book of pen-portraits of men and women involved in Irish science and technology is a step in redressing the neglected state of the history of Irish science. It should assist in bridging the gap between the earlier Irish scientific tradition and the present scientific activities of our universities and research institutes, which underpin the phenomenal and continuing growth in Ireland's high-technology industry.

Professor T. David Spearman,
President, Royal Irish Academy

Mr Patrick J. Molloy
Chairman, Enterprise Ireland

CONTENTS

CONTENTS

Page

Entries in Alphabetical Order

Entries in Alphabetical Order

Authors

Author	Subject
Anderson, Robert	James Apjohn
Archer, Jean	William Fitton
	Patrick Ganly
	George Du Noyer
Attis, David	John Tyndall
	William Thomson
	Joseph Larmor
	John Townsend
Barry, Noel	Robert Murphy
Bennett, Jim	Thomas Romney Robinson
	Henry Hennessy
	Edward Hull
Bowler, Peter	Robert Scharff
	Ernest MacBride
Breathnach, Caoimhghín	Henry Martin
	Joseph Barcroft
	Edward Conway
	Denis Burkitt
Brown, Martin	Hans Sloane
	William Traill
Burnett, John	Walter Noel Hartley
Butler, John	John Dreyer
	Ernst Öpik
	Eric Lindsay
Buttimer, Anne	Estyn Evans
Casey, Michael	Nicholas Callan
Chesney, Helena	John Templeton
	Ellen Hutchins
	William Thompson
	Mary Ball
	Matilda Knowles
Childs, Peter	James Muspratt
Collins, Timothy	John Philip Holland
	Robert Lloyd Praeger
Cox, Ronald	Robert Mallet
	James Thomson
	Bindon Blood Stoney
	Osborne Reynolds
Davis, William	Peter Woulfe
	Richard Lovell Edgeworth
	William Higgins
	William MacNevin
	Richard Chenevix
Davis, William (continued)	Aeneas Coffey
	Dionysius Lardner
	Robert Kane
	Thomas Andrews
	Maxwell Simpson
	William Kirby Sullivan
	John Mallet
	Cornelius O'Sullivan
	James Emerson Reynolds
	Sydney Young
	Vincent Barry
	John Bell
Davison, David	Parsons, Mary
Elliott, Ian	Thomas Maclear
	William Wilson
	Andrew Crommelin
	Ralph Sampson
	William McCrea
	Patrick Wayman
Finch, Eric	Frederick Trouton
	Ernest Walton
Gabbey, Alan	William Molyneux
	William Rowan Hamilton
Garvin, Wilbert	James Murray
	James Martin
Gee, Brian	Edward Clarke
Glass, Ian	Thomas Grubb
	Howard Grubb
Gow, Roderick	Henry Smith
Guiry, Michael	Ellen Hutchins
Hall, Robert	Johannes Scottus Eriugena
Harmey, Matthew	Phyllis Clinch
Harry, Owen	Mary Ward
Hayes, Michael	William Orr
Herries Davies, Gordon	Richard Griffith
	Edward Cooper
	Joseph Jukes
	Thomas Oldham
Houghton, Raymond	George Berkeley
James, Kenneth	William Cole
Johnston, Roy	John Desmond Bernal
Landy, Sheila	Francis Beaufort

Authors

Author	Entries
Landy, Sheila	Geroge Francis FitzGerald
Lyons, J.B.	Thomas Molyneux
	Sylvester O'Halloran
	Philip Crampton
	William Stokes
	Dorothy Price
McConnell, David	William Hayes
McConnell, James	Arthur Conway
	Erwin Schrödinger
	Walter Heitler
MacHale, Desmond	George Boole
McKenna-Lawlor, Susan	
	Robert Ball
	Agnes Clerke
	Margaret Huggins
Mollan, Charles	Introduction
	Kathleen Lonsdale
Molloy, Patrick	Preface
Monaghan, Nigel	Frederick M'Coy
Montgomery, Alfred	
	John Dunlop
	Harry Ferguson
Moriarty, Christopher	
	Augustin Hiberniæ
	Roderic O'Flaherty
	William Spotswood Green
	Anne Massy
Morrison-Low, Alison	
	William Petty
	Samuel Yeates & Family
	James Joseph Hicks
Nelson, Charles	Richard Turner
	David Moore
	William Harvey
	Augustine Henry
O'Brien, Eoin	Robert Graves
	Dominic Corrigan
O'Connor, Sé	Thomas Preston
O'Connor, Thomas	Alexander Anderson
O'Hara, James	Humphrey Lloyd
	George Johnstone Stoney
O'Rawe, Des	James Murray
	James Martin
Porter, Neil	Cormac O Ceallaigh
Scaife, Brendan	James McConnell
Scaife, Garrett	William Parsons
	Laurence Parsons
	Charles Parsons
Scannell, Mary	Henry Hart
	Edwin Butler
Scott, Tony	J.J. Nolan
	P.J. Nolan
	James Drumm
Somerfield, Adrian	John Joly
	William O'Leary
	William Gosset
Spearman, David	Preface
	James MacCullagh
	George Salmon
	Samuel Haughton
	John Synge
Spencer, John	Robert Geary
Thorburn Burns, Duncan	
	Robert Boyle
	John Rutty
	Joseph Black
	Richard Kirwan
	Charles Cameron
Wallace, Patrick	Introductory Essay
Wayman, Patrick	John Brinkley
	Edmund Whittaker
Weaire, Denis	William Thomson
	Thomas Preston
	Thomas Lyle
Webb, David	Henry Horatio Dixon
White, James	Thomas Johnston
	Matilda Knowles
	Joseph Doyle
Winder, Frank	Arthur Stelfox
Wood, Alastair	George Stokes
Wyse Jackson, Patrick	
	William Hamilton
	Charles Giesecke
	Henry Medlicott
	Richard Oldham
	Francis Mitchell
Wyse Jackson, Peter	
	David Webb

NOTES ON MEDALS

It will be seen from the biographies in this book that many Irish people have won medals for their scientific work. It may be helpful to give some details of those awards which have been won most frequently – from the Royal Society, the Royal Dublin Society and the Royal Irish Academy.

The Copley Medal of the Royal Society

The Copley Medal 'which has long been regarded as the highest scientific distinction that the Royal Society can bestow' originated from a legacy of £100 from Sir Godfrey Copley, Bart. FRS in 1709. In 1736 it was resolved that a medal should be awarded to the author of the most important scientific discovery or contribution to science by experiment or otherwise. In 1831 it was agreed that the medal should be awarded to the living author of such philosophical research, either published or communicated to the Society, as may appear to Council as deserving of that honour. The medal was redesigned in 1942, and this premier award of the Society normally alternates between the physical and biological sciences.

The Royal Medals of the Royal Society

The Royal Medals were founded by King George IV, the proposal to found them being conveyed in a letter from Sir Robert Peel to Sir Humphry Davy dated December 3, 1825. They were at first awarded for the most important discoveries completed and made known to the Society in the year preceding the day of their award, but this was soon changed to within five years. Following some other changes in regulations, it was resolved in 1850 that two medals should be awarded each year for the two most important contributions to the advancement of Natural Knowledge published in Her Majesty's (Queen Victoria's) dominions within a period of not more than ten years and not less than one year before the date of the award. One was to be awarded in 'each of the two great divisions of Natural Knowledge' (physical and biological sciences). In 1965 Queen Elizabeth II added a third medal to be awarded annually for distinguished contributions to the applied sciences.

The Boyle Medal of the Royal Dublin Society

Named after Robert Boyle (1627–1691), and the highest scientific distinction which the Society can bestow, the Boyle Medal was awarded by the Society between 1899 and 1996 to 32 distinguished researchers in either pure or applied science. Up until 1961, it was awarded for work reported in the scientific publications of the Society, but this was then extended to any research of exceptional merit carried out in Ireland. Previously it had been awarded specifically for either pure or applied science, but this distinction was also dropped. Following co-operation with the newspaper, *The Irish Times*, the conditions were changed for the 1999 centenary of the award. The RDS Irish Times Boyle Medal is now awarded every two years, and alternates between a scientist based in Ireland, and an Irish born scientist working abroad.

The Cunningham Medal of the Royal Irish Academy

In 1789 Timothy Cunningham, a barrister of Gray's Inn, left £1000 to the Academy to be spent in awarding premiums for 'the improvement of natural knowledge and other subjects'. At irregular intervals, the Academy set subjects and awarded a medal to the best essayist. In 1817 it was decided that the medal might be awarded annually for the best essay read in each of the three sessions, but this was only occasionally acted upon. In 1838 the decision was made that a medal should be awarded each year in science and the humanities, but this also was only occasionally acted upon. In 1878, it was agreed that the Cunningham bequest could be used to fund both medals and publications – 'Cunningham Fund Memoirs'. Between 1796 and 1885, 36 medals were awarded. The distinction was re-instated in 1989 for Professor Frank Mitchell, and has not been awarded since that date.

INTRODUCTION

Following the publication of *Some People and Places in Irish Science and Technology*[1] in 1985 and its sequel, *More People and Places in Irish Science and Technology*[2], in 1990, the original intention of the National Committee for the History and Philosophy of Science of the Royal Irish Academy was to publish a third volume featuring biographies of Irish born scientists who had carried out all or most of their major work abroad. These had purposely been omitted from the earlier volumes (although a few had crept in). It had been agreed that the title of the third volume should **not** be *Yet More People and Places in Irish Science and Technology!*

For a variety of reasons there was considerable delay in progressing Volume 3. In the meanwhile, both positive and negative things had happened. On the positive side, more work had been carried out in the history of Irish science, which had thrown up new names worthy of inclusion, or additional information about the entries previously published. On the negative side, some eminent Irish scientists had died in the 1990s, and they deserved inclusion, but some of them did not fit into the proposed criterion of the third volume in that they had carried out much of their major work here. Several of these had indeed been authors of articles in the earlier volumes – David Webb (1912–1994), Frank Mitchell (1912–1997), James McConnell (1915–1999) and Patrick Wayman (1927–1998). So, it was decided that a new composite volume should be produced, updating the people entries from the first two volumes and including many new ones. The 'places' category has been omitted in the new volume, although quite a few of the places of the earlier volumes have been acknowledged by highlighting instead the people who made them famous. Our new volume includes biographies of colleagues who have passed away since 1990 – our four authors already mentioned, as well as John Synge (1897–1995), Ernest Walton (1903–1995), William McCrea (1904–1999), Denis Burkitt (1911–1993), and Cormac O Ceallaigh (1912–1996). As was the original intention for Volume 3, it also includes a sizeable collection of people of Irish birth whose families had moved away or who took advantage of opportunities which other countries presented, and who therefore carried out their major work abroad. These include some very eminent names – like Robert Boyle, George Gabriel Stokes, John Tyndall, Lord Kelvin, Charles Parsons, J.D. Bernal and Kathleen Lonsdale. While claiming or reclaiming these as our own, we must acknowledge with gratitude the opportunities which they were given outside our country – opportunities of which they took advantage with such distinction.

Apart from Dr Patrick Wallace's introductory essay, *Origins and Originals—The First 10,000 Years*, the entries are in the form of either single or double pages, roughly depending on the eminence of the person being featured. However, it will be conceded that this is not a terribly rigorous distinction – there are people who have one page but who may be considered to deserve two, and *vice versa*. However, despite limitations such as this, we hope this new volume will give a very clear demonstration that Ireland has a history of scientific and technological innovation of which we can be proud. It is not a tradition which has been much acknowledged in our recent history – but it can help explain how Ireland has entered so successfully into the technological age at the beginning of the twenty-first century. We have a notable tradition, and we have opportunities to build on this to the advantage of our country and the world.

Finally, our best thanks are due to the Royal Irish Academy and to Enterprise Ireland, who have made publication possible, and, of course, to our authors for their contributions and for their patience during the long gestation period of this volume.

Charles Mollan
Blackrock, Co. Dublin
April 2001

References

1. Charles Mollan, William Davis and Brendan Finucane (eds): *Some People and Places in Irish Science and Technology*, Royal Irish Academy and National Board for Science and Technology, Dublin, pp. ix + 107, 1985.

2. Charles Mollan, William Davis and Brendan Finucane (eds): *More People and Places in Irish Science and Technology*, Royal Irish Academy and EOLAS, Dublin, pp. 107, 1990.

In many of the articles reference is made, without further elaboration, to the following well-known sources:

Dictionary of National Biography (Oxford University Press – multiple volumes and dates, now most conveniently available as a CD ROM: a new version is in preparation).

Dictionary of Scientific Biography (16 volumes bound in 8, Chales Scribner's Sons, New York, 1981).

ORIGINS AND ORIGINALS – THE FIRST TEN THOUSAND YEARS

Patrick F. Wallace

Ongoing specialised work on all aspects of our material culture, including on items made of organic substances like wood, leather, textile and bone, will better help future evaluations of Ireland's possible early contributions to invention and technical innovation. The assumption that inventions once made were shared or passed on has always to be questioned in the context of ancient evidence. And good though some of our evidence is by survival standards in other more industrialised European countries, which were often subjected to deep ploughing, mining and other damages and not blessed (from the archaeological standpoint) in having bogs and other water-logged sites which preserved quantities of ancient organic objects in comparatively intact conditions, it has to be remembered that even here the evidence is often uneven, patchy and broken in nature.

Comparisons of material remains should be driven by the need to understand a place and its people at a particular time by looking at the influences which contributed to the sum of its technology (as well as its products and the expected relative quality of life which might accrue therefrom) rather than providing impressions for others to use in social, regional or national contests. We are not in the business of proving that the Irish had the first horizontal mill or making statements like Orson Welles in the film 'The Third Man' when he wittingly but selectively focussed on the cuckoo clock as the invention of one race compared to another which produced the renaissance. Nor is it so much a case of the Irish consciously setting out to save civilisation, as the title of a recent best-seller has it, as that in Ireland at a particular time a set of circumstances obtained which provided a context which preserved, utilised and handed on values upon which others were later to build. We should also remember here that technical and technological innovation are only part of the expression of a people, albeit one closely related to progress and even survival itself. Innovations and inventions can and did also occur in the minds of unrecorded grammarians, poets, lawyers and other thinkers often working in concert, producing sparks of genius which fuse to light the fire of a language, a literature, a legal code...a civilisation.

Rather than using the boundaries of present day nation states in a comparative assessment of culture – especially material culture – we should seek to explain why technical innovations came about at certain places and times, how they were transferred and possibly by whom. (The introduction of the concept of urbanism to Ireland in which the Scandinavians were catalysts for example is a case in point.) While it may be allowable for the historians of any former subject people to list the ancient deeds and innovations of their countrymen, we should by now be mature enough to assess Ireland's material remains in the dispassionate way that anthropologists carry out ethnographic studies on peoples in other continents who are very different from themselves and their culture. It has to be admitted though that things are somewhat different when the deeds and innovations being studied are the accomplishments of people of the same stock and place as the student. Pride does come into it but it must be restrained, informed and generous as well as understanding of places and peoples who may not have enjoyed the benefits of circumstances as favourable to invention as our ancestors may have had.

As with any list of achievers, be they in science, arts, politics, commerce or sport, we should also define exactly what we mean by 'Irish' – otherwise, rather as with the Irish born olympians who competed for the United States, combinations of pride and confusion might have us include figures such as Michael Cudahy who revolutionised the American meat packing industry or go farther and include Marconi who, like the architect Edward Luychens, had an Irish mother! There will always be some confusion and rival claims for the achievements and inventions of a country from which so many had to emigrate, especially in recent centuries. John Philip Holland, the inventor of the modern submarine, may have been working for the American navy but he was also very much a Liscannor-born Fenian!

While a history of inventions using a 'Eureka'-style approach (as in the biro was invented by Hungarian brothers of that name in New York in 1945) may be applicable to the documented achievements of industrial and post industrial societies where inventions are often driven by the published results of laboratory experiments and trials as well as by patents, such a model is not suitable for the achievements of pre-industrial societies. This is because with such societies documentation is poor or, mostly, non-existent and the archaeological or material remains so uneven, patchy or chronologically ill-defined that measurement or comparison of achievements in the way possible for industrialised societies is out of the question. We will never know exactly when the wheel came to Ireland, by whom it was brought and why. Neither will we know who made the Doogarrymore wheel or with what kind of vehicle precisely it was used. Nor will we know the name of the person or persons who made Ireland's arguably greatest artefact (after the Book of Kells) – the Ardagh Chalice, or even for whom it was made or, even most tantalisingly perhaps, where exactly it was made. Less still have we any chance of identifying the genius who first made the later Bronze Age lock rings, or the master craftsman of four thousand years ago who presided over the carving of the Addergoole dug-out canoe, or his near contemporary at Ross Island who was Ireland's first metallurgist. An idea of how little we know of Ireland's extraordinary Bronze Age is implicit in the scarcity of surviving words from that now silent era in the Gaelic of the succeeding Celtic Iron Age. *Portán* for a starfish is one of a handful of possible survivors from before two and a half thousand years ago.

Before unfolding the detail of the technological achievements of Ireland in the Stone, Bronze and Iron ages and later, two further points have to be raised. Firstly, the degree to which there was a tendency to pattern in both art and literature and how this has been seen as a peculiarly Irish genius and, secondly, there is the conscious fixation of the early historic Irish with their own vernacular which is, apparently, unique in early medieval Europe. As to whether there was a distinctively Irish way of doing things and of thinking things, A.T. Lucas's work on the earliest written sources and his knowledge of material culture including decoration led him to conclude that 'in many respects Irish art and literature are two aspects of the same thing; a fundamental compulsion to pattern in the Irish mind'. He cited the elaboration of early syllabic meters and the complex assonantal schemes of later poetry 'both of which made the thing said of far less moment than the manner of saying it' as being of the same nature as the art, and saw the much more recent development of Irish dancing with its mathematical sophistication 'utterly unlike the loose patterns and freer rythms of European folk dancing in general' as a modern example of the same tendency. He further saw in all this evidence for 'perfectionism and adroitness' in Irish art in all its manifestations, the roots of what he identified as yet another national characteristic, deference to the specialist. Early historic society itself was tightly defined with even house size and the capacity of vessels in houses of the various social grades and professions described in such detail as to betray a fixation with classification. As Lucas puts it (and I quote it at length because of its essential relevance not only to these remarks but to the subject of the present study): 'specialisation is self-evident in the great corpus of law tracts which has come down to us, in the variety and complexity of the verse metres, in the many styles of literary composition, in the long genealogical lists, in the topographical treatises, in the different kinds of craftsmen who are mentioned and in the particulars which are given of their separate skills, in the grades of musicians, and in the countless references to individuals who are described as masters of some acrobatic feat, slight-of-hand or other physical or mental proficiency.... there was no room for the amateur in Irish society'.

Irish interest in the vernacular is described by J.F. Lydon as 'unique in the West'. It resulted in monks writing down stories thereby preserving a corpus of ancient literature. They actually 'created a vernacular literature of their own – the lives of saints, voyagers, visions and versions and pagan mythology were all composed in Irish'. This early focus on the vernacular seems to underline the antiquity of a common sense of nationality (if not the idea of the nation state), and a sense of distinctiveness reinforced perhaps by not having been part of the Roman world and relieved at not being exposed to the threat of barbarian invasion (until the advent of the Vikings) from the outside, while at the same time being allowed to develop their own brand of christianity. The archaeological proof is so thin for any major invasion(s) in the first millennium BC, which includes the expected time of arrival of Celtic influences, that it seems to support the idea that there was a common shared culture and language and a relatively homogenous stock not majorly upset by any large-scale arrival of new comers.

From the dawn of history, and probably for a long time before, Ireland, the place, its people, their mind set and shared national psyche were distinctive and original in a collective rather than in a merely individualistic way. Maybe this is the difference between the respective tasks of my colleagues and myself in this publication – the objects and doings which are the products of our pre-industrial society are more shared and in the nature of communal achievements, whereas those ideas which industrial and later societies patented belong more to individuals. Essentially, we are talking about two different worlds, even in regard to art. It would have been unthinkable for the makers of the Book of Durrow or the Tara brooch to sign their work or be identified with its creation in the way the maker of the fifteenth century Lislaughtin (Ballylongford) cross is. It was not until the time of Giotto and after that European artists began to sign their work in the way Renaissance and subsequent artists did.

At first sight, Ireland in its island setting off the west coast of mainland Europe might be perceived of as not having had much opportunity to lead its neighbours in areas of technological innovation. Such an impression is further reinforced when it is remembered that this was one of the last parts of western Europe to be inhabited (as recently as 9,500 years ago), the island lacks significant mineralogical resources like tin (a vital alloy for the production of bronze), it never directly came under Roman domination and remained doggedly rural and non-urban until about the tenth century. And because of its geographical location it was one of the places most distant from south east Europe where the effects of the Neolithic revolution and the great civilisations of the Fertile Crescent and the East Mediterranean, and the technological innovations connected with bronzeworking and ironworking, were first felt. The civilising effects of the Greeks and, after them, the Romans and even of Christianity itself could hardly have originated from locations more remote from Ireland. But this is to look at geography and communications in terms of road, rail and air access and to ignore water – seas and rivers – as means of linkage and exchange. As we shall see, Ireland was not remote at all, and in some ways was witness to distinctive and original technological achievements. A naturally bountiful landscape and seashore, as well as a mild maritime climate without temperature extremes, anyway contributed to a distinctive natural genius which was able to flourish comparatively freely from the economic challenges and hardships of other countries such as, for instance, having to provide built shelter and food for animals in winter.

THE STONE AGE

For all our sense of ourselves as an ancient people, Ireland was one of the last parts of Europe to have human habitation. The island does not appear to have had any human witness to Europe's long Palaeolithic era, so we missed out on the cave paintings and figurine sculptures of 20–40,000 and more years ago. We did, however, have an active Mesolithic or Middle Stone Age hunter-gatherer era from about 9,500 years ago when a local range of flint and other stone tools appears, by and large, to have mirrored the stone tool assemblages then current elsewhere in north western Europe. Among surviving stone tool assemblages from this time in Ireland, however, polished stone axes do feature, and these appear to be technologically distinctive and different from those obtaining elsewhere, even those in Britain. Elsewhere stone axes are more usually diagnostic of the subsequent Neolithic era. Their early presence here, while not suggestive of Irish invention, underlines a distinctiveness of local manufacture and product. The arrival firstly of domesticated animals and then of the knowledge of crops during the long Neolithic or New Stone Age, which began less than six thousand years ago, saw the evolution of settled farming communities whose stone tombs rank among the finest of their type and time in Europe. The earliest types – the single large cist burials, court tombs and the portal tombs, with their often large lintels and capstones balanced on portals, imply great engineering and organisational skills. The mostly later but still Neolithic passage tombs were often built in groups and, in eastern Ireland, had kerbstones, orthostats, lintels and entrance stones often covered in a glorious hieratic art. These represent the finest engineering and artistic buildings of their time in northern and western Europe, to say nothing of the knowledge of astronomy that informed the choices of their situation, orientation and precision planning, as in the case of the famous light box at Newgrange, Co. Meath, through which the first sun on mid winter's day spills down a passage, apparently to light the back wall of the furthest burial chamber. Erected about five thousand

Figure 1–Knowth macehead

Figure 2–Sun disc

years ago, the Boyne Valley passage tombs may be the culmination rather than the start of a movement; if so, they represent the high point of European Stone Age building and, with the corbelled roofs (rows of stones stepped one above another) of their domed chambers, they impress with their engineering as well as their megalithic art and implied astronomical observation. These are European originals if not in concept certainly in execution and scale.

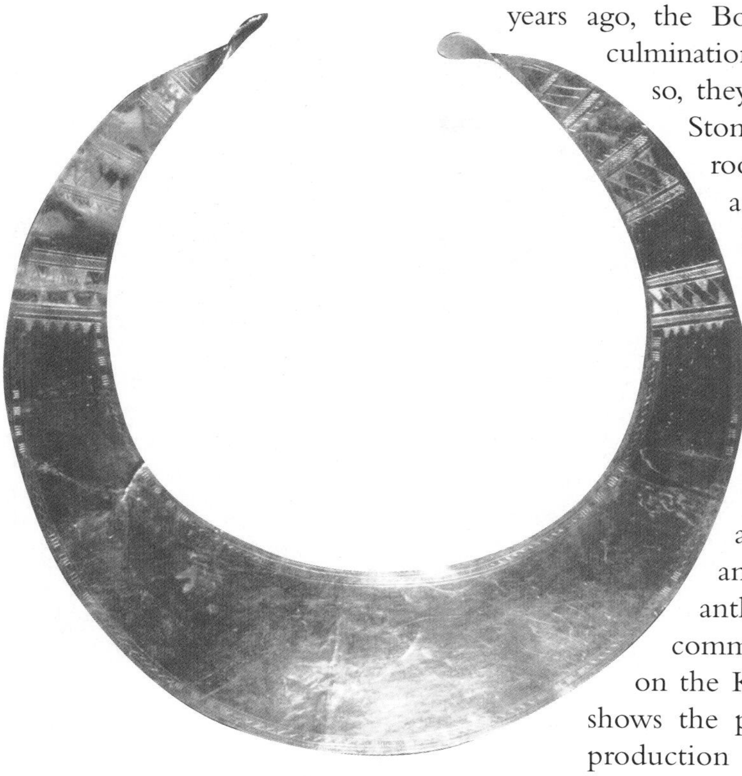

Figure 3–Lunula

Of course there was much more to the Neolithic than building, engineering and astronomy. There is evidence for pottery and the working of wood, bone and antler in what were essentially farming communities. The refinement of the carving on the Knowth, Co. Meath macehead (*Figure 1*) shows the peak which handcraft attained, and the production of polished axes in pocellanite and serpentine may constitute Ireland's first tangible evidence for serious export. The extensive field wall pattern of the Céide Fields in Co. Mayo suggests farm ownership and the hierarchically owned organisation of landscape which was to develop further in the ensuing Bronze Age.

THE BRONZE AGE

As a prelude to the development of the full evolution of bronzeworking, there appears to have been a period of intense copper prospecting in the south west. We can now say that Ireland's technology gradually changed from stone to metal from about 2400 BC onwards, as the earliest mining evidence of that date from Ross Island, Co. Kerry, shows. Whether our earliest metallurgists are one and the same as the so-called Beaker people remains to be conclusively demonstrated. We can however say that the earliest Copper Age mine in North West Europe and the first Beaker associated copper mine in Europe are represented at this site in Co. Kerry and, while we are not looking at the earliest copper mining in northern or western Europe, we are witnessing one of the earliest examples and seeing a range of products which in terms of their accomplishment, technology and variety are the equal of any then being produced anywhere.

Despite Ireland's lack of tin resources and the implied need to have frequent and safe links with the nearest outside source, which is taken to be Cornwall, a vigorous Early Bronze Age based on the continued exploitation of rich native copper resources grew. The production of impressive (and eventually decorated) bronze axeheads and halberds in one-piece, stone moulds is paralleled by the early working of sheet gold. Earrings, so-called sun-discs (*Figure 2*) and lunulae (*Figure 3*) were all produced at this time, that is, about 4,000 years ago. Lunulae are neck ornaments and are so widely distributed here that they have to be seen as a type which was invented on the island. Although relatively simple to make, the balanced layout of their mostly incised geometric ornament, and the repoussé technique used in the production of similar ornament on the contemporary sun discs, show a technical confidence and competence in design which are impossible to match in neighbouring countries at this time. There is also some evidence for the export abroad (or at least the carriage for some reason other than trade) of Irish made bronzes and even gold ornaments in the centuries after 2000 BC.

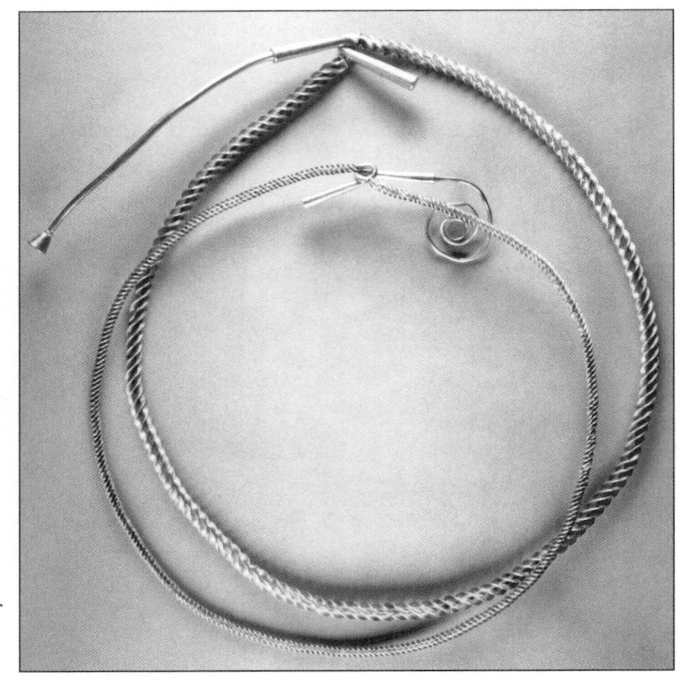
Figure 4–Tara torc

The adaptation of twisted bronze ornaments or torcs of east Mediterranean and north and central European origin to Ireland and their digestion by native goldsmiths who, from about 1200 BC onwards, responded by producing gold torcs in a range of sizes and purposes in a series of ways (including finishing and twisting) to replicate usually cast originals, adds up to one of the most distinctive and glorious achievements of Ireland's long prehistory. The great torcs, such as those from Tara, Co. Meath (*Figure 4*) are amongst the most impressive ornaments of their time and type in prehistoric Europe.

As the Bronze Age wore on, the application of gold foil to bronze and lead backings, most notably the sunflower pins from Ballytegan, Co. Laois (*Figure 5*) and the Bog of Allen, Co. Kildare bulla; the production of great sheet gold collars, including the Gleninsheen, Co. Clare, Gurteenreagh, Co. Clare, Ardcrony, Co. Tipperary and Tory Hill, Co. Limerick specimens; dress fasteners, bracelets, collars, cuff-links, boxes and particularly the intricate hair or lock rings (*Figure 6*) constitute a

Figure 5–Ballytegan gold pinheads

Figure 6–Gurteenreagh Lock rings

spectacular range of achievements in metalwork which neither in their type, variety or sheer numbers are paralleled elsewhere. The lock rings must rank among the most intricately difficult metalworking achievements of any western or north European country of their time (about 800 BC). They are biconicals with slits through which it is presumed locks of hair were threaded to be engaged in a central vertical tube, all in gold. The biconicals consist of pairs of what appear to be gold sheets with grooved surfaces. In fact, the sheets were made of dozens of gold wires soldered together and hammered flat to look like sheets. Why the makers did not simply produce the sheet effect by cutting and grooving real sheets rather than going to so much trouble cannot be explained. What we can say is that in terms of technical dexterity and skill there is nothing to match these in north or west Europe at this time.

The skill of the contemporary bronzeworker did not lag far behind: the use of the *cire perdue* or lost wax and, possibly, the lost lead processes in the production of bronze rings in staples in which one clay or sand mould has to be made around another to enable the staple and ring to be produced in one pour is an example of the skill to which Irish bronze casters could reach. The cast bronze end-blow and side-blow horns are similarly indicative of skill at casting and finishing, as are some of the spearheads which were probably for show and parade rather than actual battle. In the latter case, liquid bronze has to be poured into the mould so that it fills an extremely narrow and elongated cavity in one pour. The sheet bronze workers, whether making shields from single pieces, or cauldrons and buckets from several riveted pieces, also had few equals in the Europe of their

Figure 7–Castlederg cauldron

day. In this regard, the Castlederg, Co. Tyrone cauldron (*Figure 7*) is not only aesthetically beautiful and suggestive of the grandeur of this heroic era in Ireland, it is practically sound also with rows of rivets with heads to retain and distribute the heat during cooking; it is also an intricate piece of technology incorporating the best sheet metalwork of its time as well as the implication of intricate casting in its staples and their suspension rings.

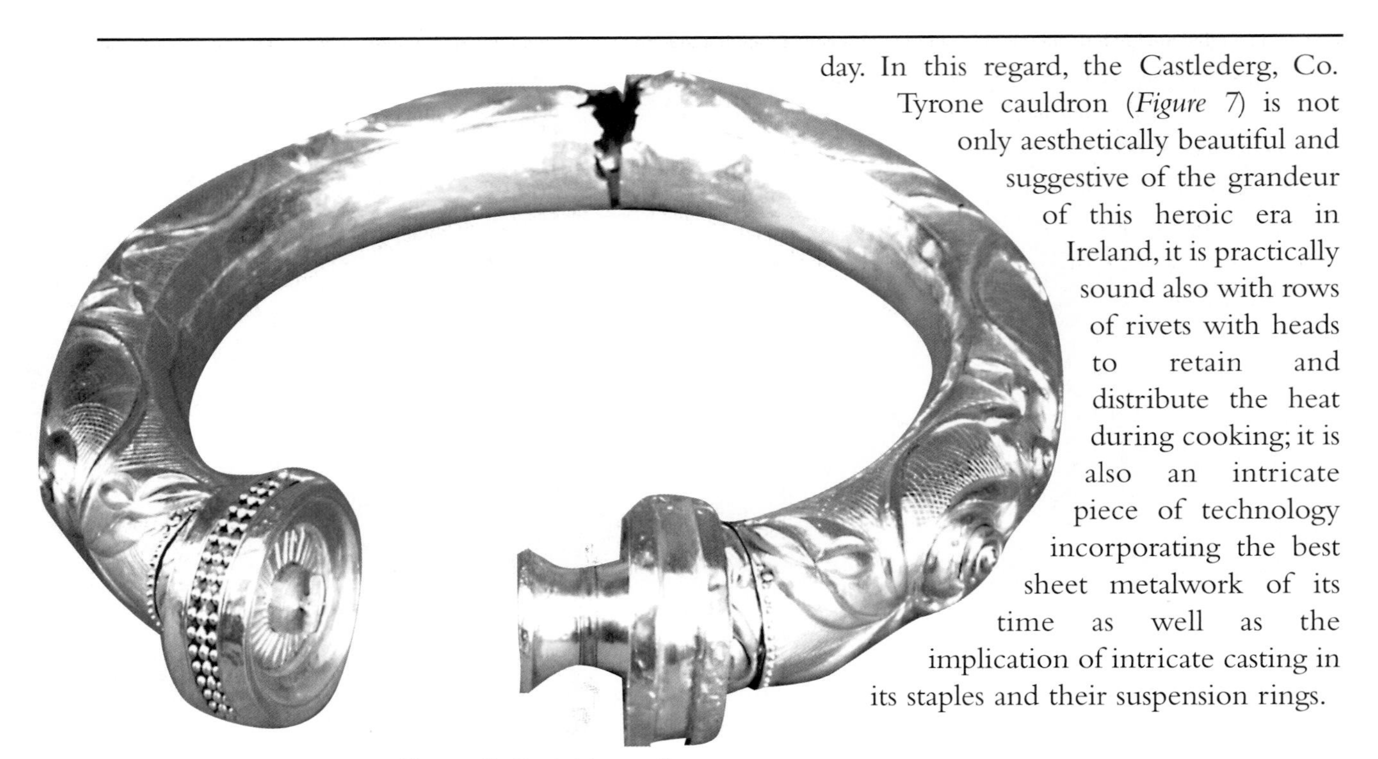

Figure 8–Broighter collar

THE IRON AGE

Ireland in the late Bronze Age had already witnessed an heroic and aristocratic age not dissimilar to that of the continental Celts of the Early Iron Age as described by classical writers. Irish Iron Age archaeological artefacts and monuments, while reflective of strong British and continental Celtic influences, in their totality seem to be different from the material cultural remains from any of the other regional variations of what is termed Celtic Europe. Chronological reconciliation of Celtic material cultural remains to possible waves of Celtic language implantation may be some way off, though examination of relevant material remains – particularly the beautifully ornamented series of bronzes – are instructive. They suggest that the Irish picked and chose object types and bronze casting techniques, and particularly the insular variety of the Celtic La Tène ornament style in which they were to excel.

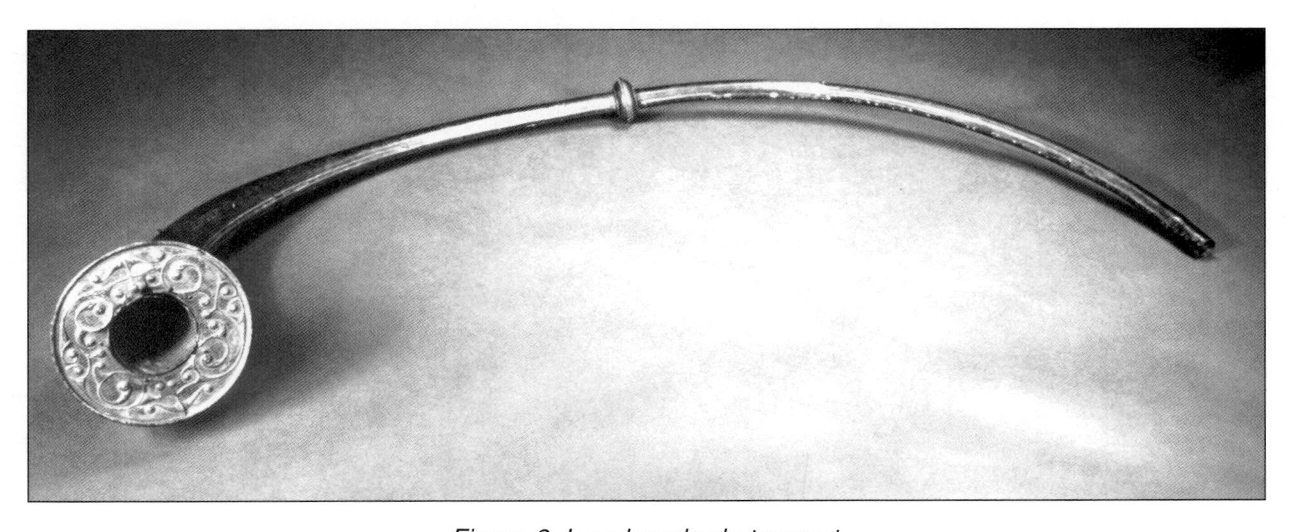

Figure 9–Loughnashade trumpet

The workmanship on the first century BC Broighter, Co. Derry collar (*Figure 8*) shows the level of skill attained by the Irish goldsmith who, apart from displaying techniques like granulation not known to his Late Bronze Age predecessors, also lavished the lively La Tène art style of continental origin to sheets of gold in which the design was arranged so as to allow for the subsequent rolling of the sheets. The dexterity of the contemporary bronze sheet-worker can be seen in the flange of a ceremonial Celtic trumpet from Loughnashade, Co. Armagh (*Figure 9*), while the quality of casting is evident in the Attymon, Co. Galway horsebits (*Figure 10*) and the ability to cut away bronze to produce sharp openwork features in the Cornalaragh, Co. Monaghan (*Figure 11*) and other boxes. In the century or two either side of the birth of Christ, La Tène ornament – its layout and the number of variations which could be wrung out of a handful of classically derived ideas (scrolls, commas, bosses, voids and the ubiquitous triskele, that is three curved limbs radiating from a centre) – engaged the metalworker who, while producing distinctive designs in a barbarian tradition, choose to ignore the mainline artistic influences (including representational and iconic approaches) though not the technology of the Romans which was then engulfing much of the rest of western Europe. Quality, artistic, prestige objects including ornaments like horsebits were cast and chisel finished in bronze in the east and north of the island, especially in the centuries after Christ. At the dawn of history, as the Roman empire was drawing to a close, there is archaeological evidence in the east of the island, and on the Central Plain, for armed and mobile cattle lords with a weakness for British influenced jewellery and decorated horse trappings, who are the probable ancestors of great dynasties who were to figure prominently later on.

Figure 10–Attymon horsebits

Ireland's metalsmiths and other artists appear to have taken what they wanted and what they valued from the Roman world. Enamel work and glass production seem to have impressed them, as did a range of pin and dress fastener types and other ornaments which were soon adapted. The adaptation of Germanic and British derived interlace was eventually to have a huge impact on the Irish art repertoire, where it coexisted alongside the indigenous triskele and

Figure 11–Cornalaragh box

Figure 12–Ardagh chalice, base

other Celtic forms, most notably in illuminated manuscripts like the Book of Durrow and on the reverse face of the eighth century Tara brooch, as well as on the roundel under the cup of the Ardagh chalice (*Figure 12*).

The introduction of iron technology to Ireland is generally attributed to the Celts. Iron appears to have been a late introduction to Ireland and to have been slow to make any significant impact. The absence of iron from the Iron Age record cannot be put down merely to its poor preservation qualities and its notorious propensity to disintegrate in most soil conditions. While the existence of sharp-edged iron tools can be implied from surviving woodwork of quality finish and while iron weapons, especially swords of sub-Roman derivation, are in evidence, the relative paucity of iron has to be acknowledged. It may be that the desire to compensate for this relative absence of iron is responsible for the often all too easy acceptance of iron objects as of the Iron Age when in fact they are of much later dates. The Lough Gara, Co. Sligo adze with its welded-on, bent reinforcement plate, some of the socks and coulters and especially the Drumlane, Co. Cavan cauldron with its distinctively early medieval twisted iron suspension rings and the method of their being welded to the rims, seem to be cases in point. Iron-working prowess in Ireland may have been due more to Scandinavian (Viking) than to any direct Celtic or Romano British influence, the Lagore, Co. Meath collar notwithstanding.

THE EARLY MEDIEVAL PERIOD

From the seventh century, by which time Ireland had enthusiastically embraced christianity, a golden age of art, learning and missionary endeavour sprung up around proliferating numbers of monasteries. The achievements of this era can also be seen in ornamental metalwork, manuscript illumination, stone sculpture and architecture, as well as in a somewhat less tangible but no less significant way in contemporary developments in literature, language and grammar, scientific description especially geography, liturgy and devotional practices. It is also evident in industrial production and in carpentry and engineering as at water and tidal mills.

Figure 13–Derrynaflan paten, filigree detail

Filigree, pressed metal foil work, knitted wire,

chisel carving, granulation and much else add up to the compendium of metalworking excellence achieved by Irish metalsmiths. The early eighth century Ardagh, Co. Limerick chalice probably represents the highest achievement of the Irish metalworker. It displays a mastery in silver smithing, the application of gold wirework or filigree panels to a girdle, and of grills to glass and enamel studs, as well as of difficult techniques such as knitted silver wirework or trichonopoly, stamped silver (*pressblech*) and chisel carving (*kerbschnitt*) on the bronze stem. Back-to-back C-curves of beaded gold wire were set into tiny glass studs in the roundel on the underside of the chalice (*Figure 12*). Patterns of the pressed silver shine through the translucent blue glass studs of the ring around the base when the chalice is held aloft. Most impressive of all is the balance and proportion of the overall product and the tasteful ordering of the materials used in its ornament and assembly.

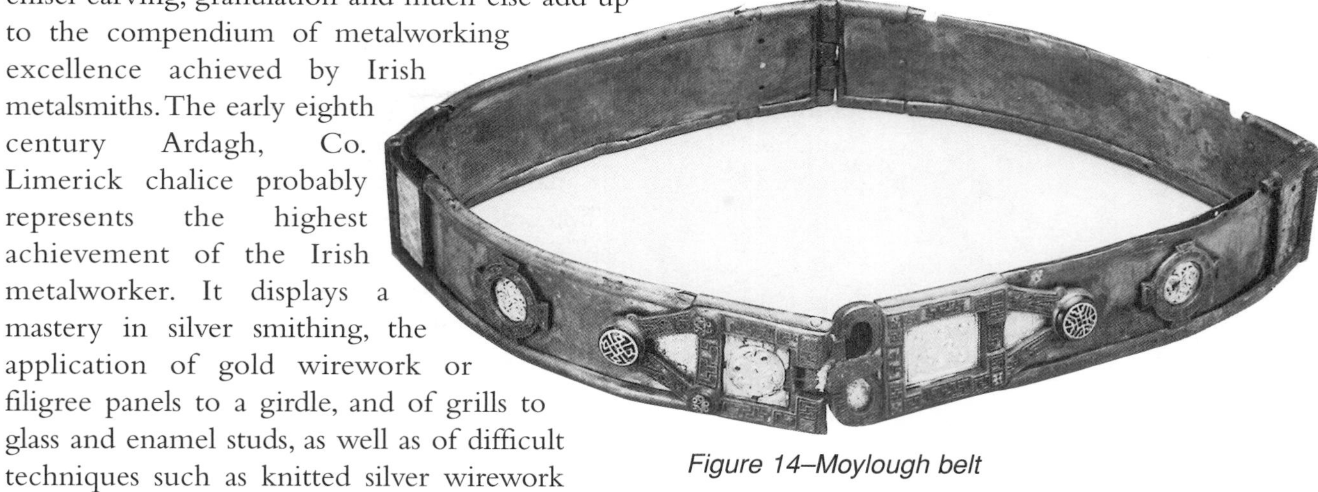

Figure 14–Moylough belt

The contemporary Derrynaflan, Co. Tipperary paten (*Figure 13*) features the finest known combinations of gold filigree and enamel studs and has an extensive run of pressed foil on its supporting stand. The latter technique is also beautifully evident on the earlier Moylough, Co. Sligo belt (*Figure 14*). The gilt silver ring of the early eighth century Tara, Co. Meath brooch has gold filigree animals set on gold foil on the terminals of its front face as well as on the main panel of the pinhead face while, at the back (*Figure 15*), Celtic triskele devices cut away from polished silver expose a copper underlayer making for an exquisite contrast. The amber studs on the front have granulated gold at their centres and the chain hung from the side features a pair of tiny human face masks in cast blue glass. And these are but a sample of the jeweller's techniques which were developed and perfected in the Ireland of their time.

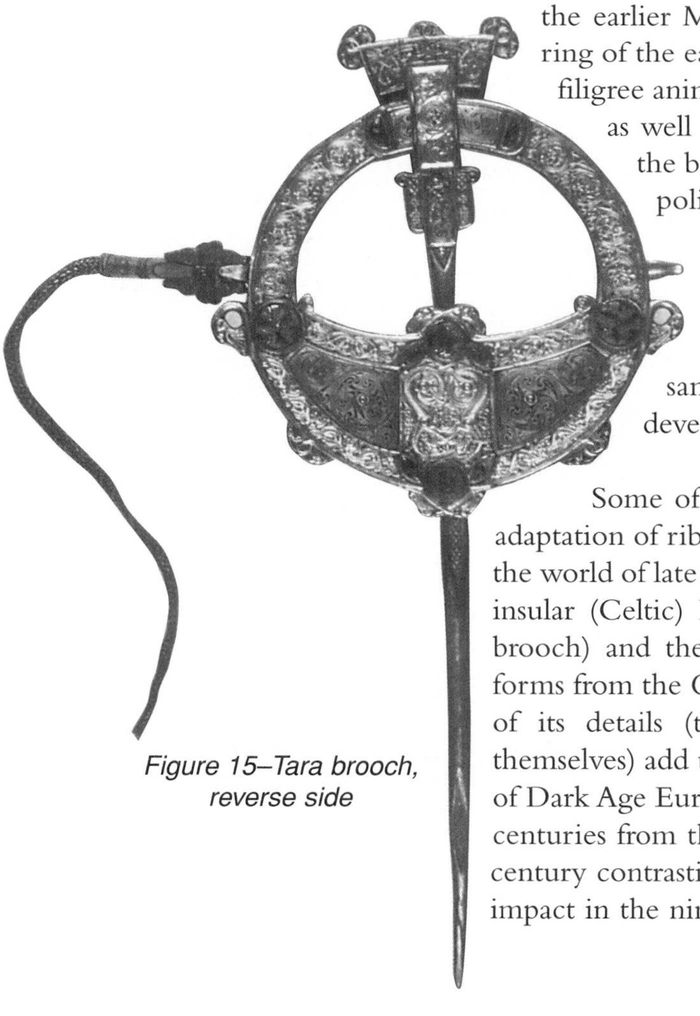

Figure 15–Tara brooch, reverse side

Some of the inspiration, particularly the use of enamel, the adaptation of ribbon interlace and some of the ornament came from the world of late Antiquity but the successful marriage of these to the insular (Celtic) La Tène repertoire (as on the reverse of the Tara brooch) and the slightly later adaptation of zoomorphic interlace forms from the Germanic world in its totality rather than in any one of its details (though they are all impressive and originals in themselves) add up to one of the most distinctive achievements in all of Dark Age Europe. The metalwork went on being produced in the centuries from the seventh, the originality and vigour of the eighth century contrasting with something of an easing off after the Viking impact in the ninth.

The Vikings brought silver to Ireland in great amounts and the tenth century saw a new range of silver ornament types and techniques produced by Irish metalworkers. Eventually, Hiberno-Norse Dublin itself became a major metalwork producing centre. One of the most distinctively Irish pieces of the non-ferrous metalworker's equipment or tool chest was the mainly bone/antler motif – or trial piece – which really comes into its own in the eleventh century workshops of Dublin, where many have turned up. In the later eleventh and early twelfth century all this was to culminate in a veritable metalworking rejuvenation which was part of an overall renaissance associated with church reform. The Cross of Cong, Co. Mayo (*Figure 16*), probably made in Tuam or Roscommon, is the best known product of this time.

Whatever their origins, the carved stone crosses of Ireland, the so called high crosses, have no equal outside Ireland and the areas of northern Britain influenced by the Celtic church except perhaps for the Pictish crosses of eastern Scotland and, slightly later, the Anglo Saxon wheel crosses of Northumbria and northern England. Claims have been made *ex silentio* for continental ivories and wall paintings having been inspirational in the genesis of high crosses but these are difficult to sustain. Inspection of the earliest main group (eighth century) in the Ahenny, Co. Tipperary area shows that they are really scaled up interpretations in stone of contemporary crosses made of decorated metal panels fixed to wooden crosses by large bosses such as are found on the late seventh century processional cross from Tully Lough, Co. Roscommon (*Figure 17*). Both the metal covered wooden prototype and the stone versions are Irish originals. The later and more widely distributed mainly tenth century stone crosses, with their characteristic panels of carved figures and scenes in relief, such as those at Clonmacnoise, Co. Offaly and Monasterboice, Co. Louth represent the highest achievements of the early medieval stone sculptor. Recent work on the preparation of stone for carving, on units of measurement, and on the multi-purpose wooden templates used in their production, only heighten appreciation of their execution.

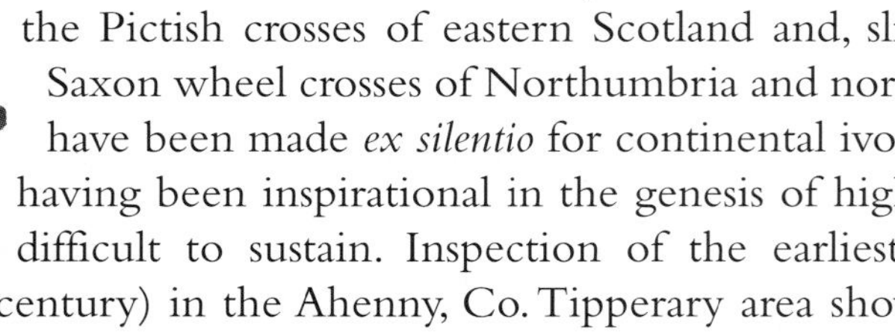

Figure 16–Cross of Cong

The earliest church architecture was probably carried out in wood. This was probably done in both the prevailing native post-and-wattle method as well as in more carpentered frame built form with possible wooden roof tegulae (tiles). Although the archaeological evidence to date only supports the existence of the simpler tradition, the existence of the sturdier carpentered tradition cannot be

Figure 17–Tully Lough cross detail

ruled out given our knowledge of the heights to which carpenters went in mill construction and the implications of both historical, graphic and three dimensional depictions of Irish churches. Most secular construction before and immediately after the introduction of Christianity was of timber and earth with sparing use of stone for revetments, pathways and souterrains, except in areas of the country where stone was the main building medium. Here, as at the great stone forts, drystone walling, bonding, cobelling and prudent use of lintels were all in evidence. Buildings in Ireland whether of post and wattle, earth or stone were invariably of round plan until about the tenth century when there is a switch to rectangular houses, although of course church buildings had to be rectangular from the start.

The corbelled cells of round plan, the so-called beehives which are associated so much with our Atlantic island monasteries and which are thought to date early (they also go on being created long after the demise of the hermitages) are undoubtedly a native invention. So are the drystone rectangular oratories such as Gallarus which were derived from the round originals and which pushed the potential of corbelling to the limit. These distinctively Irish architectural forms, particularly the rectangular oratories, deserve to be on any list of Irish innovations.

While the forms I have discussed were of drystone construction, ninth, tenth and especially eleventh and twelfth century round tower, church and oratory walling was done with mortar, mainly lime mortar which was probably an introduction from outside. While small oratories such as those on Aran, Co. Galway or at Killulta, Co. Limerick show how walls of mortared construction in small buildings have endured, twelfth century cathedral buildings like that at Glendalough, Co. Wicklow and Clonmacnoise, Co. Offaly, with their projecting antae (vertical projections at the edges of the front walls) and the typically Irish trabeate doorways (with flat tops rather than arches) show that large buildings with load bearing walls were also erected in Ireland. The trabeate doorway often come with sloped jambs, and the large lintel is a very Irish architectural detail which goes on in use well into the twelfth century when it was adapted for use in Romanesque-style buildings.

Whatever its origins, the round tower became a distinctively Irish building form and, later, a symbol of Ireland itself! It also shows that Irish masons could erect prestige buildings of great height which afford reasonable security. These and the great monastic buildings show the mason's, stone cutter's and scaffolder's skill. Possibly the most challenging of all Irish early medieval building constructions were cells with the internal corbelled and the external stone roofs as at Kells, Co. Meath and St Kevin's 'Kitchen' at Glendalough, Co. Wicklow which has a similar cavity roof with the additional complication of suspending a round tower. When the Romanesque architectural style eventually came to Ireland, late by continental and English standards (Cormac's chapel is almost a century later than Durham cathedral), only its decorative doorways and details – like gable finials, windows and chancel arches – were taken on by native builders.

Dendrochronology (tree ring dating) has provided irrefutable evidence for the early presence in Ireland of quality, well finished carpentry, especially from the engineering contexts of water mills. Vertical and horizontal water mills go back to at least the early seventh century in Ireland, the horizontal type being eventually replaced by the vertical. Vertical water wheels were employed at Littleisland, Co. Cork (AD 630), Morett, Co. Laois (AD 770) and Ardcloyne, Co. Cork (AD 789), but Littleisland also had a twin-flume horizontal wheeled mill. It is important to note, especially in the overall context of the present book, that the earliest varieties of water mill so far documented in either Europe or Asia have been found in Ireland. Not altogether surprisingly, in view of earlier comments, the earliest vernacular terms for the principal components of the horizontal-wheeled mill in any European language are in Irish.

We now know that by the seventh century mills were commonplace and the specialised craft of millwrighting was so important that even 'regional millwrighting styles had already evolved'. Ireland has also produced the earliest evidence for the use of tide mills and, according to Colin Rynne, the seventh century 'double penstock mill' at Littleisland, along with the vertical undershot specimen which was built alongside it, 'are the earliest known examples of tide mills in either Europe or Asia' (*Figure 18*). The only difficulty with this implication for the existence of quality carpentry at such an early date is that, to date, as has been remarked, there is a scarcity of evidence for iron and, when it is found, only poor quality iron woodworking tools survive in the archaeological record.

A scarcity of iron which, when it is found, seems poor and a limited range of iron tools seem to be a recurring theme in our archaeological record, which is in dire contrast to the standards achieved in other materials, notably non-ferrous metals and wood. Both the range and edge quality of our iron tools are deficient until the advent of Scandinavian influence from the later ninth century onwards. This seeming contradiction in our technological history is mirrored in the archaeological record of Hiberno–Norse Dublin where poor quality carpentry was lavished on domestic houses by a population whose numbers included shipwrights capable of making one the largest and finest warships found to date – Skuldelev 2 at Roskilde! Even after the Norman invasion we appear to have had different standards of carpentry in dockside construction – a low grade *ad hoc* approach at Wood Quay in the early thirteenth century, which I put down to the use of builders and handymen of a local tradition, and a more sturdy, better carpented tradition as the dockside in Drogheda (and, presumably, at the King's works in Dublin Castle) which by analogy with the London waterfront evidence I would suggest was of English derivation.

The assured confidence of the learned classes of Ireland as their culture and language digested the influences of a Roman world, which never dominated them but which they chose to use in a variety of ways in the vernacular, is noteworthy in any review of originality and invention. The Irish produced the earliest and most copious vernacular literature in Europe, several centuries *before Boewulf.* The learned class on coming into contact with Latin tried to adapt the Latin alphabet to their requirements and invented a script of their own known as ogham. They were the first to possess an early medieval grammatical tradition and preserved the work of the Roman grammarians. In their evolution of common law about AD 700 they created categories of subjects such as kingship and rights of kings. In the great world of learning that obtained in the Celtic monasteries, original treaties on geography and astronomy were produced by figures of European stature like Dicuil the Geographer. In Dun Scotus and in Johnannes Scottus Eriugena they produced two of the greatest scholars of the age.

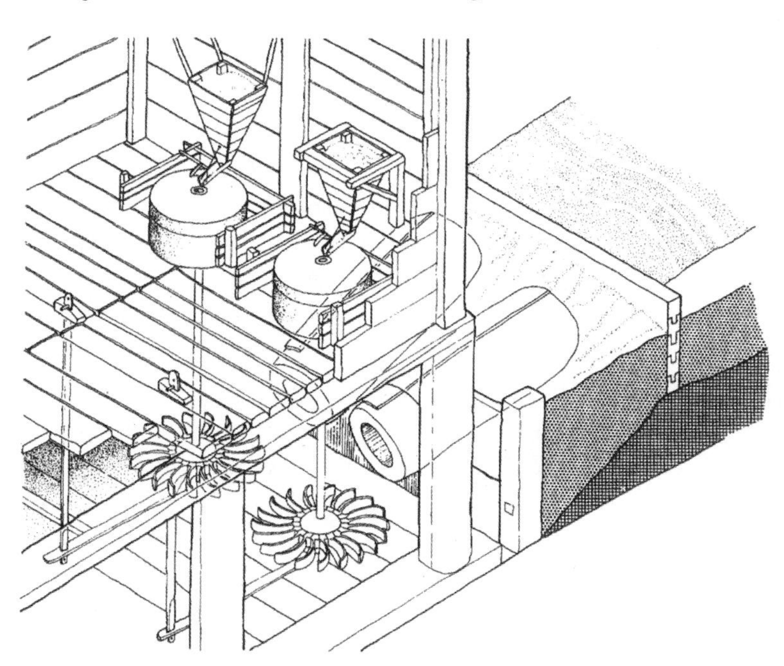

Figure 18–Reconstruction of watermill

The originality of Eriugena is unquestioned: 'It was only a metaphysician of great insight who could have found all that he needed in the incomplete and disorderly literature to which he had access, and it was only a scholar of great speculative power who could have wrought it all into so massive a system.' According to Fr John Ryan 'in the four centuries preceding his birth, and again in the two centuries after his death, his equal in intellectual strength could scarcely have been found in Europe'. Ireland's golden age witnessed the production of numerous descriptions of Christian subjects like the seven deadly sins, monastic rules, penances (they may even have invented auricular confession), alphabets of spiritual life, commentaries on the Bible as well as lives of the saints and the *Voyage of Brendan* which was consulted all over the medieval world. There were also countless poems including very early nature poems of exquisite beauty and originality. The monks in the eighth century created a new literary form in the lyrics in which they sang of 'nature in all its manifestations' as a way of honouring God. According to Professor Lydon 'nowhere else at this time did poets celebrate nature in this way and it is a tribute to the originality and independence of the Irish monastic schools that they did so'.

Irish monks ingratiated themselves with Charles the Bold and with Charlemagne and his court and exerted an influence disproportionate to their numbers on subsequent European history. It has been said that indirectly Irish monks at the court of Charlemagne influenced present day typefaces because of the development of their script into the Carolingian script which was revived in the fifteenth century by Italian humanists. As Ireland's present-day leading calligrapher Tim O'Neill puts it: 'About 1300 years ago, the Irish developed a beautiful formal script for writing books which forms the basis of the lower case or small letters we use today. These, along with Roman capitals and Arabic numerals, are all that is needed in the western world for the visual presentation of language'.

While most of the techniques involved in manuscript illustrations have their origins in the classical world and late Roman manuscripts, techniques like the multi-layering of pigments as in the Book of Durrow and most spectacularly in the Book of Kells do seem unique. Irish scribes may also have contributed to early medieval manuscript layout and design as in the use of *diminuendo* or diminishing letters at the start of paragraphs as well as in the use of innovative punctuation and abbreviation devices. In Francis John Byrne's words there were 'peculiarly Irish suspensions and abbreviations....and the Irish developed an even greater array of abbreviations, far more than were current in Europe before the twelfth century'. The Irish also developed a distinctive insular Greek script.

The last word on the golden age of Irish art just touched upon and its place in European culture should be left to the palaeographer historian Ludwig Bieler: 'Irish art....can claim a place in European civilisation entirely on its merit....whether it is manifested in metal and stone or on the illuminated pages of manuscripts over the wide field of Irish expansion from Scotland to Franconia and Italy, it is not only unique in the Middle Ages, it is also the first, and in the West the only, example of an abstract art in an articulate civilisation which was still spiritually integrated....its spirit can still be felt in the romanesque art of the eleventh and twelfth centuries....our eyes are opened again for the appreciation of Irish art, which is one of the glories of our Western heritage'.

The achievement of the Irish in this Golden Age included countless firsts in many fields. The age reached its peak in the seventh to ninth century period, reaching its apogee in the eighth century. For a variety of reasons the overall fervour of activity and accomplishment in many areas fell somewhat in the tenth century before undergoing a recovery in the earlier eleventh century, leading to a full revival in the later eleventh and twelfth centuries.

EPILOGUE

What might be termed an Irish solution to continental and English art, architectural, fashion, ceramic, military and technological influences, that is the often successful adaptation of the influences to local situations and conditions, was to be the main characteristic of Irish civilisation from the absorption of Norman-French and English influences in the twelfth and thirteenth centuries onwards to the end of the medieval centuries. In the south and west of the country these witnessed a vigorous Gaelic revival, the north having being largely untouched by the outside and the east as always more influenced by the Irish Sea and the great island neighbour. Up to the final fall of Gaelic Ireland in the early seventeenth century, even at their most productive and distinctively different, it seems to have been a case of adaptation and regurgitation rather than total innovation when it came to matters technological and artistic. Matters may have been somewhat similar in the linguistic and literary areas but even here, too, continental and English influences were being felt at the dawn of the ever shrivelling early modern world.

Acknowledgements
My thanks to Dr Edel Bhreathnach, Professor Padraig Breathnach, Sinéad McCartan, Dr William O'Brien, Dr Bernard Meehan, Felicity O'Mahony and Timothy O'Neill for their help on various points. They cannot be held responsible for any of the infelicities. My gratitude to Marie O'Rourke for her typing.

References:

A.T. Lucas: *Treasures of Ireland*, Dublin, 1973, esp. pp 66–7.
J.F. Lydon: *The Making of Ireland*, London, 1998, esp. pp 12–13 and 23.
F.J. Byrne: Introduction, in Timothy O'Neill, *The Irish Hand*, Dublin, 1984.
Timothy O'Neill: *The Irish Hand*, Dublin, 1984.
John Ryan: *Ireland from AD 800 to AD 1600*, Dublin, no date, esp. pp 42–6.
Ludwig Bieler: *Ireland Harbinger of the Middle Ages*, Oxford, 1963. esp. p. 144.

Born/Died/Addresses:

Augustin's writing shows that he was living and working in Ireland in 655. He was probably a monk, had either visited the sea coast or lived close to it, and was certainly the only scientist of his time in Ireland whose work has been preserved.

Augustin is known only as the writer of the treatise *De Mirabilibus Sacrae Scripturae.* In it he examines each of the miracles recorded in the Bible and shows how they can all be explained as unusual extensions of the laws of nature rather than as instances of the Creator breaching his own rules. For example, in the context of Noah's Flood, he remarked that we are all familiar with the rise and fall of the tide. The Flood was a particularly high and protracted tide.

No Christian of his time would have questioned the truth of the scriptures and therefore Augustin begins by accepting the stories as they are given. But this leads him to the problem: after all the animals in Ireland had been destroyed by the flood, how did they make their way back when the waters subsided?

He gives in his treatise a list of these animals, the earliest inventory of the Irish fauna to be written down. It includes wolves, deer, forest pigs, foxes, badgers, hares and squirrels and also the two aquatic species, otter and seal. The evidence is very clear that he made his observations at first hand. In this he stands in contrast to his contemporary scholars who depended on written sources.

Augustin concluded that Ireland and Britain had been joined to the Continent and that they had become islands as a result of erosion by the sea.

Again he quotes a familiar example: the rock stacks on the sea coast which have obviously been cut off by wave action. If it can happen on a small scale, it is likely to happen on a large scale and the animals must have made their way back to Ireland before the break with Britain took place.

Augustin's theories were not acceptable, since Christians believed that the dry land had been created as an entity and could not be changed. In making direct observations in nature and using them to explain problems, his approach was truly scientific.

Further reading:

R.F. Scharff: The Earliest Irish Naturalist, *The Irish Naturalist*, **30**, 128–132, 1921.
C. Moriarty: The Early Naturalists, pp 71–90 in J.W. Foster (ed.), *Nature in Ireland: A Scientific and Cultural History*, Lilliput, Dublin, 1997.

Christopher Moriarty, Marine Institute Laboratories, Abbotstown, Dublin 15.

Born: Ireland, first quarter (first decade?) of Ninth Century.
Died: France or England(?), last quarter of Ninth Century.

Addresses:
Childhood and youth: Ireland
Maturity: France, from before about 850; Carolingian court (under Charles the Bald who reigned from 840–877); Palace School of Laon. Latter years: France or England, after 877

Eriugena (the 'Irish-born') certainly came from Ireland, but where or of what stock is unknown. He almost certainly attended a church school in Ireland, and would have been educated in Christian classical learning, probably including some Greek. Irish scholars were very prominent on the Continent and, for whatever reasons, Eriugena left Ireland and went to the Frankish realms, gaining the patronage of Charles the Bald. He had perhaps the best theoretical mind of his time in regions west of India and China. The chief contender, far broader in his interests but perhaps less acute, was al-Kindi, who flourished at the Arab court in Baghdad.

With Martin, another Irishman, Eriugena headed the Carolingian Palace School at Laon, and re-instituted the largely forgotten study of the liberal arts (grammar, logic, rhetoric, geometry, arithmetic, astronomy, music). Eriugena regarded those disciplines as necessary preparation for philosophy and theology. The main text was Martianus Capella, *On the Marriage of Philology and Mercury*, with Boethius' *Consolation of Philosophy* for the more promising students. The production of teaching commentaries was established, and a serious attempt was made to require the learning of Greek. This liberal arts curriculum (without Greek) became the model for the schools of Europe until the twelfth century. His polemical *On Divine Predestination* (about 850–851), a carefully reasoned interpretation of Augustine, brought renown to Eriugena. It revealed a good philosophical mind and also a complete mastery of Greek. Charles the Bald requested him to translate into Latin the writings by pseudo-Dionysius the Areopagite (Fifth Century), and one by Maximus 'Confessor' (Sixth Century), major works of neoplatonising Greek Christian theology which introduced this thought to the Latin West. These and further translations led Eriugena to his greatest achievement, the *Periphyseon* [On Natures], or *De Divisione Naturae* (written 860–862). This was not a work of natural philosophy or natural science, but a theological treatise synthesising Eastern and Western (Augustinian) Neoplatonism in an individual way. In the *Periphyseon*, the natural world is not only something created by God's will but essentially a manifestation of God's nature in a lower realm (a theophany), and it thus resembles a neoplatonic emanation. Everything is from God and returns to God, and the natures are four (or really three, or even two): that which creates but is uncreated (God); that which is created and which creates (like the neoplatonic universal intellect, an adaptation of the Platonic ideas); that which is created and does not create (the world perceived by the senses); and that which is uncreated and does not create (again, God). Eriugena combined rational argumentation (philosophical/theological/scientific) with interpretation through symbol and metaphor. His thought often influenced later Scholastics.

Further reading:

Dictionary of Scientific Biography.
Dermot Moran: *The Philosophy of John Scottus Eriugena,* Cambridge, 1989.

Robert Hall, The Queen's University of Belfast

Born: Romsey, Hampshire, 27 May 1623.
Died: 16 December 1687.

Portrait, from Hiberniae Delineatio, *1683 (courtesy of the National Library of Ireland)*

Family:
Son of a cloth-merchant, Anthony Petty, and his wife Frances Denby.
In 1667 he married an Irish heiress, Elizabeth, widow of Sir Maurice Fenton, daughter of the regicide Sir Hardress Waller: they had five children of whom three survived.

Distinctions:
Petty was one of the founders of the Royal Society, and knighted at its incorporation on 22 April 1662; he twice refused a peerage.

A child prodigy, at the age of 15 William Petty was shipwrecked in France, but managed to turn disaster into opportunity by teaching English and navigation; for the next few years he travelled the Continent, making influential friends and generally developing his considerable intellectual abilities. He returned to London in 1647/8, where he became involved with the circle which became the Royal Society. He moved with them to Oxford, where, aged 28, he became Vice-principal of Brasenose College, and Professor of Anatomy.

In 1651 he was granted leave to go to Ireland as physician to the Commonwealth army there. This army was still awaiting reward for having quelled earlier rebellion, and a small, inadequate survey had been begun of forfeited Irish estates. Petty boldly proposed to survey the country in thirteen months, for a modest fee, and was awarded the contract in 1654. Because the results were set down as maps, with a scale of about eight inches to the mile – which lasted for centuries as the legal basis for Irish land deeds – the enterprise was known as 'the Down Survey'. He published this in 1685 as *Hiberniae Delineatio.*

The Restoration of the Stuart monarchy meant that Petty, like all other Cromwellian supporters, was ruined. He retired to London, where his intellectual abilities impressed the King. He was knighted at the incorporation of the Royal Society in 1662. He published in the *Philosophical Transactions* a paper concerning a double-keeled boat, and there were others on 'political arithmetick'.

He hated extremism in religion and managed to avoid involvement in the Popish Plot. He retained the personal goodwill of the new king, James II, while deeply disappointed in his disastrous policy in Ireland. Petty died in 1687.

Further reading:
Sir Irvine Masson and A.J. Youngson: Sir William Petty, F.R.S. (1623–1687) in *The Royal Society: Its Origins and Founders,* edited by Sir Harold Hartley, pp. 79–90, London 1960.
J.H. Andrews: Sir William Petty: a Tercentenary Reassessment, *Map Collector,* 34–39, 1987.

A.D. Morrison-Low, National Museums of Scotland, Edinburgh.

4 ROBERT BOYLE Chemist, Natural and Moral Philosopher

Born: Lismore Castle, Co Waterford, 25 January 1627.
Died: Pall Mall, London, 31 December 1691.

Family:
Fourteenth child, seventh son, of Richard Boyle, first Earl of Cork. He never married.

Distinctions:
Founder Member, Royal Society London, 1660; Governor of the Society for the Propagation of the Gospel in New England, 1661/2–1689.

Addresses:

1627–1635	Lismore Castle, Co Waterford
1635–1638	Eton College
1636–	London and Stalbridge
1638–1644	Grand Tour of Europe
1645–1655	Manor of Stalbridge, Dorset
1656–1668	Oxford, house on High Street, next to University College, now demolished
1668–1691	At his sister's house, Pall Mall, London, next door but one to Nell Gwyn

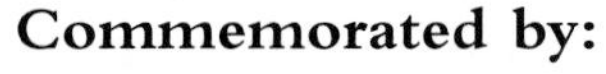

Commemorated by:
Boyle Lecture (Sermons), 1691/2– ; Boyle Medal of Royal Dublin Society, 1899– ; Robert Boyle Gold Medal of Analytical Division of the Royal Society of Chemistry, 1982– ; Boyle-Higgins Gold Medal of the Institute of Chemistry of Ireland, 1990– .

Robert's early education was by tutors at home then, at the age of eight, at Eton College, during which time holidays were spent at Stalbridge Manor in Dorset. In 1638 he was sent under the charge of Issac Marcombes on a Grand Tour of Europe. They got, *via* France and Switzerland, as far south as Italy. Two years were spent in Geneva, where much of his time was spent on mathematics and language studies.

Upon return to England, mid-1644, he lived with his widowed sister in London before moving to Stalbridge Manor, inherited from his father. Here he sat out the Civil War and early Commonwealth times and started on his largely self-taught career of chemical experimentation. He was back and forth to London and intimate with the 'Invisible College' – a circle of friends – the embryo of the Royal Society. He was back in Ireland in 1652–1654 dealing with family estate business. In a letter from Ireland to Fredrick Clodius he used the term 'chemical analysis' in the sense it has since been used by chemists. Boyle moved to Oxford in 1656 where he worked, privately with paid assistants such as Robert Hooke, to provide experimental proofs of the corpuscularian, mechanical theory of nature. Boyle's first scientific publication after four years work in Oxford was *New Experiments Physical and Mechanical Touching the Spring of Air.....*(1660). A few criticisms turned up amongst the generous praise by most scientists. Franciscus Linus, in particular, offered an entirely different explanation of why mercury stays up in a barometer tube. Boyle decided that the best way to refute Linus was by experiment rather than argument by letters in journals. From simple experiments Boyle discovered the reciprocal relationship between pressure and volume which now bears his name.

Less well known today are Boyle's moralistic, theological and utopian writings which account for about a third of his published work. One of the most interesting, his first devotional work, was *The Style of the Scriptures.....* (1661). It was a remarkable forerunner of modern higher criticism often said to have begun 100 years later with Jean Astruc (1753). Boyle compared the Gospels in a scientific manner, commented on various incongruities, but emphasised their essential harmony. He supported financially the translation of the Bible into Irish. His scientific writings have three distinct components. The first or traditional is a result of his reading a great variety of authors to which he provides references in the modern manner; the second or experimental is based on the outcome of his own research or that of his paid assistants. The third component is a mass of information and misinformation acquired as a result of conversation or correspondence. His style is diffuse, rambling, apologetic, deprecatory and not to the modern taste. However, in the midst of prolixity and verbage are some clear statements and illuminating opinions – e.g. on heat 'the nature of it seems to consist of namely, if not only, in that mechanical affectation we call local motion'. He studied a range of topics in physics, medicine and biology, but his main contributions were to Chemistry. These can be divided into six main sections:

1) The validity of fire assay and the nature of elements described in *The Sceptical Chemist.....*(1661). In this and other works he was attempting to provide a philosophical basis for chemistry and to set educated chemists above unlearned ones in the public estimation.
2) The nature of combustion.
3) Solution chemistry, indicators and selective reactions applied to for example *Mineral Waters* (1684/1685). Boyle achieved the earliest quantitative colorimetric reaction on record, that of iron in Tunbridge Wells water.
4) Specific gravity measurement and its applications in analysis.
5) Clinical chemistry including the analysis of urine and blood.
6) Luminescence studies and the manufacture of phosphorus from urine.

Many of the studies are inter-related.

Robert Boyle died on 31 December 1691 eight days after his sister. Both were buried in St Martin-in-the-Fields, London. When the old church was demolished in 1721 no record was kept of the disposal of remains. The present church, built 1722–1724, has no memorial of Robert or his sister nor does it contain their remains.

Further Reading

R.E.W. Maddison: *The Life of the Honourable Robert Boyle*, Taylor and Francis, London, 1969.

J.F. Fulton: *A Bibliography of the Honourable Robert Boyle*, Second Edition, Clarendon Press, Oxford, 1961.

D. Thorburn Burns: Robert Boyle (1627–1691): A Foundation Stone of Analytical Chemistry in the British Isles: Part I, Life and Thought, *Analytical Proceedings*, **19**, 224, 1982; Part II. Literary Style Contribution to Analytical Science, *Analytical Proceedings*, **19**, 228, 1982; Part III, American and Dutch Connections, *Analytical Proceedings*, **22**, 253, 1985; Part IV, Determination of Iron in Tunbridge Waters, *Analytical Proceedings*, **23**, 75, 1986; Part V, Hungarian Mines and Mineral Waters, *Analytical Proceedings*, **23**, 349, 1986; Part VI, Luminescence, *Analytical Proceedings*, **28**, 362, 1991.

D. Thorburn Burns: A Man of God and Gases, *Chemistry & Industry*, 921, 1991.

Duncan Thorburn Burns, Department of Analytical Chemistry, The Queen's University of Belfast.

Born: Moycullen, Co. Galway, 1629.
Died: Parke, Co. Galway, 1718.

Family: One son, Michael.

Addresses:

1629–1649	Moycullen Castle	1649–1653	Sligo
1653–1718	Parke		

Roderic O'Flaherty was head of a major clan and owner of a very large estate to the west of Lough Corrib. He was educated at Alexander Lynch's school in the city of Galway. This was a famous establishment, attracting scholars from all parts of Ireland and O'Flaherty distinguished himself in his studies of the Classics and History.

As an Irish native and a Catholic, O'Flaherty lost all his land, and therefore his wealth, after the execution of King Charles I. He fled to Sligo before his twentieth birthday. At the Restoration of Charles II, he had hoped to regain his territory, but received only one acre for every fifty he had owned previously. He lived on a meagre income at Parke, a few miles to the west of Galway between Furbo and Spiddle. In spite of his reduced circumstances, he continued his studies and made a major contribution to the knowledge of Irish history and antiquities, culminating in a seminal work in Latin entitled *Ogygia seu Rerum Hibernicarum Chronologia*, published in London in 1685. Like many seventeenth century writers, he did not exclude the mysterious and the supernatural from his study of natural phenomena.

His great scientific treatise *Chorographical Description of West or h-Iar Connaught* was written in English in 1684 for the Dublin Philosophical Society but was not published in his lifetime. It describes the topography and natural history of the O'Flaherty country, from Lough Corrib to the west coast.

This work is unique, as the first scientific account of the fauna by an Irishman since that written by Augustin, a thousand years earlier. It contains very important information on the introduction of freshwater fishes to Ireland and gives the first account of seabirds and marine life. The greater part of the treatise is based on observations made at first hand. But O'Flaherty also sets down records made by other observers, making it clear that detailed knowledge of the fauna was widespread in the seventeenth century. The work includes an account of the first experimental study of animal behaviour recorded in Ireland. The experiment was made to determine whether the salmon returned to the river of its birth. Confirmation of the salmon's powers of homing laid the foundation for modern conservation of the species.

Further reading:

Roderic O'Flaherty: *West or h-Iar Connaught*, Kenny's Bookshop, Galway (1978 reprint, with Introduction by W.J. Hogan, of J. Hardiman's edition of 1846).
C. Moriarty: The Early Naturalists, pp 71–90 in J.W. Foster (ed.), *Nature in Ireland: A Scientific and Cultural History*, Lilliput, Dublin, 1997.

Christopher Moriarty, Marine Institute, Laboratories, Abbotstown, Dublin 15.

Born: Dublin, 17 April 1656.
Died: Dublin, 11 October 1698.

Family:
Married: Lucy Domville in 1678. She died in 1691. Children: One daughter and two sons, of whom only Samuel survived infancy. His son, Samuel Molyneux (1689–1728), was a distinguished astronomer, a Fellow of the Royal Society (1712), a Member of the English Parliaments of 1715, 1726, and 1727, a Member of the Irish Parliament of 1727, a Lord of the Admiralty (1727), and Secretary to the Prince of Wales, 1714–1727.

Addresses:

1656–1674	Fishamble Street, Dublin
1674–1677	London (studied law at Middle Temple)
1677–1689	Dublin and Castle Dillon, Co. Armagh
1689–1690	North Gate, Chester
1690–1698	Dublin and Castle Dillon

Distinctions:
Fellow of the Royal Society 1685; Chief Engineer and Surveyor General of the King's Buildings and works in Ireland (shared with William Robinson) 1684–1688; Member for City of Dublin, then University of Dublin, in the Irish Parliament 1690, 1692; LLD *honoris causa,* Trinity College Dublin 1693.

Scientifically speaking, seventeenth-century Ireland lagged behind Britain and the Continent. Largely unaffected by the Renaissance, Ireland had almost no scientific tradition, and the transformations in science elsewhere were not reflected in Irish intellectual life until well into the second half of the seventeenth-century. The new developments after 1650, principally in natural history and practical mathematics (surveying and cartography), created the climate in which was born and nurtured William Molyneux, in his day the single most influential figure in Irish science.

Molyneux's importance lay less in scientific discoveries than in an important institutional innovation. Inspired by the example of the Royal Society of London (founded 1660), and by his reading of Francis Bacon (1561–1626), who saw science as a co-operative enterprise with utilitarian ends, Molyneux founded and was the first secretary of the Dublin Philosophical Society, which formally constituted itself on 7 February 1684 under the official name 'The Dublin Society for the Improving of Naturall Knowledge, Mathematicks, and Mechanicks'. Despite its chequered history – its demise in 1708 followed the failure of Samuel Molyneux to revive interest after periods of inactivity – the Society helped create an Irish scientific tradition through its widely-ranging scientific work, astronomy and optics being the areas within which William Molyneux contributed most actively. In 1685 he designed a sundial-mounted telescope, of which he published an account in *Sciothericum Telescopicum* (Dublin, 1686). His popular *Dioptrica Nova* (London, 1692) was the first optical treatise to be published in English. His first publication was philosophical: *Six Metaphysical Meditations* (London, 1680), a translation of the famous *Meditationes* (1641) of René Descartes (1596–1650).

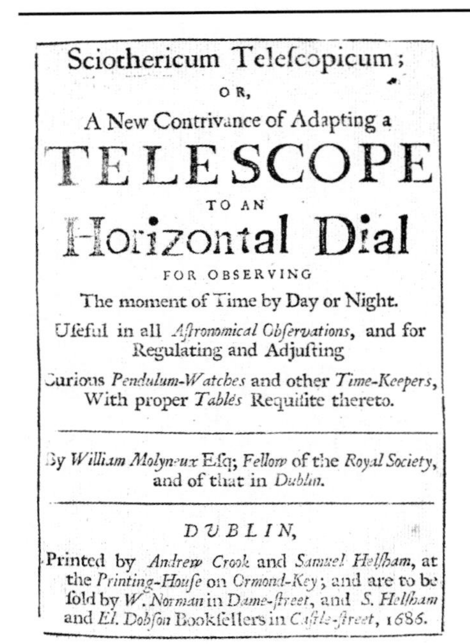
Sciothericum Teleſcopicum;
OR,
A New Contrivance of Adapting a
TELESCOPE
TO AN
Horizontal Dial
FOR OBSERVING
The moment of Time by Day or Night.
Uſeful in all *Aſtronomical Obſervations*, and for Regulating and Adjuſting
Curious *Pendulum-Watches* and other *Time-Keepers*, With proper *Tables* Requiſite thereto.

By *William Molyneux* Eſq; *Fellow* of the *Royal Society*, and of that in *Dublin*.

DUBLIN,
Printed by *Andrew Crook* and *Samuel Helſham*, at the *Printing-Houſe* on *Ormond-Key*; and are to be ſold by *W. Norman* in *Dame-ſtreet*, and *S. Helſham* and *El. Dobſon* Bookſellers in *Caſtle-ſtreet*, 1686.

Title page of Sciothericum Telescopicum *(Dublin 1686)*

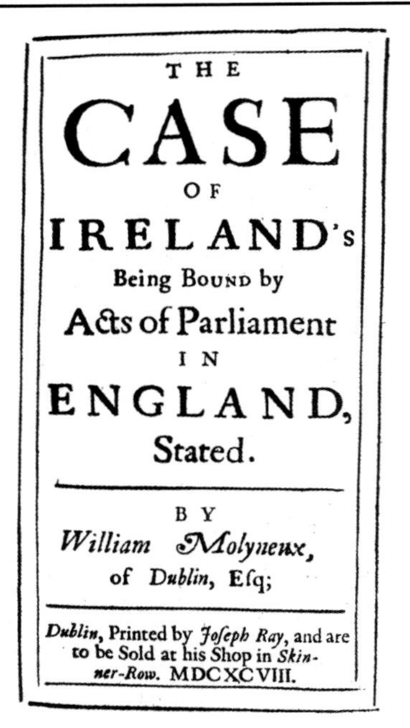
THE
CASE
OF
IRELAND's
Being Bound by
Acts of Parliament
IN
ENGLAND,
Stated.

BY
William Molyneux,
of *Dublin*, Eſq;

Dublin, Printed by *Joſeph Ray*, and are to be Sold at his Shop in *Skinner-Row*. MDCXCVIII.

Title page of The Case of Ireland Stated *(Dublin 1698)*

Illustrations are taken, with kind permission, from William Molyneux of Dublin, *by J.G. Simms, Irish Academic Press, 1982.*

Molyneux originated the celebrated 'Molyneux Problem', which he proposed to the philosopher John Locke (1632–1704) in 1693 (an earlier version dating from 1688). A man born blind has learned to distinguish between a globe and a cube through touch. Suppose he suddenly gains his sight, and is shown a globe and a cube. Could he tell which is which without touching them? Those who held that all our knowledge comes from experience of the external world, i.e. the 'empiricists' (which included Molyneux and Locke), said no; those who held that some elements in our knowledge are innate and independent of experience, i.e. the 'rationalists', said yes, claiming that we are born with the ability to recognise geometrical shapes. Molyneux's Problem might have been suggested to him by the poignant circumstance that in January 1679 his wife became incurably blind.

In the 1690s Molyneux became involved in politics and economic matters, and published *The Case of Ireland's Being Bound by Acts of Parliament in England Stated* (Dublin, 1698), his most popular work, in which he argued for Ireland's autonomy under the English crown.

Further reading:

K. Theodore Hoppen: The Royal Society and Ireland: William Molyneux, F.R.S. (1656–1698), *Notes and Records of the Royal Society,* **18**, 125–135, 1963.
K. Theodore Hoppen: *The Common Scientist in the Seventeenth Century: A Study of the Dublin Philosophical Society 1683–1708,* Routledge & Kegan Paul, London, 1970. A full-scale study of science in Ireland in the second half of the seventeenth century, in which Molyneux figures prominently.
J.W. Davis: The Molyneux Problem, *Journal of the History of Ideas,* **21**, 392–408, 1960.
J.G. Simms: *William Molyneux of Dublin,* Irish Academic Press, Dublin, 1982.
Menno Lievers: The Molyneux Problem, *Journal of The History of Philosophy*, **30**, 399–416, 1992.

Alan Gabbey, Department of Philosophy, Barnard College, Columbia University, New York.

HANS SLOANE Physician and Collector

Born: Killyleagh, Co. Down, 16 April 1660.
Died: London, 11 January 1753.

Courtesy of the British Museum

Family:
Father: Alexander Sloane, a tax collector for the Hamilton family who had come to Co. Down from Scotland in 1603; Mother: Sarah Hickes.
Married (1695): Elizabeth Rose, daughter of John Langley and widow of Fulk Rose of Jamaica.
Children: one son and three daughters.

Addresses:

1660–1669	Killyleagh
1669–1683	London
1683–1684	Paris and Montpellier
1684–1687	London
1687–1689	Jamaica
1689–1753	4 Bloomsbury Place, London

Distinctions:
Baronet 1716; President, Royal College of Physicians 1719–1735; President, Royal Society 1727–1741.

Young Hans spent much of his time looking at plants and gathering specimens from around Strangford Lough. He noticed that many local people chewed dulse, which was thought to prevent or cure scurvy – now known to be a disease caused by lack of vitamin C. When he was nineteen, Sloane went to London to study medicine and botany. He learned to prepare and use chemicals from plants as medicines, and spent time studying in the Physic Garden, where plants from all over the world were cultivated for making medicines. He also realised that our health depends very much on the food we eat. In 1683 he went, for a short time, to study botany and medicine at the University of Paris, then to Montpellier, considered the best medical university in Europe. He spent a year there, visiting many sites of rare plants. He graduated at the nearby Huguenot University of Orange, as Protestants were not allowed to graduate from Paris or Montpellier.

For four years he was a junior doctor helping Thomas Sydenham – who held that botany was for greengrocers, and anatomy for butchers, and felt that examining patients carefully was a much more useful way of investigating and curing diseases than knowing about plants and animals. But Sloane learned from Sydenham that, by careful observation and recording, it was possible to classify the symptoms of a disease. He was able, for example, to list the differences between measles and scarlet fever. Following the process of classification, diseases could be identified and effective medicines prescribed.

In 1687, Sloane went to Jamaica, as physician to the Duke of Albemarle who had been appointed Governor there. He was keen to explore Jamaica, making notes on his expeditions, and recording many activities of the local people, their diseases and the food they ate, like snakes and caterpillars, rats and turtles. He dried and kept many plants, and employed an artist to draw pictures of things like fish, insects, birds and fruit that could not be preserved. Returning to London, he brought home many specimens, including about eight hundred plants, many of which had not been seen in

England before. Although he was physician to rich people like Queen Anne and Samuel Pepys, he tended many poor people free. He was interested in the welfare of the poor of London, and tried to make sure that the food and medicines they used were as clean and pure as possible.

In the eighteenth century many people died of smallpox. People who survived, like Sloane himself, were often marked for life by pock marks on their faces. The Princess of Wales, hearing that in Turkey people who were deliberately given a mild attack of smallpox never had the full disease, asked Sloane to find out if it would work. Her face had been scarred by smallpox, and she wanted to protect her children from it.

Sloane was not sure whether such inoculation would work safely, so he first tried it on six condemned prisoners from Newgate jail, and on six orphan children. Even when they deliberately lived near people with smallpox, none of the convicts or orphans caught the disease. As his experiment was successful, the children of the Princess were then inoculated, and the royal example was followed by many others. After this, the number of smallpox cases in London was much less than in Paris, where inoculation was not used.

Sloane was elected a member of the Royal Society in 1685 when he was only 24. He was President of the Society from 1727, after the death of Sir Isaac Newton, until 1741 when he resigned because of failing health.

In 1716 he was created a baronet by King George I. He was also president of the Royal College of Physicians, which was the main society in London for medical doctors. He helped produce *The London Pharmacopoeia* – a list of medicines that doctors at that time thought were safe to use.

When he died at the age of ninety-two, there were about 200,000 specimens in his collection: stones, crystals, eggs, books and papers, mathematical and scientific instruments, coins and medals, butterflies, insects and shells, small dried plants and large stuffed animals, and many more objects collected over nearly seventy years. He wanted these to be kept together. But he also wanted his family to benefit from the sale of his specimens. So he had agreed that, after he died, the collection should be offered to King George II for £20,000, even though it was worth very much more.

The Government ran a lottery which raised about £95,000. Some of this was used to buy Sloane's collection and other specimens, the rest to buy and furnish a large house in Bloomsbury. The British Museum opened there on 15 January 1759, just six years after Sir Hans Sloane died.

Further reading:

W.W.D. Thomson: Some Aspects of the Life and Times of Sir Hans Sloane, *Ulster Medical Journal*, Belfast, 1938.

W.R Sloan: *Sir Hans Sloane, Legend and Lineage*, W.R. Sloan, Helen's Bay, 1981.

Arthur MacGregor (ed.): *Sir Hans Sloane, Founding Father of the British Museum*, British Museum Press, London, 1994.

Martin Brown: *Hans Sloane*, The Blackstaff Press, Belfast, 1995.

Martin Brown, Bangor, Co. Down.

Born: Dublin, 14 April 1661.
Died: 19 October 1733.

Statue in Armagh Cathedral

Family:
Youngest son of Samuel Molyneux, Master gunner for Ireland, and his wife Margaret Dowdall.
Married: Catherine Howard of Shelton, Co. Wicklow, 1693
Children: Several

Addresses:
1690–1711 Thomas Court, Dublin
1711–1733 Peter Street, Dublin

Distinctions:
Fellow of the Royal Society 1686; President, King's and Queen's College of Physicians of Ireland 1702, 1709, 1713, 1720; Professor of Medicine, Dublin University, 1711–33; State Physician 1715–30; Physician-General of the army 1718–33.

The scion of a family that held public office for generations in Ireland, Thomas Molyneux attended Dr Henry Rider's school before entering Trinity College Dublin in 1676. Having graduated MA and MB in 1683, he went abroad to complete his education. In London, where he arrived about 15 May 1683, he took lodgings 'at the sign of the Flower-de-luce, over against St Dunstan's Church in Fleet Street'. The young doctor visited Gresham College on a day when the Royal Society was meeting and he was permitted to sit in and listen to the discourse. 'At this meeting', he informed his brother, 'I had the opportunity of seeing several noted men as Mr Evelyn, Mr Hooke, Mr Isaac Newton...' While in London he also met Robert Boyle, Sir William Petty and John Flamsteed, who was to be the first Astronomer Royal.

Molyneux visited Windsor and Eton College where he found the rooms were kept 'very nastily stinking when you come in to them'. He also went to Oxford and Cambridge and, in July, moved to Leyden, where he met John Locke with whom he later corresponded. In August 1684 his first scientific article, which discussed 'the dissolution and swimming of heavy bodies in Menstruums far lighter than themselves', was published in *Nouvelles de la République des Lettres* and, in December 1684, he sent his first contribution (an account of a 'prodigious os frontis') to the *Philosophical Transactions*. He spent some time in Paris in 1685 and was elected a fellow of the Royal Society in the following year.

On his return to Dublin in 1687, Thomas Molyneux proceeded to his M.D. and set up in practice at Thomas Court but, because of the unsettled political situation after the accession of James II, he and many Protestants left Ireland temporarily. For a time he practised in Chester but, after the victory of William of Orange at the Boyne, he returned to the Irish capital and before long had attained remarkable professional success. He has, indeed, been called the 'father of Irish medicine'.

Thomas and Catherine Molyneux had a large family. They also took charge of a nephew Samuel (a future astronomer), when William Molyneux died in 1698. The latter ('the single most influential figure in Irish science' in his day, according to Alan Gabbey) had held office as secretary and treasurer of the Dublin Philosophical Society. He was the first to demonstrate microscopically the circulation of the blood in reptiles. He represented Dublin University in the Irish parliament of 1692. Thomas Molyneux was also a member of the Dublin Philosophical Society, which wished to establish experimental science on a proper footing in Ireland. Like his brother, he contributed through moral support, enthusiasm and example rather than by major discoveries. Being an antiquary, geologist and physician, Thomas Molyneux's contributions to the society were correspondingly varied and included an over-credulous account of an operation to remove an ivory bodkin from the bladder of an ingenuous young woman who said she had accidentally swallowed it, a detailed account of the Giant's Causeway, and a paper on 'the late general coughs and colds in Ireland'. Some of his contributions were republished in the *Philosophical Transactions.* He also wrote an account of coal-mining in Ireland.

His acquaintances included Roderick O'Flaherty (1629–1718), and a picture from the early eighteenth century worth keeping in mind is that of Molyneux cantering along a road in Connemara on 21 April 1709 to visit the old historian, at the mercy of an icy wind off the Atlantic which cut through his rain-sodden greatcoat. 'I never saw so strangely stony and wild a country', he reftected later. He and O'Flaherty shared a common interest in scholarship and aptly represented their Anglo-lrish and Irish heritages, one belonging to a privileged, the other to an oppressed class. Both, of course, would have owned to a sense of nationalism. Molyneux's brother, William, had had his book *The Case of Ireland Stated* burned by the common hangman; O'Flaherty was the author of *A Chorographical Description of West or h-Iar Connaught.*

Thomas Molyneux's career was in many ways an ideal one: a classical scholar, a man of scientific temperament, a successful practitioner, he wrote on such diverse topics as the Irish round towers, the Irish elk and the ancient Greek and Roman lyres. He represented Ratoath in the Irish parliament from 1695 to 1699 and was one of the founders of the Royal Dublin Society.

Molyneux held important offices and, in 1730, he was created a baronet. He amassed considerable wealth – he boasted that he had spent more than Dr Richard Steevens ever made – and in 1711 built a mansion in Peter Street.

Sir Thomas Molyneux died in 1733. His burial place is disputed – Armagh Catheral (where there is a statue of him by Roubiliac) according to the *Dictionary of National Biography* and St Audoen's church according to A.M. Fraser. Lady Molyneux lived on for some years in the Peter Street mansion, which in a later period served as the Molynuex Asylum for Blind Women.

Further reading:

Sir William Wilde: *Dublin University Magazine,* **18**, 305–27, 470–89, 604–18, 744–63, 1841.
A.M. Fraser: The Molyneux Family, *Dublin Historical Record,* **16**, 9–15, 1960.
K.T. Hoppen: *The Common Scientist in the Seventeenth Century,* London, 1970.
A. Gabbey: William Molyneux, p23-24 of this volume.

J. B. Lyons, Department of the History of Medicine, Royal College of Surgeons in Ireland, Dublin.

9 GEORGE BERKELEY Bishop of Cloyne, Philosopher

Dean Berkeley and his Entourage, The Bermuda Group (Dublin), artist John Smibert (courtesy of the National Gallery of Ireland)

Born: Co. Kilkenny, 12 March 1685.
Died: Oxford, 14 January 1753.

Family:
Married: Anne Forster.
Children: Four sons and three daughters.

Addresses:
1685?–1700 Dysart Castle, Thomastown, Co. Kilkenny
1700–1728 Trinity College Dublin, London and continental travel
1729–1732 Whitehall, Aquidnick Island, near Newport, Rhode Island
1733–1753 Cloyne, Co. Cork

George Berkeley was born in County Kilkenny and grew up at Dysart Castle, Thomastown. He received his early education at Kilkenny College, Swift's school, where he started a lifelong friendship with Thomas Prior, who became one of his major correspondents. He entered Trinity College Dublin in 1700, took fellowship in 1704, and began working on his major ideas in what have become known as his *Philosophical Commentaries*. By 1713 he had published *An Essay Towards a New Theory of Vision* (1709), *A Treatise Concerning the Principles of Human Knowledge, Part 1* (1710) and *Three Dialogues between Hylas and Philonous* (1713). Any one of these would assure him continued fame. The three were a magnificent output. His essential idea, 'To be is to be perceived' earned him the title of 'Immaterialist Philosopher'. While he became immediately famous, Berkeley was not really understood. He was saying that what we know of things is the product of our sensory experience with them. Things have existence only as they are perceived. However things do not vanish when not being perceived by one of 'us'. Rather, they have existence in that an omniscient and omnipotent God is always 'there' to perceive them.

During the following decade he spent extensive time in London, where he was presented at court through his friendship with Jonathan Swift and other literary and social figures. Considered by Pope to 'possess every virtue under Heaven', Berkeley was well regarded in London. His friends included Richard Steel, the essayist, the Earl of Pembroke, former Lord Lieutenant of Ireland, Addison, the playwright, Pope, Gray, and Parnell, the poets.

Twice he toured Europe, crossing the Alps in midwinter and travelling through France and the major cities of Italy to Sicily. We know of his travels through his correspondence with Thomas Prior and his patron, Sir John Perceval. During this period he wrote *De Moto*, an important contribution to physics that some say anticipates Einstein. By 1723, having taken orders, Berkeley, impatient with the world as he had experienced it, resolved to found a college in the New World. He was appointed Dean of Derry in 1724. He worked to develop support for his Bermuda scheme, obtaining pledge of a grant from the English Parliament in the sum of £20,000. By the end of 1728, and reinforced by a beneficence in the will of Swift's friend Hester von Homburgh (Vanessa), he chartered a ship, gathered his followers and sailed to America with all of the equipment for his college. He arrived in Newport, Rhode Island, on 23 January 1729, after having first stopped in Virginia, where he visited the College of William and Mary.

Just before his departure he married Anne Forster, daughter of John Forster, Speaker of the Irish House of Commons. Newport was to be the base for his establishment of a utopian community in Bermuda, the centrepiece of which would be St Paul's, to serve sons of settlers and native Americans from the mainland who would be educated to MA standard. He remained in Rhode Island nearly three years awaiting the promised funding from Parliament. While in Newport he preached in Trinity church and on the 'continent' at the Old Narragansett church, visiting with his friends, the Reverend James MacSparran and Attorney General Daniel Updike. He wrote *Alciphron*, a dialogic restatement of his basic principles.

Through his friend and admirer, the Reverend Samuel Johnson of Connecticut, he outlined the plan for what is now Columbia University. As he left Rhode Island he gave his books and his property, Whitehall, to Yale University in Connecticut. He provided Harvard University with a set of the classics. He is often considered to be the father of higher education in America. Berkeley, California, is named for him, as is the University there.

Upon his return to Ireland, he was made Bishop of Cloyne in Co. Cork. He remained there for twenty years serving all the people of Cloyne. He became known for his advocacy of tar water as a universal cure. His *Querist*, written during this period, continues to be read and admired for its penetrating analysis of the economic and political conditions in eighteenth century Ireland. Yeats, de Valera and Jack Lynch read and admired him. His *Siris* is a strange and not fully understood treatise in which he promotes tar water for its curative powers. He died peacefully in January 1753 in Oxford, where he had proceeded to oversee the education of his son.

Berkeley is a major figure in western philosophical thought. His theses, written in clear and elegant prose, have continued to attract attention and have proved difficult to refute. He is remarkable for his breadth of interest and his seminal contribution across the spectrum of ideas. He is still read for his insights in mathematics, physics, psychology of perception, art and architecture, medicine, economics and politics.

Further Reading:

George Berkeley: *The Works of George Berkeley*, A.A. Luce and T.E. Jessop (eds), Thomas Nelson and Sons, Edinburgh, 1948.
David Berman: *George Berkeley, Idealism and the Man*, Clarendon Press, Oxford, 1994.
Raymond W. Houghton: *The World of George Berkeley*, The Irish Heritage Series No. 53, Easons, Dublin.
Raymond W. Houghton, David Berman and Maureen T. Lapan: *George Berkeley*, Wolfhound Press, Dublin, 1986.
A.A. Luce: *The Life of George Berkeley*, Thomas Nelson and Sons Ltd, London, 1949.
I.C. Tipton: *Berkeley – The Philosophy of Immaterialism*, Methuen and Co. Ltd, London, 1974.

Raymond W. Houghton, North Kingston, Rhode Island 02852, USA.

10 JOHN RUTTY Physician and Chemist 1697–1775

Born: Wiltshire, 25 December 1697.

Died: Dublin, 26 April 1775 – buried in the Society of Friends burying ground, York Street, site of present Royal College of Surgeons, Dublin. He never married.

Addresses:

1722–1723	University of Leyden, graduated MD
1723	Moved to Dublin
1755–1775	Resided in Pill Lane.
	Later moved to corner of Boot Lane and St Mary's Lane

John Rutty was born in Wiltshire of Quaker parents. After education in Dublin, London and Leyden, where he graduated MD in 1723, he settled in Dublin as a physician. He continued as a student of medicine and related sciences such as chemistry all his life, in addition to his reading of spiritual books. He lived sparingly and often gave his services free to the poor. John Wesley noted that Rutty was 'held in high repute for his professional skills, as a writer, naturalist and highly spiritual member of the Society of Friends'.

Rutty published a major text *Experimental and Medical History of Mineral Waters of Ireland* (Dublin 1757) and *A Methodical Synopsis of Mineral Waters* (London 1757). This brought him into dispute with Dr Charles Lucas (1713–1771) also of Dublin who had earlier published an *Essay on Waters* in three volumes (London 1756). The dispute raged for at least three years. Lucas was also in dispute with Dr Oliver and others of Bath on mineral water analyses. Like all chemists of the period, Rutty had problems with speciation of elements, the nature of some elements, and with the properties of organic matter. Lucas was misled by the phlogiston theory of Stahl, then at the height of its fame. Lucas is now more remembered as an Irish patriot, parliamentarian and supporter of free holders of Dublin.

Rutty published four papers on mineral waters in the *Proceedings of the Royal Society*; Tracts on *The Analysis of Milk* (1762), the *Weather and Seasons, and Diseases in Dublin for 40 Years* (1770); and a major two-volume work *A Natural History of the County of Dublin* (1772). His last work, in Latin, which occupied him for 40 years, was *Materia Medica Antiqua et Nova* (Rotterdam 1775). The other side of Rutty's life was as a mystic, and student of spiritual books. He was an active and devoted member of the Society of Friends all his life. He maintained a *Spiritual Diary* from 1753–1774 which was published after his death. This diary was referred to in Boswell's *Life of Johnson*. In 1751 Rutty published *A History of the Rise and Progress of the People called Quakers in Ireland, from 1653–1751*, a continuation of a book originally written by Thomas Wright of Cork in 1700: this appeared in at least four editions up to 1811. John Wesley records in his *Diary* that on 6 April 1775 he visited 'that venerable man Dr Rutty, just tottering over his grave; but still clear in his understanding'. Earlier, in 1748 and 1749, he had been a patient of Dr Rutty.

Further Reading:

J. Osborne: Memoir of John Rutty, *Dublin Quarterly Journal of Medical Science*, **3**, 555, 1847.

W.K. Sullivan: ....Short Notice of Irish Chemists ... prior to 1800, *Dublin Quarterly Journal of Medical Science*, **8**, 465-495, 1849.

W.T.S. Sharpless: Dr. John Rutty of Dublin, *Annals of Medical History*, **10**, 249, 1928.

Duncan Thorburn Burns, Department of Analytical Chemistry, The Queen's University of Belfast.

11 PETER WOULFE Chemist and Mineralogist 1727–1803

Born: Co. Limerick 1727.
Died: London 1803.

Addresses:
Clerkenwell, London; Paris

Distinctions:
Fellow of The Royal Society 1767; awarded its Copley Medal in 1768; chosen to give the Society's first Bakerian Lecture in 1776.

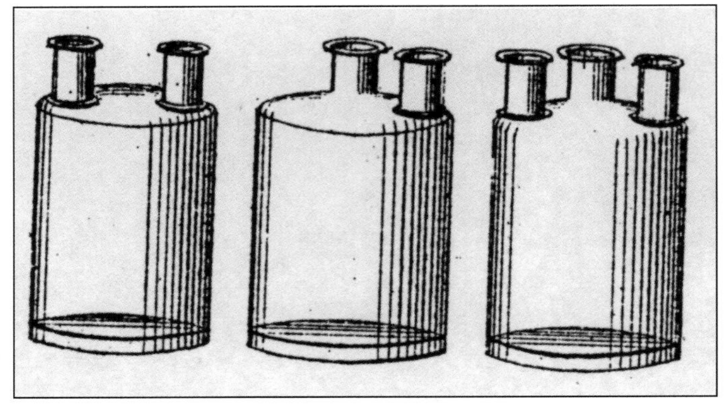

Variations of Woulfe's bottle

Peter Woulfe is a shadowy figure about whom we know little, except through his published papers. He seems to have worked mainly in England and to have found plenty of employment as a mineralogist. He investigated the Cornish tin deposits. In 1779 he was the first to show that the mineral dolomite was not just a mixture of magnesium and calcium carbonates, but a double salt. He was an intimate friend of Joseph Priestley, who valued his competent expertise. His name lives on in the history of chemistry because of his invention of Woulfe's Bottle, a two-necked flask, which, it is said, he invented as a trap to catch vapours produced during heating or distilling operations which 'might be hurtful to the lungs'. This piece of glassware remains a standard piece of laboratory equipment, if in slightly modified form.

He carried out an investigation into various ways of preparing mosaic gold (a form of tin sulphide), by heating together mixtures of tin, sulphur and sal-ammoniac (ammonium chloride). He produced the most beautiful specimens by heating five parts of tin sulphide with eight parts of mercury chloride. This form of stannic sulphide, gold-coloured and highly crystalline, shows great resistance to attack by acids. Its density is, however, only about a quarter of that of gold. Mosaic gold was known to the alchemists as *Aurum mosaicum* and its preparation must often have led some to think that they had managed to perform the transmutation of base metals into gold, that being the goal of many who took up the study of chemistry. Even in Woulfe's day, claims to produce gold artificially were still being made.

He was recognised as a chemist of sufficient merit to be appointed to a commission, (along with Irish chemists Richard Kirwan and Bryan Higgins), to examine the claims of a Dr Price of Guildford, who presented King George III with samples of 'gold' that he had obtained by 'transmutation'. Price however declined to demonstrate how he had achieved this miracle. Like Kirwan, Woulfe appears to have become reclusive and a little eccentric in his late years. He died alone in London in 1803.

Further Reading:

Dictionary of Scientific Biography.
P. Woulfe: Nature of Some Mineral Substances, *Philosophical Transactions of the Royal Society,* **66**, 605–623, 1776.
W. A. Campbell: Peter Woulfe and his Bottle, *Chemistry and Industry*, 1182–3, 1957.

William J. Davis, University Chemical Laboratory, Trinity College, Dublin.

12 JOSEPH BLACK Chemist

Born: Bordeaux, 16 April 1728.
Died: Edinburgh, 10 November 1799 – buried in Grey Friars Kirk Yard.

Engraving by John Kay, 1787

Family:
Son of John and Margaret (née Gordon) Black; never married.

Addresses:
1728–1740 Bordeaux
1740–1744 Latin School, Belfast
1744–1748 University of Glasgow
1748–1754 University of Edinburgh
1756–1766 Lecturer, University of Glasgow
1766–1799 Professor, University of Edinburgh

Distinction:
Fellow Royal Society of Edinburgh 1783.

Joseph Black was born in Bordeaux where his father, a Belfast man of Scots descent, was in the wine trade. He was educated by his Scottish Mother until the age of twelve, when he was sent to Belfast to become a pupil in the old Latin School. This school, established by the Earl of Donegall in 1666, stood at the corner of Ann Street and Church Lane, which was then called School-house Lane. It remained in existence until the Belfast Academy was established in 1786.

In 1744 Joseph Black entered the University of Glasgow. His matriculation entry reads 'Josephus Black filius natu quartus Joannis Black Mecatoris in Urbe Bordeaux in Gallia, ex urbe de Belfast in Hibernia'. He read the general arts curriculum for three years, then, after being pressed by his father to choose a profession, studied medicine and chemistry under William Cullen.

In 1752 Black went to Edinburgh to complete his medical studies and graduated MD in 1754. His now famous thesis 'On the acid humour arising from food, and magnesia alba' deals mainly with chemical experiments on acidity and an explanation and proof of Black's doctrine of the relationship between mild and caustic alkalis. In June 1755 he read a paper to the Philosophical Society of Edinburgh, a revised account of the chemical experiments in his thesis together with further work. His paper was published in the following year, reprinted 1779, 1782, 1898 and reissued in 1963. This was Black's most significant chemical publication. He showed that the change from chalk to lime consists of the withdrawal of fixed air (carbon dioxide) i.e.:

$$CaCO_3 \rightarrow CaO + CO_2$$
$$CaO + H_2O + K_2CO_3 \rightarrow CaCO_3 + 2KOH$$

Black's theory was opposed by J.F. Meyer but in due course confirmed by N.J.E. Jacquin. A detailed account of the controversy was given by Lavoisier in his *Essays Physical and Chemical* (1774). Black, as recognised by Lavoisier, thus laid a foundation stone of the revolution in chemistry based on quantitation. The letters from Lavoisier to Black were published at the behest of Thomas Andrews in the Report of the 41st Meeting of the British Association held in Edinburgh in 1871. His equipment was simple: the balance and other apparatus used in this and other early work are in the Playfair Collection housed in the Royal Museum of Scotland.

Black succeeded Cullen in the Glasgow lectureship in 1756. Whilst at Glasgow, he developed his second important line of research, that on latent heats and specific heats. This important work was never published independently, but discussed in his chemical lectures, and recorded in manuscript form by students and in the edited edition of his lectures, published posthumously by Robison. James Watt was, at the time of Black's lectureship, instrument maker to the University of Glasgow, and his experiments on the improvement of the steam engine were greatly assisted by Black's discovery of latent heat. Watt owed much to Black's advice and their extensive correspondence is extant.

Black left Glasgow in 1766 and again succeeded Cullen, this time in Edinburgh. It has been stated that Black did no significant research work after being appointed to the Edinburgh Chair in 1766. This is not true. His paper 'Analysis of the Waters of Some Hot Springs in Iceland' exhibited fine analytical work of a quality far ahead of his time. It was also innovative. He was the first to observe the interference of carbon dioxide in alkali titration, and an indicator error: he corrected the latter with a blank and was the first to use a back titration, and titration by weight.

After 1766 Black took an increasing interest in the rapidly developing chemical-based industries in Scotland, and above all in teaching. He achieved his most widespread fame in his lifetime for his teaching, and specifically for his use of lecture demonstrations. The lecture contents were kept up to date as can be judged from the contemporary student lecture notes: over 90 sets survive over the session 1766–1767 through to 1796–1797.

It is of interest to note that the world's first Chemical Society met in Edinburgh. The list of members, dated 1785, in Black's handwriting, contained 59 members of whom 58 were students attending Black's classes 1783–1787. It included Thomas Beddoes (founder of the Pneumatic Institute at Clifton). The first volume of the proceedings *Dissertations Read Before the Chemical Society Instituted in the Beginning of 1785*, was presented to the Royal Irish Academy by Sir William Betham, 26 January 1846, and lay unnoticed in its library for more than a century. It attracted the attention of Professor P.J. McLaughlin of St Patrick's College, Maynooth, Ireland. Subsequently this volume was presented to the Edinburgh University Chemical Society by the Council of the Royal Irish Academy.

Further Reading

D. Thorburn Burns: Dark Ages to Enlightenment, Alchemy to Analysis, in D. Littlejohn and D. Thorburn Burns (eds), *Euroanalysis VII, Reviews on Analytical Chemistry*, Royal Society of Chemistry, London, 1994.

J.G. Fyffe and R.G.W Anderson: *Joseph Black a Bibliography*, Science Museum, London, 1992.

J. Kendall: The First Chemical Journal, *Nature*, **159**, 869, 1947.

Duncan Thorburn Burns, Department of Analytical Chemistry, The Queen's University of Belfast.

SYLVESTER O'HALLORAN Ophthalmic Surgeon and Antiquarian

Born: Caherdavin, Co. Clare, 31 December 1728.
Died: Limerick, 11 August 1807.

Family:
Third son of a well to do farmer, Michael O'Halloran and his wife Mary McDonnell.
Married: Mary O'Casey, 1752.
Children: Three sons and a daughter. Michael, the eldest son, was killed by a fall from a horse in 1782, leaving no issue; Catherine also died young; John served in Colonel Brown's regiment of American Loyalists and in 1787 was granted three hundred acres on Long Island; Joseph (b.1763) earned a knighthood for long and meritorious service with the army in India. At least two of Sir Joseph O'Halloran's sons settled in Australia – Thomas Shuldham O'Halloran was elected member of the first Parliament of South Australia; William Littlejohn O'Halloran became Auditor General of Western Australia.

Sylvester O'Halloran attended school in Limerick before setting out to study surgery in Leyden, Paris and London. Early in his career he displayed an experimental flair and, being particularly interested in ophthalmology, investigated the blood supply of the eye. 'The method I took for this (he wrote) was, by hanging a young dog by his hind feet to a post: in about ten moments after, I could perceive the Blood vessels of the Eye became more manifest . . .'.

When he returned to Limerick to start practice, O'Halloran had a Treatise on Cataract ready for publication. A contemporary has described him graphically: 'The tall, thin doctor in his quaint French dress, with his gold-headed cane, beautiful Parisian wig and cocked hat . . .'. He presented a strikingly elegant figure in the provincial city and before long had an increasing number of patients. 'I could produce an instance in this Town (he wrote) of a Woman who had a cataractous Eye . . . which Taylor had declared *incurable*, which I nevertheless restored her the Use of the Twentieth of March 1749'.

O'Halloran's critical analysis of the operative treatment of cataract was presented to the Royal Society in 1752. His *New Method of Amputation* was published in 1765. This book included an appendix 'Proposals for the Advancement of Surgery in Ireland' which may have influenced the Dublin surgeons who in 1784 founded the Royal College of Surgeons in Ireland. He was one of the chief founders of the County Limerick Infirmary and his important book, *External Injuries of the Head*, contained a plea for caution in the use of the trepan (an instrument for cutting out small pieces of bone, especially from the skull).

His principal diversion was the study of Irish history; he was the author of *Ierne Defended* and *A General History of Ireland*. A prolific letter writer, his correspondents included Charles O'Conor of Bellangare and Edmund Burke.

An important political event, the Act of Union, overshadowed Sylvester O'Halloran's later years, and he took the chair at a protest meeting of the Roman Catholics of the city of Limerick on 23 January 1800. In 1802 he was among those who formed the Limerick Medical Society and in the following year an edition in three volumes of his historical writings was published.

But old age advanced relentlessly. The Reverend J. Hall, who visited Limerick in 1807, wrote in his *Tour of Ireland*: 'I found Dr. O'Halloran the celebrated antiquarian . . . old, infirm and confined to his chair.' He died on 11 August 1807.

Further reading:

J.B. Lyons: Sylvester O'Halloran (1728–1807), *Irish Journal of Medical Science*, 217–32, 279–88, 1963.
S. O'Halloran: The Letters of Sylvester O'Halloran, edited by J.B. Lyons, *North Munster Antiquarian Journal*, **9**, 25-49, 168–81, 1962–3.
J.B. Lyons: Sylvester O'Halloran's 'Treatise on the Air', *Irish Medical Journal*, **1**, 37–39, 1983.

A N

APPENDIX,

C O N T A I N I N G

Propoſals for the ADVANCEMENT of SURGERY in IRELAND;

With a retroſpective View of the ANTIENT STATE of P H Y S I C amongſt us.

P R E S E N T E D

With Great DEFERENCE and high ESTEEM to

L U C I U S O'B R I E N, Eſq;

REPRESENTATIVE in P A R L I A M E N T for the

BOROUGH of E N N I S.

'T HOUGH it be univerſally admitted, that the profeſſion of ſurgery is of the greateſt utility to the ſtate in time of war, and to the public at all times; and as ſcarce a man from the prince to the peaſant, but muſt fall under the hands of ſurgery, at ſome period or other of his life, it muſt neceſſari-
ly

Sylvester O'Halloran's Proposals for the Advancement of Surgery in Ireland *(1765) may have influenced the foundation of the Royal College of Surgeons in Ireland*

J. B. Lyons, Department of the History of Medicine, Royal College of Surgeons, Dublin.

14 RICHARD KIRWAN Chemist, Mineralogist, Geologist and Meteorologist

Born: Cloughballymore, Co. Galway, 1733.
Died: Dublin, 1 June 1812.

Family:
Second son of Martin Kirwan of Cregg Castle.
Married: Anne Blake 1757 (died 1765). Anne was the daughter of Sir Thomas Blake of Menlough Castle.
Children: Two daughters – Maria Theresa and Eliza.

Addresses:

1733–1741	Cregg Castle, near Corrandulla
1741–1750	Cloughballymore, near Kilcolgan
1757–1765	Menlough Castle
1777–1787	11 Newman Street, London
1787–1812	6 Cavendish Row, Dublin

Richd. Kirwan Esqr. LL.D. F.R.S. &c. &c.

Distinctions:
Founder Member of the Royal Irish Academy 1785, President 1799; Fellow of the Royal Society 1780; Copley Medal of the Royal Society 1782; LLD Dublin University 1794; Fellow of the Royal Society of Edinburgh 1796; Gold Medal Royal Dublin Society (for services in acquiring and arranging the Leskean cabinet of minerals) 1794. In addition he was an Honorary Member of the Academies of Stockholm, Uppsala, Berlin, Dijon, Philadelphia, the Mineralogical Society of Jena and the Manchester Society.

Richard Kirwan belongs to the chemical age when the foundations of stoichiometry or quantitative chemistry were laid. As luck would have it no reaction or basic law bears his name but historians of chemistry such as Proust and Klaproth rank him with famous names in Europe. He corresponded with all the major chemists in Europe – Bergman, Scheele, Chaptal, Klaproth and Lavoisier – and was frequently cited by them as an authority on his subject.

Richard was originally destined for the priesthood. His early education was in Ireland but was completed at Poitiers, France. In 1754 he went to Paris to enter the Jesuit noviciate. His elder brother Patrick was killed following a duel. Richard then abandoned his noviciate and returned to Ireland in 1755. He lived most of his married life at Menlough Castle, where he fitted up a laboratory and amassed a library. He renounced Catholic beliefs in 1764 as a prerequisite to being called to the Irish bar in 1766, but practised for only two years. During the next nine years he increased his knowledge of science and languages. Three years (*circa* 1769–1772) were spent in London, and in 1777 he returned to London where he stayed for ten years. His house became a well-known meeting place for the distinguished, particularly those in science. He was awarded the Copley Medal by the Royal Society (1782) for his contributions to 'chemical affinity'. His results contributed to the formulation of the Richter's law of reciprocal proportions. Kirwan was the first to describe the titrimetric determination of iron with ferrocyanide.

His *Elements of Mineralogy* (1784, 2nd edition 1794 – *figure 1*) was the first systematic work on the subject in English. Its significance was such that it was translated and published in Paris within a year. Kirwan's *An Essay on the Analysis of Mineral Waters* (1799 – *figure 2*) is an excellent account of qualitative and quantitative analysis of the period and was written when he had returned and settled in Ireland. He returned to Ireland in 1787, was active in Irish science and technology, and was President of the Royal Irish Academy from 1799 until his death. Most of his papers were published in the *Proceedings* of the Academy (38 papers 1788–1808), with some in the *Transactions of the Dublin Society* (6 papers), and his earlier work (6 papers, 1781–86) is in the *Philosophical Transactions of the Royal Society*. Many of his papers appear in translation in European Journals.

An Essay on Phlogiston and the Constitution of Acids (1787, 2nd Edition 1789, reprinted 1968 – *figure 3*) is possibly his best known work. Phlogiston was in the eighteenth century supposed to be a component of all combustible bodies. When they burned phlogiston was lost: metal → calx + phlogiston. In the *old* theory phlogiston is equivalent to minus oxygen: later it was sometimes assumed to be hydrogen. His *Essay* was translated into French (by Madame Lavoisier) in 1788 with notes refuting Kirwan's views by leading 'antiphlogistonists' including Lavoisier. He was later converted to this view by his fellow Irishman, William Higgins.

Kirwan's contributions to meteorology *via* weather pattern recognition are only now being fully appreciated. He was interested in the application of science and wrote informatively on coal mining, manures and bleaching. Late in life he was eccentric. For example he refused to remove his hat even at court and became interested in logic and metaphysics. He died in 1812 starving a cold.

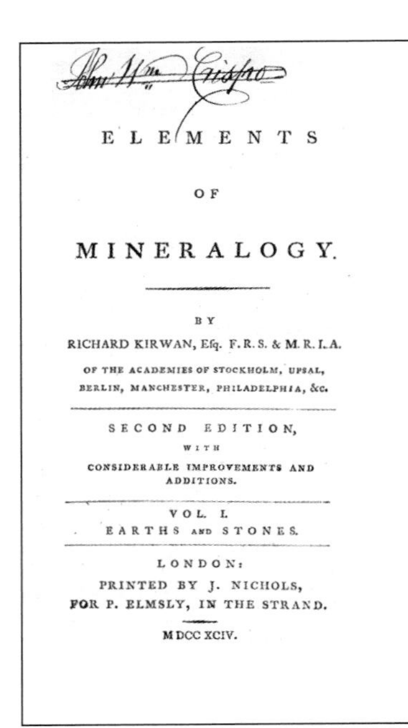
ELEMENTS

OF

MINERALOGY.

BY

RICHARD KIRWAN, Efq. F. R. S. & M. R. I. A.

OF THE ACADEMIES OF STOCKHOLM, UPSAL, BERLIN, MANCHESTER, PHILADELPHIA, &c.

SECOND EDITION,

WITH

CONSIDERABLE IMPROVEMENTS AND ADDITIONS.

VOL. I.

EARTHS AND STONES.

LONDON:

PRINTED BY J. NICHOLS, FOR P. ELMSLY, IN THE STRAND.

MDCCXCIV.

Figure 1

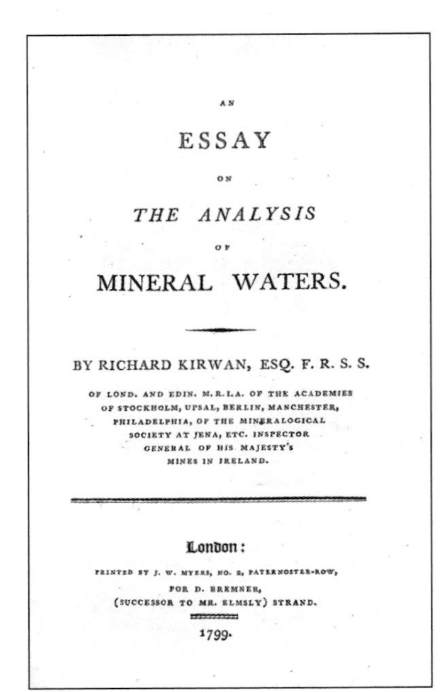
AN

ESSAY

ON

THE ANALYSIS

OF

MINERAL WATERS.

BY RICHARD KIRWAN, ESQ. F. R. S. S.

OF LOND. AND EDIN. M.R.I.A. OF THE ACADEMIES OF STOCKHOLM, UPSAL, BERLIN, MANCHESTER, PHILADELPHIA, OF THE MINERALOGICAL SOCIETY AT JENA, ETC. INSPECTOR GENERAL OF HIS MAJESTY'S MINES IN IRELAND.

London:

PRINTED BY J. W. MYERS, NO. 2, PATERNOSTER-ROW, FOR D. BREMNER, (SUCCESSOR TO MR. ELMSLY) STRAND.

1799.

Figure 2

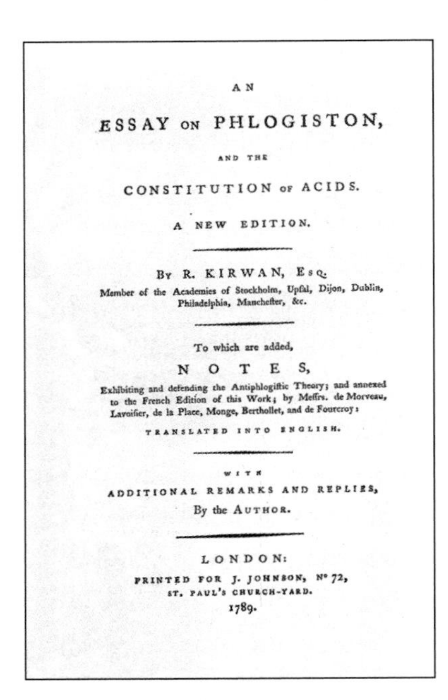
AN

ESSAY ON PHLOGISTON,

AND THE

CONSTITUTION OF ACIDS.

A NEW EDITION.

By R. KIRWAN, Esq.

Member of the Academies of Stockholm, Upfal, Dijon, Dublin, Philadelphia, Manchefter, &c.

To which are added,

NOTES,

Exhibiting and defending the Antiphlogiftic Theory; and annexed to the French Edition of this Work; by Meffrs. de Morveau, Lavoifier, de la Place, Monge, Berthollet, and de Fourcroy:

TRANSLATED INTO ENGLISH.

WITH

ADDITIONAL REMARKS AND REPLIES,

By the AUTHOR.

LONDON:

PRINTED FOR J. JOHNSON, N° 72, ST. PAUL'S CHURCH-YARD.

1789.

Figure 3

Further reading:

Dr Pickells: *Proceedings of the Royal Irish Academy*, **4**, 481–484, 1850.

Dictionary of Scientific Biography.

D. Thorburn Burns, Irish Contributions to European Analytical Chemistry, in D.M. Carroll, D.T. Burns, D.A. Brown and D.A. MacDaeid (eds), *Euroanalysis III, Reviews on Analytical Chemistry, Applied Science*, London, 1979.

Duncan Thorburn Burns, Department of Analytical Chemistry, The Queen's University of Belfast.

15 RICHARD LOVELL EDGEWORTH Inventor and Educationalist

Born: Bath, England, 1744.
Died: Edgeworthstown, Co. Longford, 17 June 1817.

Family:
Son of Richard Edgeworth and Jane Lovell.
Married: 1. Anna Maria Elers, daughter of Paul Elers of Oxford: 2. Honora Sneyd of Lichfield: 3. Elizabeth Sneyd (sister of Honora): 4. Frances Anne Beaufort, daughter of Daniel Augustus Beaufort, rector of Navan, Co. Meath, and sister of Francis Beauford. Fathered 22 children, the most famous of whom was Maria Edgeworth, the author and novelist.

Addresses:

1744–1773	England;	1776–1782	Northfield, Hertfordshire
1773–1776, 1782–1817	Edgeworthstown, Co. Longford		

Distinctions:
Founder member of the Lunar Society 1766; Original member of the Royal Irish Academy 1785; MP for St John's Town, Longford 1798; Member of the Parliamentary Commission on Irish Education 1806.

Not perhaps a genius, but certainly ingenious, Richard Lovell Edgeworth was the seventh son of a reasonably wealthy Anglo-Irish landowner, whose family had lived in Ireland since the sixteenth century. Born in Bath, where his mother had gone because of her delicate health, at the age of three Richard returned with his family to their 600-acre estate in Co. Longford.

He took to learning from an early age and entered Trinity College Dublin at the age of sixteen as a Fellow Commoner, being allowed to dine with the Fellows. Richard seems to have inherited some of his male antecedents' rather boisterous, independent thinking nature, which included contracting many marriages. Near the end of his first year in Trinity, in which he had been a far from serious student, his father moved him to Corpus Christi College, Oxford. Here Richard took his studies a little more seriously, but became romantically entangled with Anna Elers, daughter of a family friend, whom he married. He was a father before he was twenty: he soon realised that the marriage was a mistake, for his wife had no interest in the intellectual curiosities that were her husband's passion. Up to the time she died, bearing their fourth child in 1773, Richard seems to have spent much of his time away from home in the company of English scientific entrepreneurs. These included Thomas Beddoes, Matthew Boulton, Erasmus Darwin, John Smeaton, James Watt, and Josiah Wedgeworth. It was with some of these that Edgeworth helped to found the Lunar Society at Birmingham in 1766. This scientific discussion group was so named because it met when the Moon was full, as this helped its members to see their way home after their deliberations.

At this time the Industrial Revolution was in full swing, and much thought and effort was being put into the improvement of mechanical devices. The growth of industry had led to a movement of labour away from the land, and those connected with agriculture were much interested in labour-saving devices to maintain or improve efficiency on the farm. As a progressive landlord, Edgeworth did much in this area, inventing machines for turnip-cutting, loading machines, and land reclamation. He was particularly interested in improved methods of transport, inventing a velocipede (a large cylinder inside which for every step taken the wheel moved forward by six feet), a carriage with a sail to provide the driving force, improved springing for horse-drawn carriages, and a portable railway and carriages. In practice, some of his inventions ended in disasters, mainly because of the poor conditions of the roads. In time he turned his attention to road-building: not only did he carry out practical building in the Longford area, his book *An Essay on the Construction of Roads and Carriages*, published in 1813, predates any other book on modern road-building. He advocated a construction in which successive layers of stones of decreasing size was placed one on top of another. The top surface was sealed with a layer of sharp gravel, which, by inserting itself into the crevices in the lower layer, bound the roadway together. This method was superior to that used by John Macadam, who disagreed with this binding method. Paradoxically it is Macadam's name which lives on in connection with road-making, and not Edgeworth's.

The Edgeworth Family 1787 – pastel by Adam Buck (1745–1781)

It was Edgeworth who suggested to Thomas Romney Robinson, who married one of his daughters, the idea of the anemometer for measuring wind speed. Another of Edgeworth's inventions was a series of semaphore signalling towers that allowed messages to be quickly relayed over long distances. Originally conceived when Edgeworth was studying law in 1766, with a view to transmitting the results of horse races from Newmarket to London, he successfully put it into action thirty years later. Now the need for the fast relay of news arose from the fear of a French invasion on the west-coast of Ireland. Working with Francis Beaufort, a portable system was set up which relayed messages from Galway to Dublin in eight minutes. However, as time went on and the threat of invasion receded, the Government abandoned Edgeworth's novel invention.

Edgeworth was also interested in education, and espoused the ideas put forward in J.J. Rousseau's *Emile*. With his daughter Maria, he wrote the three-volume work, *Practical Education*, published in 1798. While much of the writing was done by Maria, many of the ideas were those Richard had been turning over in his mind for many years. It became the most important book on education produced in the eighteenth century and remained so well into the second half of the nineteenth. It was to be the first of a number of writings on education that were to continue until Edgeworth's death in 1817.

Further Reading:

Desmond Clarke: *The Ingenious Mr Edgeworth*, Oldbourne, London, 1965.
Maria Edgeworth (ed.): *R.L. Edgeworth, His Life and Times*, London, 1844.

William J. Davis, University Chemical Laboratory, Trinity College, Dublin.

16 WILLIAM HAMILTON Cleric and Geologist 1755–1797

Born: Derry, 16 December 1755.
Died: Sharon, Co. Londonderry, 2 March 1797.

From an original Profile Shade in the Possession of the Family.

Family:
Son of John Hamilton, merchant.
Married: his widow and nine children were provided for after his death by a grant from the House of Commons.

Addresses:
1771–1790 Trinity College Dublin
1790–1797 Clondevaddog, Co. Donegal

Distinctions:
Fellow of Trinity College Dublin 1779; Founder Member of the Royal Irish Academy.

Hamilton had a distinguished university career at Trinity College, Dublin; he graduated in 1776 and was elected to Fellowship three years later. He founded the Palaeosophers which amalgamated with the Neosophers to form the nucleus of the Royal Irish Academy. In 1777 Hamilton founded the Dublin University Museum which soon acquired fine mineralogical and geological collections.

In 1740 Susannah Drury was awarded a premium of £25 by the Dublin Society for her paintings of the Giant's Causeway. Engravings of these were widely distributed across Europe, at a time when the true nature of basalt was under debate – some felt basalt was precipitated from water while the Frenchman Nicholas Desmarest argued for the first time that it was the product of volcanic activity and, on the basis of the engravings, concluded that the Giant's Causeway was the product of volcanic activity. Hamilton's *Letters Concerning the Northern Coast of the County of Antrim* (1786; German edition 1787), in which he followed Desmarest's opinion, did much to advance the volcanic theory for the origin of basalt. The 1822 edition included a memoir on its author.

He also wrote on the principles and state of democracy in post-revolutionary France and its inferences for democracy in the British Isles (1792). In the *Transactions of the Royal Irish Academy* he examined experimental methods for determining the Earth's surface temperature (1788), and wrote on Ireland's climate (1794).

In 1790 he resigned his Fellowship and moved to the remote Donegal parish of Clondevaddog where he also served as the magistrate. Several years later he became the focus of ill-feeling; his rectory was attacked in February 1797 and he had to employ bodyguards. The following month, while visiting a Mr Waller, the house was attacked by a mob and Mrs Waller was killed. Terrified servants ejected Hamilton who was promptly killed. He is buried in Derry Cathedral.

Further reading:
Dictionary of National Biography.
Memoir in 1822 edition of *Letters Concerning the Northern Coast of the County of Antrim.*

Patrick N. Wyse Jackson, Department of Geology, Trinity College, Dublin.

Born: Ausburg, Bavaria, Germany, 6 April 1761.
Died: Dublin, 5 March 1833.

Courtesy of the Royal Dublin Society

Family:
Son of George Melchior Metzler and Sibylla Magadalena Götz. He was unmarried.

Addresses:

1780s–1794	Vienna
1794–1806	Frieberg, Constantinople, Italy, Copenhagen
1806–1813	Greenland
1813–1833	14 George's Place (now Hardwicke Place), Dublin

Distinctions:
Knight Commander of the Royal Danish Order of Danebrog 1816; Member of the Royal Irish Academy 1816.

Born Johann Georg Metzler, he assumed the name Karl Ludwig Giesecke (anglicised after 1813). Following divinity, law and mineralogy studies at Göttingen, he embarked on various careers, including those of musician, actor, diplomat, and mineralogist. He appeared in the first performance of Mozart's *The Magic Flute* in 1791. After further study at Frieberg in 1794, he served as an Austrian diplomat in Constantinople and later in Italy, but was invalided out of the service. He became a mineral collector and dealer in Copenhagen, from where he travelled to the Faeroe Islands and Greenland to assess their mineral wealth and trading potential. He spent seven years in Greenland between 1806 and 1813 where he assembled mineral collections and observed the traditions of the indigenous people, whose artefacts he collected. He discovered the new mineral species Arfvedsonite, Sapphirine, and Sodalite, and located the source of the rare mineral Cryolite. His first Greenland collection, worth over £5000, was captured as a prize-of-war while bound for Copenhagen, and sold to the Scottish collector Thomas Allen for £40. Undaunted Giesecke assembled another collection, later destroyed in Copenhagen in 1807. In 1813 Giesecke met Allan in Scotland who suggested he apply for the job as Professor of Mineralogy at the Dublin Society. Giesecke was appointed in January 1814 and began to learn English in order to deliver public lectures. He donated 415 Greenland mineral specimens to the Society's valuable collections, which included the Leskean collection acquired in 1792. Giesecke enlarged the holdings of Irish minerals through a series of collecting trips made around Ireland in the 1820s, and published a catalogue of the collection in 1832. Smaller collections of his specimens are in museums in Trinity College Dublin, Vienna, and Copenhagen.

Further reading:
Dictionary of National Biography: missing persons.
P.N. Wyse Jackson: Sir Charles Lewis Giesecke (1761–1833) and Greenland: a recently discovered mineral collection in Trinity College, Dublin, *Irish Journal of Earth Sciences*, **15**, 161–168 (and papers referenced therein), 1996.

Patrick N. Wyse Jackson, Department of Geology, Trinity College, Dublin.

18 WILLIAM HIGGINS Chemist

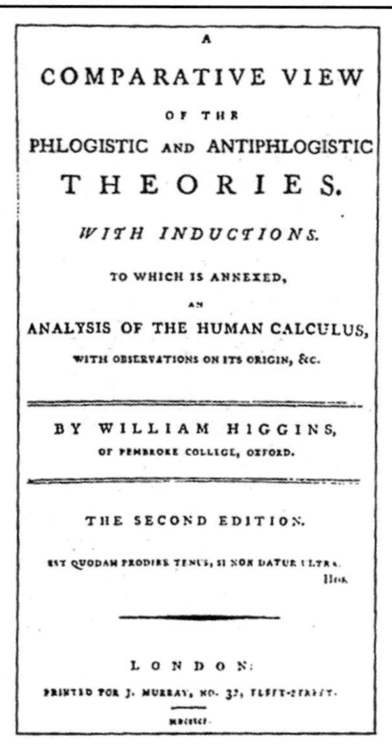

A

COMPARATIVE VIEW

OF THE

PHLOGISTIC AND ANTIPHLOGISTIC

THEORIES.

WITH INDUCTIONS.

TO WHICH IS ANNEXED,

AN

ANALYSIS OF THE HUMAN CALCULUS,

WITH OBSERVATIONS ON ITS ORIGIN, &c.

BY WILLIAM HIGGINS,

OF PEMBROKE COLLEGE, OXFORD.

THE SECOND EDITION.

EST QUODAM PRODIRE TENUS, SI NON DATUR ULTRA.

HOR.

LONDON:

PRINTED FOR J. MURRAY, NO. 32, FLEET-STREET.

[illegible]

Title page of Higgins
Comparative View
The unclear date reads
MDCCXCI–1791

Born: Colloney, Co. Sligo, 1763.
Died: Dublin 1825.

Family: Higgins' father, Thomas, and his grandfather, Bryan, were medical doctors, the traditional profession of the Higgins family since the seventeenth century. William Higgins never married.

Addresses:
6 Grafton Street, Dublin
3 College Green, Dublin
71 (now 75) Grafton Street, Dublin

Distinctions:
Member of the Royal Irish Academy 1794; Member of the Royal Society 1806; Founder Member of the Kirwanian Society 1812; Professor of Chemistry at the (Royal) Dublin Society 1791–1825.

Most school students know that the English meteorologist, John Dalton, was the first man to rediscover the atom and so created the climate in which chemistry flourished and grew into the subject we know today. But was he? An Irishman has a strong claim to the discovery which has always been attributed to Dalton. His name is William Higgins.

The real value of scientific hypothesis lies in the way it allows a body of observations to be organised into meaningful relationships with each other. It is only of secondary importance whether the theory is 'right' or 'wrong'; the further development of science decides this point. For the hundred years before Higgins the unifying principle in chemistry was the 'something' which made its appearance visible in the flame and was called Phlogiston. It was supposed to be released when combustion of a substance such as a metal takes place. Only when it was realised that the gain in weight which occurs when metals are burned is too important a fact to ignore, was the phlogiston idea challenged. Lavoisier's experiments, which showed that oxygen was absorbed whenever phlogiston was supposed to appear, slowly led to the demise of the theory. Higgins contributed largely to its downfall.

Higgins appears to have been apprenticed to his wealthy uncle Bryan Higgins (1737–1820), a medical doctor in London, who also ran a school of practical chemistry in Greek Street, Soho. It was there that William joined him in the early 1780s, before going up to Oxford in 1786.

He was assistant to the Professor there, but left without a degree in the middle of 1788. We can only speculate on the reason for his departure. He had abandoned the phlogiston theory even before he had gone up to Oxford and it may be that he found the established figures there too rigid for his innovative turn of mind. A short time after leaving he published, at the age of 26, the book on which his claim to fame is based. It was called *The Comparative View of the Phlogistic and Antiphlogistic Theories* (1789), and it was very critical of the writings of one of the diehards of the old theory, the distinguished Galway chemist, Richard Kirwan.

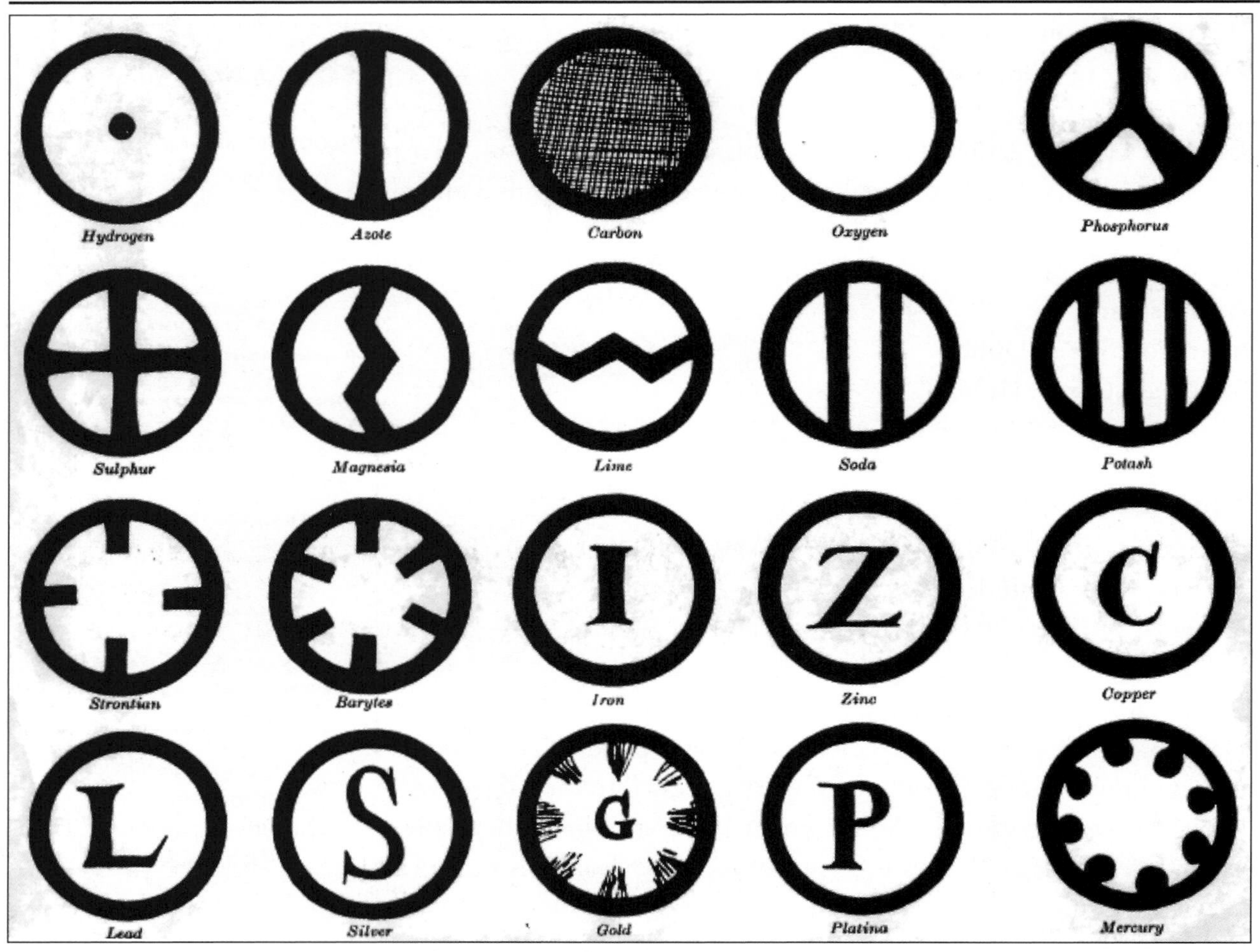

Dalton's symbols for the elements

The book was obviously a success. Five hundred copies were sold and a second edition was published in 1791. In that year Higgins was appointed Chemist at the newly established Apothecaries' Hall in Dublin and was soon occupying 'the small back room for the use of our Chemist' which was temporarily provided in Mary Street until the new Hall was completed. In 1795 he was appointed Professor of Chemistry at the Dublin Society's new laboratories in Hawkins Street. Here he was quick to organise a course on Experimental Chemistry.

Higgins is something of an enigma. His was a 'difficult' personality, for he was never far from controversy and strained relationships. Having proposed some very imaginative ideas, he did not push on with their development. His subsequent work concerned minor industrial applications of chemistry, such as his work on bleaching for the Irish Linen Board. It cannot be disputed that his *Comparative View* suggests the existence of atoms and the attractions between them, although it was Dalton who conceived the idea of individual atomic weights. After 1808, when Dalton published his *A New System of Chemical Philosophy,* much of Higgins' energy seems to have been directed towards claiming that 'he was first'. He died a disappointed man in 1825.

Further reading:

J.R. Partington and T.S. Wheeler: *William Higgins*, Pergamon, Oxford, 1960. This includes a reprinting of Higgins' *Comparative View* and his paper on Atomic Theory published in 1814.

William J. Davis, University Chemical Laboratory, Trinity College, Dublin.

Born: Ballynahowne, near Aughrim, Co. Galway, 21 March 1763.
Died: 12 July 1841.

Family:
Eldest son of James and Rosa MacNeven. A land-owning family that had held large estates in the North of Ireland, the MacNevens were moved to Galway under Cromwell's resettlement programme ('to Hell or Connaught').

Distinctions:
MD degree Vienna 1783; Member of the Royal Irish Academy 1785; admitted MD *(ad eundem)* Columbia College, New York 1807; Professor of Obstetrics in the New York College of Physicians and Surgeons 1808; Professor of Chemistry 1811; founder member of the New York Athenæum 1824.

Unable to have an education at home in Ireland because of the Penal Laws, William James MacNeven was sent, at the age of 11, to live in Prague with his paternal uncle, William O'Kelly MacNeven, who was Physician to the Empress Marie Theresa. After a sound classical education, MacNeven studied medicine at Prague and Vienna, where he obtained his MD degree in 1783. In the following year he started practice as a doctor in Dublin. Concerned to improve the lot of his co-religionists, MacNeven soon became involved in politics. He joined the Society of the United Irishmen in 1796, and was one of the Leinster Directory when he was arrested in March 1798 with the other leaders. After the executions began he was one of those who negotiated the 'treaty' which led to their cessation. Imprisoned, first in Kilmainham Jail in Dublin and later in Fort George in Scotland, in 1803 he was released, and he emigrated to New York in 1805.

His career in America was a most successful one, both as a medical doctor and as a chemist. His interests were principally centred round chemistry, especially the practical aspects of the subject. In 1811 he was appointed Professor of Chemistry at the New York College. He was to become an acknowledged expert in laboratory design and management, and many Americans then just beginning to set up laboratories sought his advice. A brilliant teacher, he was a powerful advocate of Dalton's new Atomic Theory and The Law of Definite Proportions. He published a textbook of chemistry, which was widely used. A 40-foot monument commemorating MacNeven's achievements and humanity stands in St Paul's churchyard in Lower Broadway.

Further Reading:
J.P. Murray: W.J. MacNeven: Father of American Chemistry, *Irish Medical Journal*, **79** (no. 9), 260–263, 1986.
William J. Davis: William James MacNeven: Chemist and United Irishman, in P.J. Wyse-Jackson (ed), *Science and Engineering in Ireland in 1798*, Royal Irish Academy, Dublin, 2000.

William J. Davis, University Chemical Laboratory, Trinity College, Dublin.

JOHN BRINKLEY Astronomer

1766–1835

Born: Woodbridge, Suffolk, December(?) 1766.
Died: Dublin, September 1835.

Rev. John Brinkley, F.R.S., enrobed as Bishop of Cloyne

Family:
Married: Esther Weld, daughter of Matthew Weld of Dublin, 1792.
Children: John (b.1793) and Matthew (b.1797).

Addresses:
1790–1827 Dunsink Observatory, Dublin
1827–1835 Bishop's Palace, Cloyne, Co. Cork

Distinctions:
Senior wrangler, mathematical tripos, Cambridge 1788; Fellow of Gonville and Caius College, Cambridge 1789–90; Andrews' Professor of Astronomy 1790–1827; Ordained Priest in the Church of England, Lincoln 1791; Created first Royal Astronomer of Ireland 1792; Fellow of the Royal Society, London 1803; Doctor of Divinity, University of Dublin 1806; Cunningham Medal Royal Irish Academy 1818; Copley Medal Royal Society 1822; President Royal Irish Academy 1822–35; Bishop of Cloyne, Co. Cork, 1826–35; President Royal Astronomical Society 1831–34.

Brinkley was chosen by Provost Hely Hutchinson to succeed Henry Ussher, the first Andrews' Professor of Astronomy in Dublin University, in 1790, on the advice of Nevil Maskelyne, Astronomer Royal of England. Although he maintained contact with the Royal Society and the Royal Astronomical Society, and has been the only Irish resident to serve as President of the Royal Astronomical Society, Brinkley did all his astronomical work at Dunsink Observatory.

One of the principal problems which Brinkley attacked was that of refraction in the atmosphere, which had been one of Henry Ussher's intended investigations. He measured the constants of aberration and nutation due to the motion of the earth, and he made it his life's work to try to determine the trigonometrical parallax of bright stars. This last part of his work was not successful, owing to instrumental problems, and it involved him in a celebrated controversy with John Pond, Maskelyne's successor at Greenwich Royal Observatory in London. Nevertheless, Brinkley was reckoned to have devised powerful procedures for careful treatment of observational material.

Brinkley made contributions to the mathematics of astronomy and of statistical methods, being one of the first to use least-squares methods to interpret observational readings in practical astronomy, and he did much to introduce 'continental' notation into calculus in Ireland and England.

At the same time as working as an astronomer, Brinkley became an authority in ecclesiastical law in the Church of Ireland. He was eventually created bishop of Cloyne in 1826.

Further reading:

R.S. Ball: *Great Astronomers,* Isbister, 233–46, 1895.

Patrick Wayman, 1927–1998.

JOHN TEMPLETON Botanist and Naturalist

Born: Belfast, 1766.
Died: Belfast, 15 December 1825.

Cranmore, Belfast, home of John Templeton

Family:
Married Katherine Johnston of Seymour Hill near Belfast (1799). One son, four daughters. His son, Robert Templeton, became a surgeon with the Royal Artillery and served in Ceylon and the Crimea.

Addresses:

1766–1794 Bridge Street, Belfast
The family's country residence was Orange Grove near Belfast
1794–1825 Orange Grove was renamed Cranmore by Templeton after his father's death

Distinctions:
Associate of the Linnean Society of London 1794; Vice-President of Belfast Literary Society 1803; first Honorary Member of Belfast Natural History Society 1821.

Accurate observations about the plants and animals, rocks and minerals found in the north of Ireland had seldom been made before John Templeton's time. However, as the study of natural history gained popularity from the mid-eighteenth century, it opened up new avenues of self-expression for many. In addition, the practical applications of scientific studies were encouraged by societies, such as the Dublin Society, as a means of improving agricultural practices, communications, livestock breeding and investigating useful manufacturing methods.

John Templeton was caught up in the mood of the era and, by the age of twenty, the secure family business enabled him to devote much time to the study of botany and to carry out research on the cultivation and improvement of plants. He laid out an experimental horticultural garden at Cranmore, and there successfully cultivated many foreign trees, shrubs and flowers, with many being grown from seeds sent him by other botanists. By 1800, it was said that his knowledge of natural history 'rivalled that of any individual in Europe'. Templeton continued to study, note and illustrate not only flowering plants, but also algae, fungi, lichens, mosses, geology, meteorology and many aspects of zoology, especially birds, fish and molluscs. His intended book on the natural history of Ireland was never completed, but the copiously illustrated manuscripts of his *Hibernian Flora* and *Hibernian Zoology* were an inspiration for others.

John Templeton was educated, until the age of sixteen, at the renowned school of David Manson in Belfast, before joining the family business. However, he did not enjoy good health and took up natural history studies. He had considerable literary taste and was closely involved with the liberal philosophy and intellectual progress then developing in Belfast. He was a founder of the Belfast Academical Institution, a collegiate school based on non-sectarian principles. Templeton's kind and generous nature led him to give freely assistance, encouragement and information to others.

Lough Neagh char drawn by John Templeton – it became extinct about 1825

On visits to London he met leading English botanists, including James Edward Smith, James Sowerby, Lewis Dillwyn, Dawson Turner and Robert Brown, who included his work in their publications. He became an Associate of the Linnean Society of London on 17 June 1794. Sir Joseph Banks reputedly made him an offer of a substantial salary and a grant of land to investigate Australian botany about 1800. Templeton, however, chose to remain in Ireland and Brown, who went to Australia instead, named a new genus of Australian legumes *Templetonia* after him.

Templeton's interest in cultivation of plants led him to investigate methods of weed control and the naturalisation of plants, which he published in 1802 in the *Transactions of the Royal Irish Academy*. In 1803, the Dublin Society published his description of a new species of rose (*Rosa hibernica*), the first of his numerous contributions of new species and new Irish records.

A significant part of Templeton's collection, which formed the basis of Belfast Natural History Society Museum, which opened in 1831, still survives in the Ulster Museum, Belfast, along with manuscript copies of the *Hibernian Flora*, *Hibernian Zoology* and his *Journal* dating from 1806–1825. Another small part of the Flora manuscript, with drawings of fungi and lichens, is in the Natural History Museum, London. He also donated material to the Dublin Society, mainly seeds for their Botanic Garden and birds for their Museum. Templeton's authoritative observations served as an inspiration for many Irish naturalists such as William Thompson.

Further reading:

T.D. Hincks: Memoir of the Late John Templeton Esq., *Magazine of Natural History* **1**, 403–406, 1828; **2**, 305–310, 1829 – a biography written within a few years of his death.

Helena Chesney, 35 Colenso Parade, Belfast, BT9 5AN.

FRANCIS BEAUFORT Hydrographer and Nautical Scientist

Born: Navan, Co. Meath 1774.
Died: Hove, near Brighton, England 1857.

Family:
Francis Beaufort was the second son of the Reverend Daniel Beaufort, Rector of Collon and Navan, and himself a distinguished topographer who was responsible for the best map of Ireland before the Ordnance Survey.
Married: 1. Alicia Magdalena Wilson 1812–1834. 2. Honora Edgeworth 1838–1858.
Children: Four sons and three daughters: his youngest daughter Emily travelled extensively in the Middle East where she founded organisations and hospitals to provide medical aid to the poor. His second son, Francis Lestock, became Attorney General of Bengal.

Addresses:
1774–1776 Flower Hill, Navan, Co. Meath
Collon, Co. Louth
Mountrath, Co. Laois
Portman Square, London

Distinctions:
Fellow of the Royal Society 1814; Fellow of the Royal Astronomical Society 1829; Member of the Royal Irish Academy 1832; DCL Oxford 1839; Knight Commander of the Bath 1848.

It comes as something of a surprise to many people to learn that the British Navy's greatest hydrographer and map-maker was an Irishman. Francis Beaufort is best remembered as the originator of the table which classifies the velocity and force of winds at sea – the Beaufort Scale. Beaufort also originated a system whereby the weather's various states, both afloat and ashore, are indicated by letters of the alphabet. Both tables remain in general use among British and American meteorologists, while the Beaufort Scale has achieved world wide significance.

Equally if not more important was his life-long devotion to the production of accurate and reliable navigational charts – a direct result of being ship-wrecked at the age of 15 through lack of suitable maps. Owing to his efforts the dangers of uncharted shores, which all seamen faced, were greatly diminished.

At the age of 13 Beaufort began his nautical career with the British Navy and later served in the Napoleonic wars. He spent five months in 1788 studying astronomy and meteorology under Rev. Henry Ussher, the first Andrews Professor of Astronomy of the University of Dublin and the first Director of Dunsink Observatory. While recovering from severe wounds in battle with the Spanish (1803), he helped his father-in-law Richard Lovell Edgeworth, the Irish inventor and father of novelist Maria Edgeworth, establish a semaphore system which proved capable of transmitting messages from Dublin to Galway in eight minutes. After seeing more active service, Beaufort concluded his sea-going career with the rank of Captain in 1812.

For many years he was engaged in constructing charts based on coastal surveys he had made while at sea, and in 1829 he was appointed Hydrographer of the Admiralty, a post he held for 26 years. During this time he set himself the mammoth task of planning detailed surveys of all uncharted

The Arctic Council meeting in 1851 to discuss the search for Sir John Franklin, lost in the North West Passage – Sir Francis Beaufort seated centre (painting by Stephen Pearce – courtesy National Portrait Gallery, London)

coasts, both at home and abroad. As a result, the Hydrographic Office, previously no more than a map room, became a highly respected scientific department producing charts noted for their accuracy and detail. In 1855 Beaufort, by then a Rear Admiral, retired from service, just two years before his death.

He was author of *Karamania or a Brief Description of the South Coast of Asia Minor* (1817) which was the chief book of travels in its day, and was for many years engaged in preparing the extensive *Atlas* published by the Society for the Diffusion of Knowledge.

The Beaufort Sea, near the current Alaskan oil fields, is named after Sir Francis Beaufort. The Beaufort prize, awarded annually to the best student of navigation at the Royal Navy College, Dartmouth, is a fitting memorial to the British Navy's greatest scientist, and to one who made the seas safer for all by providing a unique and invaluable set of navigational maps.

Further Reading:

Alfred Friendly: *Beaufort of the Admiralty,* Hutchinson, London, 1977.

Sheila Landy, Science Library, The Queen's University of Belfast.

Born: Ballycommon, King's County (now Leix) 1774.
Died: 5 April 1830.

Family:
Married La Comtess de Ronault

Addresses:
Ireland, England and France

Distinctions:
Fellow of the Royal Society 1801; awarded its Copley Medal 1803; Member of the Royal Irish Academy.

Chenevix's family were Huguenots who came to Ireland after their explusion from France in 1685. Richard attended the University of Glasgow and probably came under the influence of another Irishman, the professor of chemistry, Joseph Black. He soon developed skills as an analytical chemist. He was attracted back to the country of his origins by the new and exciting chemistry being developed there by French chemists led by Antoine Lavoisier. He wrote an important paper on the new chemical nomenclature. He was not put off by the political disturbances then taking place in France, which led to Lavoisier's execution by the guillotine. He was to remain in France for most of the rest of his life.

In 1801 he published his first paper on the analysis of an ore of the metal lead. His prowess as an analyst was recognised by the award of the Copley Medal from the Royal Society two years later. By this time his publications had risen to over a dozen papers. About this time the discovery of a new element named Palladium or 'New Silver' was announced in an anonymous handbill which was circulated in London. The metal was offered for sale at one shilling per grain. Feeling that the announcement was a fraud, Chenevix purchased the entire stock (332 grains, about 21.5 grammes) for 15 guineas and began an investigation to prove that the metal was not an element but an alloy of Platinum and Mercury. He sent his views to the Royal Society, where William Wollaston, who, unknown to everyone, was the actual discoverer of the new element, was a Junior Secretary. It was the entrepreneurial Wollaston who, in fact, had discovered the new element in the process of producing very pure Platinum, a process he wished to keep secret for commercial reasons. He failed to alert Chenevix of this fact and made no comment when the paper was read before the Society and subsequently published. At this time, immediately following John Dalton's publication of his Atomic Theory, there was much debate about how exactly an element should be defined. It was finally established that Chenevix's claim that it could be synthesised by combining Platinum with another metal was erroneous. However not too much damage appears to have been done to his reputation. He had focused attention on an important aspect in the development of chemical science. Wollaston, who made £26,000 from his sales of platinum, and Chenevix even remained friends.

Further Reading:

Dictionary of Scientific Biography.
I.E. Cottingham: Palladium or 'New Silver' – no stranger to scientific controversy, *Platinum Metals Review*, **35,** 141–151, 1991.
M.E. Usselman: The Wollaston/Chenevix Controversy over the Elemental Nature of Palladium, *Annals of Science*, **35,** 551–579, 1978. This is probably the best account of the controversy.

William J. Davis, University Chemical Laboratory, Trinity College, Dublin.

Born: Dublin, 7 June 1777.
Died: Dublin, 10 June 1858.

Bust in Royal College of Surgeons in Ireland

Family:
Son of Anne (née Verner) and John Crampton, surgeon-dentist.
Married: Selina Hamilton Cannon 12 May 1802.
Children: two sons and four daughters. The elder son, John Fiennes, became British ambassador to Russia.

Addresses:

1777–16	William Street, Dublin
1813–24	Dawson Street
About 1813	14 Merrion Square North
	Lough Bray House, Co. Wicklow

Distinctions:
MD Glasgow 1800; Member of the Royal College of Surgeons in Ireland 1801; President, Royal College of Surgeons 1811, 1820, 1844 and 1855; Surgeon-General 1813; Fellow of the Royal Society; A founder and later president of the Royal Zoological Society; Created a baronet 1839.

'Who is he?': Sir Philip Crampton's memorial fountain bust in Brunswick Street stirred Leopold Bloom's curiosity. The answer (not supplied in *Ulysses*) is that Crampton was a leading Dublin surgeon. The memorial was removed some years ago and today the street where it stood is called Pearse Street.

At fourteen, Philip Crampton was apprenticed to a surgeon, Solomon Richards of York Street, in due course studying at the College of Surgeons (then in Mercer Street) and the Meath Hospital. On 25 September 1798 he took the 'letters testimonial' of the College, having held a brief army attachment under the command of Sir John Moore.

Crampton was musical. He liked to play duets with a friend, Theobald Wolfe Tone, and was so engaged when a message reached the young revolutionary that the authorities had been informed of his membership of the United Irishmen. One of the founders of the College of Surgeons, Surgeon William Dease, who died by his own hand in 1798, was rumoured to have feared arrest for a similar proscribed association. Be that as it may, the tragic cloud's silver lining shone down on the newly-qualified surgeon, who was appointed to the vacancy created by Dease's death. He was to hold the appointment as surgeon to the Meath Hospital for almost sixty years.

He was elected MRCSI (equivalent to the present college fellowship) in 1801 and a few months later became a member of the Court of Assistants. In 1802 he married Selina Hamilton Cannon, daughter of an army officer. She was a noted beauty but was destined to sustain extensive burns, which were disfiguring and fatal. The Cramptons' Dawson Street residence was equipped with a dissecting-room and lecture theatre. It stood opposite the Richmond Tavern and the story persists that he earned an over-night success by boldly operating to dislodge a lump of meat from a suffocating waiter's windpipe. It is far more likely that he attained his reputation gradually and less dramatically, for he was described as being 'sagacious in diagnosis, ready in resources, dextrous in the use of instruments and sympathetic in his treatment of patients'. He was also appointed surgeon to the Westmoreland Lock Hospital for Venereal Diseases in 1806.

As a teacher, Crampton made no attempt to conceal from his pupils the realities they faced. 'Now, I am sorry to say', he addressed them at the opening of the academic year, 'that the path of clinical observation is neither short nor pleasant; in truth nothing but a full and entire conviction that it is the only path (not the best), but the only path which leads to professional success, could ever induce the medical student to pursue it.'

Meath Hospital and County Infirmary, 1822

Having become Surgeon-General in 1813, he attended a levee in Dublin Castle wearing an impressive dress uniform. Someone asked who was this striking figure. 'He's the surgeon-general', was the reply. Whereupon a wit said, 'I suppose that's the general of the Lancers?'

A well-read man, Crampton was also devoted to outdoor pursuits, especially fox-hunting. Erinensis, the satiric Irish correspondent of *The Lancet,* inferred that Crampton ('the Nimrod of the County Dublin') refused an academic opportunity in the College of Surgeons, preferring 'the saddle to the professor's chair'. Notwithstanding, he was elected to the presidency of the College four times between 1811 and 1855. He was created a baronet by Queen Victoria in 1839.

As well as the house in Merrion Square, the Cramptons owned a country house near Bray in County Wicklow, from where, even when elderly, after an early morning swim he could ride into the city and amputate a limb before breakfast.

His publications include an essay on entropion (inversion of the eyelid) and a description of an avian muscle, which was named *Musculus cramptonius* in his honour. These and other meritorious articles enhanced his reputation but, sparing of praise, Erinensis places them short of the line of partition between talent and genius. Other judges, nevertheless, deemed his anatomical discovery sufficiently important to warrant his election to fellowship of the Royal Society. *Musculus cramptonius,* a tiny muscle in the eyes of birds, arises fiom a bony hoop which surrounds the cornea and facilitates fine adjustments of avian vision. He was the first Dublin surgeon to perform lithotrity, a method of removing bladder stones by crushing them with a specially designed instrument. He improved the operation then current for correcting cleft palate.

Sir Philip Crampton died at his Merrion Square residence on 10 June 1858. His directions were obeyed when he was buried in Mount Jerome Cemetery – his remains were encased in Roman cement. A few years later the memorial drinking-fountain referred to in *Ulysses* was erected by his friends and admirers as a symbol of health and usefulness.

Further reading:

Sir Charles Cameron: *History of the Royal College of Surgeons in Ireland,* Dublin, 1886.
Jessie Dobson: *Anatomical Eponyms,* Second Edition, Edinburgh, 1962.
Martin Fallon: *The Sketches of Erinensis,* London, 1979.

J. B. Lyons, Department of the History of Medicine, Royal College of Surgeons in Ireland, Dublin.

25 AENEAS COFFEY Excise Man, Distiller and Inventor 1780–1852

Born: Dublin (or perhaps Calais, France),1780.
Died: London, 30 November 1852

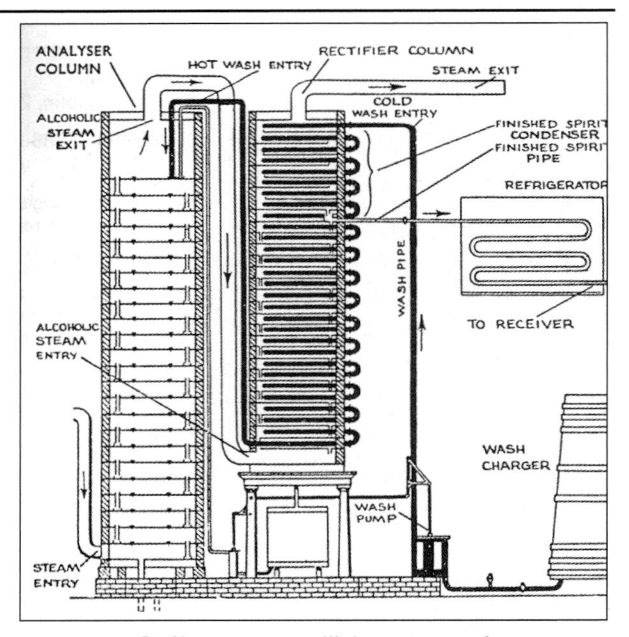

Coffey patent still (courtesy of Oxford University Press)

Family:
Aeneas Coffey was the son of Andrew Coffey, City Engineer employed by Dublin City Waterworks. He was brought up in France.
Married: Susanna Logie in 1808.
Children: 3 sons, Aeneas, William and Andrew.

Addresses:

1828	Distillery Office and Stores, 27 South King Street, Dublin
1834	Dock Distillery, Grand Canal Street, Dublin;
1837–1838	Patent Still Manufacturer, Barrow Street, Dublin;
After 1835	St Leonard's Street, Bromley, London

It occasionally happens that important discoveries are made, not as a result of deliberate, purposeful, scientific investigation, but as a 'once-off' invention, arising out of the inventor's long practical experience. Such was the case with Aeneas Coffey's Patent Still.

Born of Irish parents, Aeneas Coffey entered the Irish Excise Department in 1800. By 1809 he was Surveyor for Dublin and was transferred to Donegal in 1810 where he took charge of operations against illicit distilling. This was dangerous work and once, on the Inishowen Peninsula, he was 'beaten until he was supposed to be dead'. He became an expert on spirit distilling and was appointed Inspector General of Excise in 1818. Ten years later he retired from the Excise Service and set up his own distillery in Dublin. This continued until 1856, although long before that Coffey had moved with most of his business to London.

Until Coffey's invention, whiskey was produced by distillation from simple pot stills. This batch process was very wasteful of fuel and a very strong spirit could be obtained only by repeated distillation. By allowing the outgoing hot vapour to come into contact with a vessel containing the incoming liquid, much of its heat was exchanged before it reached the final water-cooled condenser. This method required less fuel for heating the wash and the process was capable of being used continuously rather than in the old-fashioned batch way. It had some features of the modern fractionating column and could produce a spirit containing as high as 95% alcohol with few other volatile components. Ironically it was not a success in the beverage industry, because of its efficiency in eliminating those substances which contribute most to the flavour of malt whiskey. It became the standard method for the production of industrial spirit. Coffey's invention was a significant advance in being the first heat-exchanger, an important energy saving device which still is a crucial factor in determining whether a chemical plant is to be economically viable.

Further reading:

E.J. Rothery: Aeneas Coffey, *Annals of Science*, **23**, 53–71, 1968.
J.J. Keir: Aeneas Coffey and his Patent Still, *Dublin Historical Record*, **9**, 29–36, 1946–1947.

William J. Davis, University Chemical Laboratory, Trinity College, Dublin.

Born: Dublin, January 1780.
Died: London, 13 May 1861.

Family:
Son of Nicholas Fitton.
Married: Miss James (1820).
Children: Five sons and three daughters.

Distinctions:
Fellow of the Royal Society 1815; Fellow of the Geological Society of London 1816; President of the Geological Society of London 1827–1829, Wollaston Medal 1852.

Portrait from H.B. Woodward's History of the Geological Society *(1908)*

A graduate (1799) of Dublin University, Fitton became an Edinburgh-trained medical practitioner, married 'a most amiable lady of ample fortune' and ended his days as one of Britain's foremost geologists.

Fitton's geological achievements were primarily in the field of Mesozoic stratigraphy, but he also was an important geological communicator and the first British historian of geology. He assisted the Rev. Walter Stephens (1772?–1808) with his pioneering geological explorations southwards of the Irish metropolis and edited Stephens's findings for posthumous publication as *Notes on the Mineralogy of Part of the Vicinity of Dublin* (1812). Fitton's own first geological paper (1811) was similarly titled. In Edinburgh (1808–1809), he attended Robert Jameson's famous geological lectures. He regularly contributed to the *Edinburgh Review* (1817–1841), becoming an influential commentator on geological issues of the day – his support (1841) for the contentious Devonian System, for instance, hastening its acceptance by the geological community. Fitton's own geological reputation rests on his decipherment of the geological formations which throughout southeastern England and northwestern France underlie that most distinctive of all geological 'formations' – the Cretaceous Chalk. The division of the Greensand by the Gault Clay into the Lower Greensand and the Upper Greensand was established by Fitton's painstaking research (1824–1836). This geological triad is now an important geological marker for British and Irish geologists engaged in offshore hydrocarbon exploration.

As President of the Geological Society of London, Fitton initiated the Society's annual Presidential Address and he inaugurated publication of the Society's *Proceedings*. This set an example which was followed first by the Royal Society and later by other London scientific societies.

Further Reading:

Dictionary of National Biography.
G.L. Herries Davies: *Sheets of Many Colours*, Royal Dublin Society, Dublin, 1983.
P.N. Wyse Jackson: William Henry Fitton (1780–1861) and the Wollaston Medal of 1852, *Geoscientist*, **8**, No. 10 p. 9, 1998.

Jean Archer, Nenagh, Co. Tipperary.

Sir Richard Griffith in old age holding a small version of his great geological map of Ireland

Born: Dublin, 20 September 1784.
Died: Dublin, 22 September 1878.

Family:
Married: Maria Jane Waldie (1786–1865) at Kelso in Scotland on 21 September 1812. Children: One son and three daughters constitute his officially recorded family, but there is reason to believe that there may have been another daughter, the eldest member of the family. This girl – Jane – would have been born in 1813, but at the age of sixteen she is said to have eloped with one of the family's servants – one Robert Cook – and the pair are reported to have fled to North America. Jane, it seems, was cut off from her inheritance and never again mentioned within the family. Two of the three other daughters died young. The son, George Richard Griffith, took the name Waldie-Griffith in 1865 when he inherited the Scottish estate which had belonged to his mother's family.

Addresses:
His father's estate lay at Millicent, near Clane in County Kildare, but Griffith was born at 8 Hume Street in Dublin. While working for the Bog Commissioners and for the Royal Dublin Society he took up quarters in various towns such as Ballinasloe, Castlecomer, Portarlington and Robertstown. From 1822 until 1828 he lived at Ballyellis House near Mallow in County Cork. In 1828 he returned to Dublin to live for the next fifty years at 2 Fitzwilliam Place. It was there that he died and he is buried in Dublin's Mount Jerome Cemetery.

Distinctions:
Fellow of the Royal Society of Edinburgh 1807; Honorary Member of the Geological Society of London 1808; Member of the Royal Irish Academy 1819; Honorary LLD, University of Dublin 1849; Wollaston Medal of the Geological Society of London 1854; Baronet 1858; Honorary MAI University of Dublin 1862.

Griffith had a career remarkable in its variety. At the age of fifteen he became an officer in the Royal Irish Regiment of Artillery, but he soon left the army to study engineering and geology. Between 1809 and 1813 he was an engineer with the Irish Bog Commissioners; from 1812 until 1829 he held the post of Mining Engineer to the Royal Dublin Society; and for many years after 1822 he was responsible for a programme of road and bridge building in Munster. In 1825 he became Director of the Boundary Survey; from 1830 until 1868 he was Commissioner of the General Survey and Valuation of Rateable Property in which capacity he was responsible for that valuation known throughout Ireland to this day as 'The Griffith Valuation'; and from 1846 until 1864 he was Deputy Chairman and subsequently Chairman of the Board of Works.

It is nevertheless as a geologist that Griffith is chiefly remembered among scientists and he has often been hailed as 'the father of Irish geology'. He began to study the geology of Ireland in 1809 when the Royal Dublin Society commissioned him to survey the rocks of the Leinster coalfield, and as

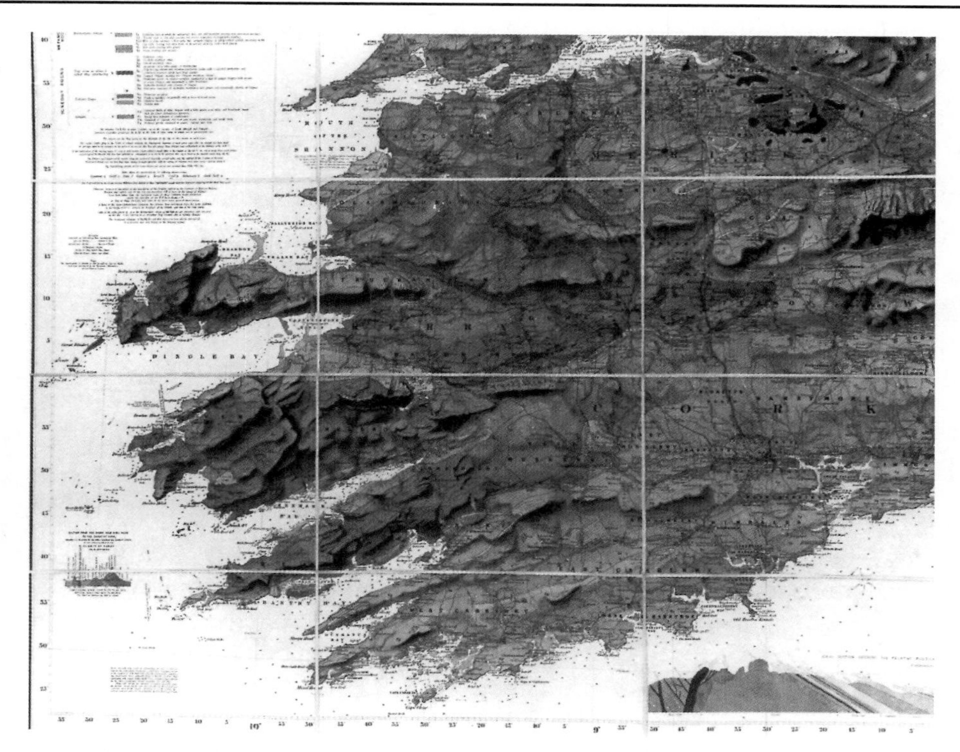

The southwestern sheet of Griffith's geological map of Ireland in its final form of 1855. The scale of the original is a quarter of an inch to one mile (1 : 253,440)

the society's Mining Engineer he later surveyed the Connaught coalfield (1814–1816), the Ulster coalfields (1816–1818), and the Munster coal district (1818–1824). Gradually, through his own field observations and through those of Patrick Ganly, one of the staff of the Valuation Office, Griffith built up a comprehensive picture of the geology of Ireland. This picture became the basis of the fine quarter-inch geological map of Ireland which Griffith published on 22 May 1839. The map, in six sheets, has an area of 147.5 x 182 cm and the geology is represented by a multitude of colour washes that were applied by hand. The colouring of just one copy of the map must have occupied an artist for many days, and now the map is a sought-after collector's item. After 1839 Griffith constantly revised his map. He personally showed one revised version of the map to Queen Victoria and Prince Albert when they visited the Great Dublin Industrial Exhibition on 31 August 1853 and two years later the map attained its final form when a new edition was prepared for display at the 1855 Paris Exposition Universelle. In its edition of 1855 Griffith's map remains to this day the finest small-scale geological map of Ireland ever produced.

Further reading:

Griffith's *Autobiography* dictated 25 August 1869. The original is in private hands but copies are in the National Library of Ireland and in the Public Record Office, Dublin.

Gordon Herries Davies and Charles Mollan (eds): *Richard Griffith 1784–1878,* Royal Dublin Society, Historical Studies in Irish Science and Technology, **1**, Dublin, 1980.

Gordon Herries Davies: *Sheets of Many Colours: the Mapping of Ireland's Rocks 1750–1890,* Royal Dublin Society Historical Studies in Irish Science and Technology, **4**, Dublin, 1983.

Gordon L. Herries Davies, Department of Geography, Trinity College, Dublin.

Born: Ballylickey House, Co. Cork, 17 March 1785.
Died: Ardnagashel, Co. Cork, 9 February 1815.

Ellen Hutchins – silhouette composed by the author (HC) from one of her father, Thomas Hutchins

Family:
Daughter of Thomas and Elinor Hutchins of Ballylickey, Bantry, Co. Cork, whose income came mainly from farming and fishing. A successful law-suit to recover land enabled Thomas to educate his family, several of whom subsequently entered the legal profession. Ellen was unmarried.

Addresses:
1785–1813 Ballylickey, Co. Cork, the family residence.

In 1813 Ellen and her mother went to Bandon in Co. Cork to receive medical attention. Elinor Hutchins died there and Ellen returned, not to Ballylickey, but to her brother Arthur's home at Ardnagashel.

In the early 1800s Ellen Hutchins' botanical contributions made her well known to a group of early naturalists and scientists, not just in Ireland, but in Britain and Europe, in an age when women were becoming accepted into scientific circles. A number of women who were unaffected by fashion and who possessed single-minded dedication pioneered female involvement in Irish natural history study. Ellen's always delicate health encouraged her to follow open-air pursuits and Whitley Stokes, a doctor, family friend and Professor of the Practice of Medicine at Trinity College, Dublin, introduced her to botany as a suitable passtime. One of his interests was the study of mosses which he also encouraged her to study, along with other non-flowering, or cryptogamic plants. In Stokes she was to have a mentor who stimulated, guided and encouraged her by introducing her to other botanists such as James Townsend Mackay, subsequently Curator of the Botanic Gardens at Trinity College. Through Mackay, she got to know English botanists such as William Jackson Hooker and Dawson Turner.

Ardnagashel in mid nineteenth century by Laura Nightengale (née Hutchins)

Her knowledge of good collecting places in the countryside and seashores around Bantry made its flora well-known to botanists, including Hooker. Ellen's records of rare, previously unrecorded species, or species new to science, led Hooker to claim '...that she finds everything'. Among the important publications to which Ellen contributed were Dillwyn's *Confervae*, published between 1802 and 1809, and Turner's *Fuci* ...., published between 1807 and 1819. Two *Hutchinsia* genera were named after her, though both are now defunct – *Hutchinsia* C. Agardh, a synonym of the seaweed *Griffithsia*, and

Hutchinsia Robert Brown (published in William Aiton's *Hortus Kewensis*), a cruciferous flowering plant. Hooker described the moss *Jungermannia hutchinsiae*, and the seaweed names *Dasya hutchinsiae* and *Cladophora hutchinsiae* are still in use to preserve her name for posterity.

Oak tree in the area known as 'Ellen's Garden' at Ballylickey, Bantry

Ellen also used her time on local beaches to make a collection of seashells, which contained the eight species of shells that she was the first to record for Ireland. She gave this collection, containing some 120 species, to her cousin, Dr Thomas Taylor, who donated it to the Dublin Society in 1814.

Like most young women of the time, she was taught to draw and paint. She became a talented artist, producing beautiful drawings of her botanical specimens, some of which were included in subsequent publications. Her valuable botanical collection of dried specimens and some 200 of her watercolour paintings were bequeathed to Dawson Turner and eventually to the Royal Botanic Gardens at Kew, London. However, duplicate specimens are to be found in herbaria in Ireland and elsewhere.

William Harvey, the eminent Irish botanist, writing in his beautifully illustrated volumes of *Phycologia Britannica* (1846–1851) about *Cladophora hutchinsiae*, stated:

...a very beautiful and strong growing species .... discovered in the year 1808 by the late Miss Hutchins of Ballylickey near Bantry, whose explorations of her neighbourhood were as unremitted as they were successful and whose name is deservedly held in grateful rememberance by botanists in all parts of the world. To her the botany of Ireland is under many obligations, particularly the Cryptogamic branch, in which field, until her time little explored, she was particularly fortunate in detecting new and beautiful objects, several of which remain the rarest species to the present day.

This is a fitting tribute to the short life of this remarkable young woman. Ellen Hutchins' all too brief life ended in her thirtieth year, but her botanical achievements ensure that her name lives on in the botanical world today.

Further reading:

H.C.G. Chesney: The young lady of the lichens, in *Stars, Shells and Bluebells*, Women in Technology and Science, Dublin, 1997.

M.E. Mitchell: Early observations on the flora of southwest Ireland. Selected letters of Ellen Hutchins and Dawson Turner, 1807–1814, *Occasional Papers, National Botanic Gardens, Dublin*, **12**, 1-124, 1999.

Helena Chesney, 35 Colenso Parade, Belfast BT9 5AN.
Michael Guiry, Department of Botany, National University of Ireland, Galway.

Born: Culnady, Co. Derry, 1788.
Died: Glasnevin, Dublin, 8 December 1871.

Family:
Married: H. Sharrock in 1809.
Children: Two sons. John Fisher Murray, became a barrister-at-law; Edward, joined his father to manufacture fluid magnesia in 1869.

Addresses:

1788–1804	Derry
1804–1807	Edinburgh
1807–1808	Dublin
1808–1831	Belfast
1831–1871	19 Upper Temple Street, Dublin

Distinctions:
Resident physician to the Lord Lieutenants of Ireland 1831–1841; Knighted 1831; Inspector of Anatomy for Ireland 1831–1871.

James Murray completed his medical studies in Edinburgh and Dublin in 1808 and practised in Belfast until 1831. To alleviate stomach pains in his patients he dissolved magnesium carbonate in water through which carbon dioxide had been bubbled. He then set up a commercial plant in Belfast to manufacture this mild palatable medicine which he called 'Milk of Magnesia' and patented the process. He shrewdly perceived that the waste heaps of silicates, bicarbonates of soda and potash at his chemical works could be used in conjunction with bones and sulphuric acid to manufacture a new fertiliser. This produced excellent grazing for cows when applied to the Point Fields, Belfast, in 1817. In 1831 Murray became resident physician to the Lord Lieutenant and Inspector of Anatomy for Ireland. He continued his interest in fertilisers and set up a company to market them, with offices in 79 Dame Street, Dublin. In 1842 he patented the manufacture of super-phosphate and published *Trials and Effects of Chemical Fertilisers with Various Experiments in Agriculture* in 1843. His business was not a prosperous concern and he sold his patent to Lawes who became the great industrialist of the super-phosphate industry.

Murray continued to strive to reduce illness. In 1849 he published *Electricity as a Cause of Cholera* and *The Relation of Galvanism to the Action of Remedies*, proposing a layer of non-conducting material underneath ground floors of houses to prevent cholera. This was translated into Italian. He also published *Heat and Humidity* and *Medical Effects of Atmospheric Pressure*. These works were very innovative at that time and he received many tributes in obituaries from the National Press.

Further reading:
W.A.L. Alford and J.W. Parkes: Sir James Murray, MD, a Pioneer in the Making of Superphosphate, *Chemistry and Industry*, 15 August 1953, pp. 852–5.
W. Garvin and D. O'Rawe: *Northern Ireland Scientists and Inventors,* Blackstaff Press, Belfast, 1993.

Wilbert Garvin and Des O'Rawe, The Queen's University of Belfast.

Trade card for Yeates & Son 1834–39 (reproduced from the Trade Card Collection in the Science Museum Library, courtesy of the Science Museum, London)

The Dublin-based Yeates family business of scientific instrument makers survived for well over a century. Although members of the Yeates family appeared in street directories from 1769, variously described as weavers, drapers and shoemakers, later there were brass founders, cutlers and surgical instrument makers. Samuel Yeates, the first to be designated 'optician', when the word meant 'instrument maker' rather than 'spectacle maker', first appeared in the directory for 1790. He moved the shop to 2 Grafton Street in 1827, and after several changes of name – both George Yeates and Stephen M. Yeates were connected with the businesss on these premises – the firm became Yeates & Son in 1865. This name still survives in neighbouring Grafton Arcade, but the business there has had no connection with the Yeates family since the 1940s. Other members of the family opened their own scientific instrument businesses for short periods in Dublin during the nineteenth century, amongst them Kendrick, William, Thomas and Horatio.

Yeates & Son made and sold a wide range of scientific instruments, and many still survive to-day. Much of their material was aimed at the growing mid-nineteenth-century market for educational material: with an increased demand for scientific apparatus in schools and colleges, there was scope for supplying reasonably priced equipment locally. From 1852 until 1910 Yeates & Son advertised as 'Instrument Makers to the University', which implies that Trinity College was a good customer. (Instruments with the firm's signature survive in the Physical Laboratory and the School of Engineering. Other institutions with either a teaching or demonstration function at this time are finding that Yeates instruments have survived until today.)

Various members of the Yeates family were of an inventive frame of mind. One of Samuel's sons, George, published a paper containing his ideas on minor improvements to the theodolite in 1845; another, Andrew, wrote about the bearings of transit instruments. Andrew went to England in 1821 to work with the pre-eminent instrument-maker Edward Troughton, then installing new apparatus at Greenwich Observatory. He eventually ran his own business at 12 Brighton Place, New Kent Road, London, between 1837 and 1873. His 'improved portable theodolite' was, however, offered for sale in the Dublin firm's 1887 catalogue, so that he must have remained on good terms with the family.

His brother George, back in Dublin, also produced papers on meteorology, while his son Stephen Mitchell Yeates, who seems to have run Yeates & Son from 1865 until his death in 1901, published a work on the barometer. Stephen became involved in some of the earliest experiments with the telephone in Ireland. Almost ten years before Alexander Graham Bell produced his system, a German, Phillip Reis, had produced a form of transmitter, and in 1865 Stephen Yeates demonstrated his modified version of this before the Dublin Philosophical Society 'when both singing and the distinct articulation of several words were heard through it, and the differences between the speakers' voices clearly heard'.

Despite a little knowledge about some of the members of the family who worked in the firm, there are a number of unanswered questions about the structure of the business – just how much they made

S.M. Yeates' improved form of G.J. Stoney's local heliostat, by Yeates & Son, Dublin c. 1880, (courtesy of the National Museums of Scotland)

themselves, how much they bought in and retailed, whether they bought in parts and then assembled them, how many people were employed, and whether they made comfortable profits or found that competition with English firms was very fierce? It would appear that in the early part of the nineteenth century much of what they sold was made by them. This is borne out in mid century by their success at trade exhibitions, where exhibitors could only show what they had actually manufactured. At the Great Exhibition of 1851 in London, Yeates & Son obtained an Honourable Mention for their surveying instruments. In 1862, at the London International Exhibition, they exhibited a large public barometer with a three foot diameter dial, designed by them in 1858 for use in agricultural areas and fishing villages where they claimed it was found 'extremely useful'. Besides exhibiting in London, they also had stands at the Cork Exhibition in 1852 and the Dublin Exhibition in 1853.

By 1887 they were able to proclaim proudly: 'Six first-class silver medals and other prizes have been awarded to Yeates & Son by the councils of the various exhibitions, both at home and abroad, for the excellence of workmanship and construction of the instruments exhibited by them. Silver medal, Inventions Exhibition, London, 1885'.

In the 1880s Yeates & Son produced a series of four trade catalogues from which some information may be gleaned. Firstly, the range of material covered was extensive: electrical apparatus, optical instruments, heat apparatus and drawing instruments, surveying and general engineering instruments. Secondly, they admitted to very few items not being of their own manufacture, yet some, such as Gramme's dynamo magnetic machine, were clearly imported. Thirdly, it is impressive that so many pieces were described as 'S.M. Yeates improved' indicating that the firm was able to produce its own version of those items. Fifty years later an observer recalled: 'Some who, like . . . Yeates, manufactured a number of various instruments have found it impracticable to do so today. Yeates made the last big attempt about 1890, but they were forced to market most of their products through the medium of English firms and they lost money. The home market is too small and because of the advance of science the variety is too great. A world market is needed, and this requires not merely highly educated scientific brains, but also large organising ability and great capital resources'.

Apparatus by Yeates & Son may be seen in St Patrick's College, Maynooth; Trinity College Dublin; and University College Dublin.

Further reading:

T. H. Mason: Dublin Opticians and Instrument Makers, *Dublin Historical Records,* **6** (4), 133–49, 1944.
A. D. Morrison-Low: The Trade in Scientific Instruments in Dublin, 1830–1921, in J.E. Burnett and A.D. Morrison-Low: '*Vulgar and Mechanick' – The Scientific and Instrument Trade in Ireland, 1650–1921,* Dublin and Edinburgh, 1989.
C. Mollan: *Irish National Inventory of Historic Scientific Instruments,* Dublin, 1995.

A.D. Morrison-Low, National Museums of Scotland, Edinburgh.

Born: Dublin, 1791.
Died: London, 1859.

Addresses:

1823–32 83 Dame Street, Dublin
1832–51 London – various addresses, notably 11 Lowther Arcade (1837–40) and 428 Strand (1840–51)

Distinctions:
Memberships of the London Electrical Society (1837–1842), Society of Arts (1838–1840) and Geological Society of Edinburgh (Honorary, 1840).

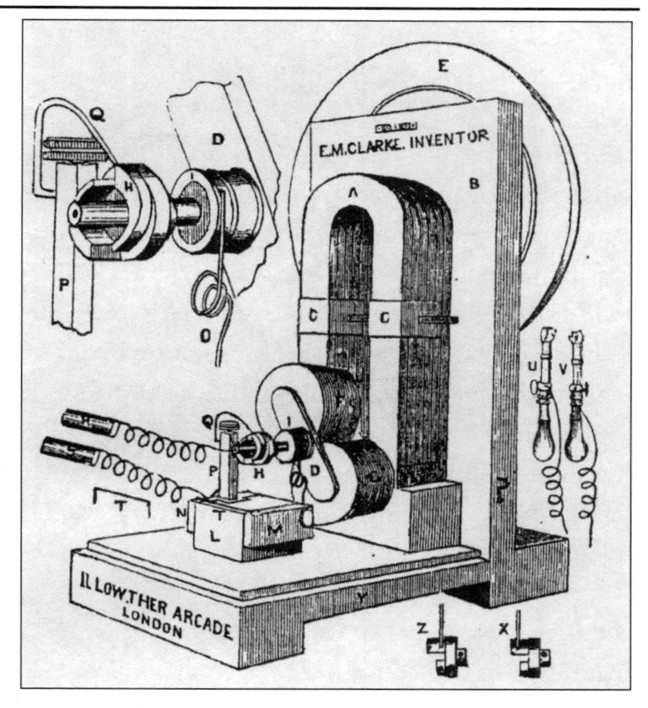

Clarke's magnetoelectric machine

A graduate of Trinity College, Dublin (BA 1815 – MA 1832) Edward Clarke seemingly trained as an instrument maker, and was active in the Dublin Mechanics' Institute in the 1820s. He left for London in 1830 to exhibit at the National Repository. There he met Francis Watkins, another instrument maker, who provided a stepping stone for a new life. His interest in electricity, particularly the new ideas that followed Faraday's discovery of electromagnetic induction in 1830, led to his demonstration of a Pixii magnetoelectric machine on Watkins's behalf. Seeing the potential of this new device for science and medicine, Clarke left Watkins to trade on his own and devote himself to the machine's improvement. He took his version to Paris where it met with much approbation at the French Academy.

His business at the Lowther Arcade 'Laboratory of Science', opposite the Adelaide Gallery of Practical Science, became a port of call for amateurs with electrical interests. It was here that William Sturgeon gave lectures on electricity which led, in 1837, to the formation of the London Electrical Society. One member – for whom Clarke made the famous Maynooth battery – was the Irish priest Nicholas Callan. During 1839 Clarke perfected a technique for projecting images using the oxyhydrogen microscope. Large versions of this instrument were sold to places of public entertainment, such as the Colosseum. Unfortunately, Clarke tended to push his claims to originality too far and so ran into several disputes. He had always been interested in the exhibition of science and, in 1851, an opportunity arose for him to join in the setting up of the Leicester Square 'Panopticon of Science and Art'. As managing director of this palace of science and entertainment he was salaried. Indeed, up to this he appears to have proved that London's pavements were made of gold. Sadly, however, the enterprise closed in 1856 due to bankruptcy. Clarke retired immediately and probably spent his remaining years a very disillusioned, if not poor, man.

Further reading:

B. Gee: E.M. Clarke, *Bulletin of the Scientific Instrument Society*, No. 58 (Sept. 1998), pp. 11–18 and No. 59 (Dec. 1998), pp. 6–13.

Brian Gee, 18 Barton Close, Landrake, Saltash, Cornwall, PL12 5BA.

32 DIONYSIUS LARDNER Physicist and Encyclopaedist

Born: Dublin, 1793.
Died: Naples, 29 April 1859.

A charicature of Dr Dionysius Lardner about the time he was Professor at University College, London

Family:
Son of a county Clare solicitor, William O'Brien Lardner.
Married: 1. Cecilia Flood, grand-daughter of the Irish parliamentarian, Henry Flood 1815. 2. Mary Heaviside, former wife of Captain Richard Heaviside 1849.
Children: Lardner had three children by his first wife and two daughters by his second. In addition he had a natural son by Anne Marie Boursiquot (née Darley) in 1820. Christened Dionysius Lardner Boursiquot, he was to become very well known as Dion Boucicault, one of the nineteenth century's most successful playwrights and author of *The Shaughraun.*

Distinctions:
Member of the Royal Irish Academy 1820; Gold Medallist of the Royal Dublin Society 1826; (First) Professor of Natural Philosophy and Astronomy, University of London 1828–31; Fellow of the Royal Society; Honorary Fellow of the Cambridge Philosophical Society; Honorary Fellow of the Statistical Society of Paris; Fellow of the Society for Promoting Useful Arts in Scotland.

Addresses:

1793–1802	88 Marlborough Street	1803–1814	195 Great Britain Street
1815–1820	12 Russell Street	1820–1822	47 Lower Gardiner Street
1823–1828	29 T.C.D.	1828–1830	3 Jermyn Street, London
After 1830	9 Great Queen Street		

Son of a solicitor, Lardner started work early in his father's Dublin office. Not liking this, he quickly abandoned the law to enter Trinity College at the age of nineteen in 1812. A brilliant student, who won almost all the available prizes in mathematics and philosophy as an undergraduate, he remained in and about the college for the next fifteen years. During this time he took the BA, MA, LLB, LLD degrees and appears to have spent much of this time as a 'grinder' of students. His obvious intention was to become a Fellow of the College and it was probably with this in view that he took holy orders, as the election of lay fellows was unusual at that time. Election depended also on vacancies arising from the retirement or death of fellows.

He was an energetic author and contributed many articles, mostly on science and education, to the *Edinburgh Review*, the literary and scientific magazine then enjoying tremendous success. (Another equally active contributor was the Edinburgh born Henry Brougham, later to become a leader in educational reform in Britain and Lord Chancellor of England.) Lardner had some reputation as a mathematician. His textbook on analytic geometry (1823) was recommended to the eighteen-year-old William Rowan Hamilton by John Brinkley, Royal Astronomer of Ireland. Lardner failed to be elected to a Fellowship in Trinity College, perhaps because the other Fellows felt that the showman side of his personality outweighed the scholarly side. As his later career was to show, he was much concerned with the use and application of science to everyday life – an aspect which was to become an important driving force throughout the coming Victorian reign. This philosophy, born of the industrial revolution, was often in conflict with the more sober and 'schoolman' approach of the older, long established, academic communities.

In 1828 Lardner moved to London to become, under the patronage of Lord Brougham – by now a powerful member of the House of Commons – the first Professor of Natural Philosophy and Astronomy in the newly founded University (College) of London. The new college was the idea of a group of radicals who felt that the time was ripe for a new style university, directed towards the teaching of useful knowledge, similar in character and style to those of France and, especially, of Germany. The pressure for its foundation came too because of the failure of the existing universities of Oxford and Cambridge to admit students other than those belonging to the established Church of England. Some of the most ardent early supporters of the new London College were Jews, Dissenters and Roman Catholics. The new University got off to a shaky start. There was much wrangling over financial matters, often with Lardner at the centre. A brilliant public lecturer, he devised demonstration equipment that was both extensive and expensive (some of it was still in use in the physics department a century later). However, his courses were not the success he expected and he resigned after five years to give himself over completely to the publishing enterprise by which he is best remembered, the production of the 133-volume *Cabinet Cyclopaedia*. Lardner early identified the clamour for knowledge of things scientific that built up both within and without the scientific community in the numerically increasing, more educated and more affluent middle-class population during the nineteenth century. Before the advent of specialist journals, the encyclopaedia was a favourite way to put facts and opinions about science before the public. The example of Diderot, the French encyclopaedist, shows how powerful an influence such publications could be. His 28-volume work was thought to be so provocative an influence on society that for some time it was suppressed by the French authorities. Not only did Lardner write many of the volumes of the *Cabinet Cyclopaedia* himself, he was able to persuade some of the leading scientists of the time, such as John Herschel, son of the astronomer who discovered Uranus, and David Brewster, a leading authority on optics, to contribute volumes.

Unlike his Irish contemporaries William Rowan Hamilton and James MacCullagh, Lardner cannot be regarded as making any really startling discoveries as a mathematician and physicist. It is through his publications (the British Library lists over a hundred works), and his lectures before such bodies as the British Association for the Advancement of Science and the Society for the Diffusion of Useful Knowledge, that his contribution as a great populariser of science must be recognised. In doing so he made himself rich.

Further reading:

N. Harte and J. North: *The World of University College, London, 1828–1978*, London, 1978.
H.H. Bellot: *University College, London, 1826–1926*, London, 1929.
R. Fawkes: *Dion Boucicault: a Biography*, London, 1979.
D. Diderot: *L'Encyclopedie,* Paris, 1751.
D. Lardner: *The Cabinet Cyclopaedia*, London, 1830–49.
D. Lardner: *The Museum of Science and Arts*, London, 1856.

William J. Davis, University Chemical Laboratory, Trinity College, Dublin.

THOMAS ROMNEY ROBINSON Astronomer and Physicist

Born: Dublin, 23 April 1793.
Died: Armagh, 28 February 1882.

Photograph of Robinson taken by Mary Countess of Rosse around 1856 (reproduced by David Davison)

Family:
Parents: Thomas Robinson and Ruth Buck.
Married: Elizabeth Isabelle Rambaut (died 1839), 1821.
Children: three, including Mary Susannah who married George Gabriel Stokes.
Married: Lucy Jane Edgeworth, daughter of Richard Lovell Edgeworth, 1843.

Addresses:

1801–1806	Belfast	1806–1823	Dublin
1823	Enniskillen	1823–1882	Armagh

Distinctions:
President of the British Association for the Advancement of Science 1849; President of the Royal Irish Academy 1851–6; Fellow of the Royal Society 1856; honorary degrees from the Universities of Dublin, Oxford and Cambridge.

The son of an English portrait painter who had moved to work in Ireland, Robinson was educated at Belfast Academy and Trinity College, Dublin, where he became successively a Scholar and Fellow and lectured in natural philosophy. He wrote a textbook for his lectures, *A System of Mechanics*, published in 1820. During his years in Dublin he was involved with different scientific institutions, such as Apothecaries' Hall, the Royal Dublin Society and the Royal Irish Academy, and he had a very wide range of interests across physics, mathematics, chemistry and natural history. No bias towards astronomy is evident in this period, and perhaps no clear commitment to a professional career in science for, after marrying in 1821, Robinson took a Bachelor of Divinity degree in 1822, and the following year accepted the College (Church of Ireland) living of Enniskillen in County Fermanagh.

However, in November 1823, he was appointed Astronomer at Armagh Observatory, and thus began the institutional association for which he is particularly remembered and which survived for more than fifty-eight years. He would die in the same post after an extraordinarily vigorous and distinguished career.

The Observatory was an institution of the Church of Ireland and, thanks to the enthusiastic patronage of the Primate, Lord John George Beresford, Robinson was able to begin by commissioning some much-needed new instruments: he bought a transit instrument and a mural circle from a leading London maker, Thomas Jones, and with these was able to pursue the programme then considered the essence of observatory work, namely a regular routine of positional measurement. But he also bought a 15-inch equatorially-mounted reflector, installed under a new dome in 1835, this time from Thomas Grubb of Dublin. This was important in two respects: it indicated Robinson's appreciation of research possibilities outside the traditional work of astrometry, and it was an important early commission for a maker who went on to establish an international reputation.

With the help of a series of assistants, Robinson maintained a rigorous and unrelenting programme of meridian observations that resulted in the publication of one of the great monuments to astronomical endeavour in Ireland, *Places of 5,345 Stars Observed from 1838 to 1854 at the Armagh Observatory*, published in 1859 at government expense. For this work, generally known as the 'Armagh Catalogue', Robinson was awarded the Royal Medal of the Royal Society in 1862. It was testimony to his conviction that making, reducing and publishing the best possible record of systematic observations was an absolute obligation of observatory astronomers, and that theoretical work, which could be done anywhere, was a secondary distraction.

Armagh Observatory already had a tradition of meteorological observation, which Robinson sought to maintain and improve. This led to his design of the instrument by which his name is best remembered today, the Robinson anemometer. The first example was occasioned by a decision to mount a lightening conductor and wind-gauge on the Observatory roof in 1839. In 1846 he announced to the Mechanical Section of the British Association meeting in Southampton his design of cup anemometer with four vertical hemispheres mounted on a cross, driven by the wind in its horizontal rotation. The example at Armagh recorded both wind speed and direction by pencil traces on a chronograph.

Robinson maintained his interest in large reflectors, being closely associated with the building programme of William Parsons, third Earl of Rosse, at Birr Castle. He was an enthusiastic champion of the observations made at Birr, more vocal in their promotion than the Earl himself. Robinson was also closely involved with the proposal for establishing a reflector in the Southern Hemisphere. Here he consistently championed Grubb as builder, and it was the Grubb design that was eventually sent to the Melbourne Observatory in 1868.

Robinson had less success with maintaining support for his own Observatory which, despite his unrelenting efforts, declined in resource in his later years as a consequence of the disestablishment of the Church of Ireland, the loss of tithe income granted by successive primates, and rent reform affecting the Observatory's holdings in land. These developments contributed to his reactionary political views, and he was a stubborn opponent of reform in Irish political institutions.

Robinson's popular reputation in his lifetime owed much to his talent as an entertaining and authoritative public speaker on scientific topics. This was particularly appreciated at the annual meetings of the British Association for the Advancement of Science. On the more academic side, he maintained the breadth of his interests and published a great many papers on a wide range of topics across physics and astronomy. In Ireland he was an effective President of the Royal Irish Academy for five years from 1851, and played a central part in acquiring the use of 19 Dawson Street from the Government for the use of the Academy.

Further reading:

J.A. Bennett: *Church, State and Astronomy in Ireland: 200 Years of Armagh Observatory*, Armagh and Belfast, Chapters 4–8, 1990.

Jim Bennett, Museum of the History of Science, University of Oxford.

JAMES MUSPRATT Industrialist chemist

Born: Dublin, 12 August 1793.
Died: Liverpool, 4 May 1886.

Engraving of James Muspratt (from J.F. Allen, loc. cit.*)*

Family:
Son of Evan and Sarah (née Mainwaring) Muspratt, both originally from England. His father and two uncles were corkcutters in Dublin.
Married: Julia Conner, 1819.
Children: Seven sons (three died in infancy) and three daughters. His eldest son, James Sheridan, was well-known as a chemist and educator, and compiled the successful *Chemistry, Theoretical, Practical and Analytical, as applied and relating to the Arts and Manufactures* (London, 2 vols, 1860). His youngest son, Edmund Knowles, was a successful industrialist, Vice-Chairman of the United Alkali Company and first Pro-Chancellor of the University of Liverpool.

Addresses:

1793–1812	Abbey Street, Dublin (works)
1812–1815	Military service overseas
1816–1822	14 Parkgate Street, Dublin (works)
1823–1834	2 Great Oxford Street, Liverpool
1834–1841	Pembroke Place, Liverpool
1841–1886	Seaforth Hall, Liverpool

After some schooling, James was apprenticed to Mr Mitcheltree, a wholesale druggist, from 1807 to 1810. His father died in 1810 and his mother in 1811. His legacy was in chancery and so he went adventuring in the Napoleonic Wars; first in Spain, and later as an officer in the Navy, from which he deserted and made his way back to Ireland in 1815.

He started making chemicals in Dublin on a small scale in 1816 and, in 1819, he went into partnership with Thomas Abbott at 14, Parkgate Street, Dublin. In 1819 he married Julia Conner and two of his sons – James Sheridan (1821) and Richard (1822) were born in Dublin.

In 1822 he made the decision to move his young family to Liverpool, judging the conditions to be better for chemical manufacture. He built a works in the Vauxhall Road and started making prussiate of potash (potassium cyanide), then sulphuric acid. (He continued to trade as Muspratt and Abbott until 1825.) The time was right to make synthetic soda, in competition with kelp and barilla, for soap-boiling. He started to make soda by the Leblanc process on a large scale in 1823. Initially, he had to give this new product away to encourage soap-boilers to switch from natural alkali but, once they saw the advantages, the alkali industry in Lancashire was launched. From this small chemical works in Liverpool grew the Lancashire alkali industry, which became the centre of the heavy chemical industry in the UK. James Muspratt was known as the 'father of the alkali industry in Great Britain'.

From 1828 to 1830 he was in partnership with another Irish émigré, Josias Gamble, in St Helens, outside Liverpool, and built an alkali works. In 1830 James left and built another alkali works at

Newton-le-Willows. Two further sons were born in Liverpool – Frederic (1825) and Edmund Knowles (1833). All four sons were educated as chemists and came in turn into the family business.

Drawing of the Vauxhall Road works around 1830 (Frontispiece to: D.W.F. Hardie & J. Davidson Pratt, A History of the Modern British Chemical Industry, *Pergamon Press, 1966 – from an engraving in the Liverpool Public Libraries)*

In 1837 James met the famous German chemist Justus von Liebig in Liverpool and they became lifelong friends. He sent all his sons to study with Liebig in Germany. He patented and tried to make (unsuccessfully) Liebig's Artificial Manure (1845–7). In 1840, he opened up the American market for British soda. In 1840 Muspratt made 15% of British alkali production. His wealth enabled him to build a new house outside Liverpool, Seaforth Hall, away from industrial pollution! The Newton works closed in 1850 and, in 1852, new works were built in Widnes and Flint.

From 1827 James Muspratt was under almost constant litigation, often successful, from individuals, landowners and corporations for air pollution. He invested money in taller chimneys and pollution prevention measures, but the Leblanc process was inherently wasteful, inefficient and dirty. Eventually James was worn out by the pressures of business and litigation and, in 1850, he went into partial retirement, retiring from 1857 when his wife died, handing over control to his sons.

From his Dublin days he had cultivated the arts and literature and his house was full of artists as well as being a place of pilgrimage for chemists visiting England. He remained at Seaforth Hall until his death, surviving three of his sons.

His long life, his wide circle of friends and acquaintances, his seminal role in the birth of Britain's heavy chemical industry, and the influence of his sons and grandsons, meant that the name of Muspratt was almost synonymous with industrial chemistry in the last century. His name survives in the Muspratt Laboratories at the University of Liverpool (named after his son Edmund Knowles). The factories are all long gone; the companies having merged into the United Alkali Company, which was in turn absorbed by Imperial Chemical Industries.

Further reading:

J.F. Allen: *Some Founders of the Chemical Industry*, London, Sherratt and Hughes, 1906.

Michael D. Stephens and Gordon W. Roderick: The Muspratts of Liverpool, *Annals of Science*, **29**, 287–311, 1972.

United Alkali Company: *Centenary of the Alkali Industry, 1823–1923*, 1923 (contains a personal memoir by his grandson Horace Muspratt).

M.L. Greatbatch: *The Muspratt Family and the Merseyside Alkali Industry*, Halton Chemical Industry Museum, Widnes.

Peter E. Childs, Department of Chemical and Environmental Sciences, University of Limerick.

35 THOMAS MACLEAR Astronomer 1794–1879

Born: Newtownstewart, Co. Tyrone, 17 March 1794.
Died: Mowbray, near Cape Town, South Africa, 14 July 1879.

Thomas Maclear about 1860 (courtesy South African Astronomical Observatory)

Family:
Eldest son of the Rev. James Thomas Maclear and his wife (née Magrath).
Married: Mary Pearse of Bedford, England, in 1825.
Children: Six daughters and four sons.

Addresses:
1823–1833 Biggleswade, Bedfordshire, England
1834–1879 Cape Town, South Africa

Distinctions:
Fellow of the Royal Society 1831, Royal Medal 1869; Knighted 1860; Lalande Prize 1867.

Rebelling against his father's wish for him to become a clergyman, Maclear studied medicine at Guy's and St Thomas's Hospitals in London. In 1815 he joined the staff of Bedford Infirmary, where he could combine medicine with a growing interest and competence in astronomy. From 1823 he had a practice at Biggleswade in Bedfordshire, where he set up a small observatory equipped with a telescope on loan from the Royal Astronomical Society.

In 1833 Maclear was appointed Her Majesty's Astronomer at the Royal Observatory, Cape of Good Hope, which had been established in 1820 to serve as the southern counterpart of Greenwich. At John Herschel's request, he assisted him in surveying the southern heavens in the same way that Herschel's father William had surveyed the northern sky. Maclear pursued his own observational programme to obtain accurate positions of stars: in particular, he observed *alpha* and *beta* Centauri and confirmed the parallax measured by Thomas Henderson. He made extensive observations of comets and double stars, and his observations of Mars in 1862 were used by others to determine the Sun's distance. Maclear's most notable achievement was his re-measurement and extension of the meridian arc of Lacaille which laid the foundation for the mapping of Africa: for this he was awarded the Lalande Prize and a Royal Medal. He showed that Lacaille's zenith measurements suggesting the Earth was not round had been affected by nearby mountains: the Earth was in fact round. Maclear also devoted himself to meteorological, magnetic and tidal observations. He helped to establish lighthouses and the transmission of time signals to Port Elizabeth and Simon's Town. He took a keen interest in the exploration of Africa and was a friend of Livingstone and Stanley. He retired to Mowbray near Cape Town, became totally blind in 1876, and died two years later. A lunar crater bears his name, as does a town in South Africa.

Further reading:
Dictionary of National Biography.
B. Warner: *Astronomers at the Royal Observatory, Cape of Good Hope*, Balkema, 1979.

Ian Elliott, Dunsink Observatory, Dublin 15.

JAMES APJOHN Chemist and Mineralogist

Born: Limerick, 1 September 1796.
Died: County Dublin, 2 June 1886.

Portrait of James Apjohn (now in Trinity College) showing his wet-bulb hygrometer

Family:
Son of Thomas Apjohn, tax collector.
Children: three sons, James Henry, Lloyd, and Richard, who became an engineer in 1872.

Addresses:

1796–1826	Sunville House, High Road, Limerick
1846–1857	Lower Baggot Street, Dublin
1858–1886	South Hill, Merrion Avenue, Blackrock, Co. Dublin

Distinctions:
Professor of Chemistry, Royal Cork Institution 1826–28; Professor of Chemistry, Royal College of Surgeons 1828–50; Professor of Applied Chemistry, Trinity College, Dublin 1841–81; Professor of Mineralogy, Trinity College, 1845–81; University Professor of Chemistry Trinity College 1850–75; Fellow of the Royal Society 1853; Fellow of the Chemical Society 1861; Vice-President of the Royal Irish Academy; Fellow of the King's and Queen's College of Physicians; Awarded the Cunningham Gold Medal of the Royal Irish Academy for essay 'A new method of investigating gaseous bodies' 1837.

James Apjohn was a competent, well-respected and long-lived chemist, who spent 50 years teaching and conducting research, mainly in physical chemistry, in Dublin. He was educated at Tipperary Grammar School and at Trinity College Dublin, to which he went in 1813 at the age of sixteen. He took his arts degree in 1817, before going on to study medicine. He received the bachelor of medicine degree in 1821 and proceeded to his doctorate sixteen years later. During his medical studies he developed a passion for experimental science and thereafter pursued researches in chemistry, his first paper being published in 1821 and his 48th in 1871 at the age of 75.

In 1824 a number of eminent physicians and surgeons established a new private medical school at Parke Street in Dublin and, in the following year, Apjohn was appointed lecturer in chemistry. Three years later he was offered the post of Professor of Chemistry in the medical school of the Royal College of Surgeons in Ireland. He quickly established a reputation for teaching and attracted students from all over the British isles. One such, William Gregory, had studied at Giessen with Justus von Liebig; Gregory was to become author of the first organic chemistry textbook in Britain and Ireland, and went on to become Professor of Chemistry at Edinburgh. Together Apjohn and Gregory analysed the natural product derived from wood, eblanine. The work eventually appeared as a joint paper in the first issue of the *Transactions of the Royal Irish Academy* in 1841.

Although Apjohn was to remain in his post at the College of Surgeons until 1850, he was additionally appointed to the Lectureship in Applied Chemistry and Mineralogy in the Engineering School at Trinity College in 1841. On the death in 1850 of Francis Barker, the Professor of Chemistry at Trinity, Apjohn was appointed to the post; he remained in it until his retirement 25 years later.

Apjohn's research interests can be seen to divide into a number of distinct topics. The specific heat of gases interested him throughout his career: his first paper on the subject was published in 1822, the last in 1853. An associated meteorological subject was of particular fascination: the determination of the dew point by the use of the wet bulb hygrometer. His paper of 1835 to the Royal Irish Academy suggested a formula for calculating the dew point: this became known as Apjohn's formula. More practically, in 1850, Apjohn and three other scientists from Trinity College were charged with the task of visiting twelve coastguard stations to give instruction on how to use the meteorological instruments that were deposited there, and to help set up tide gauges.

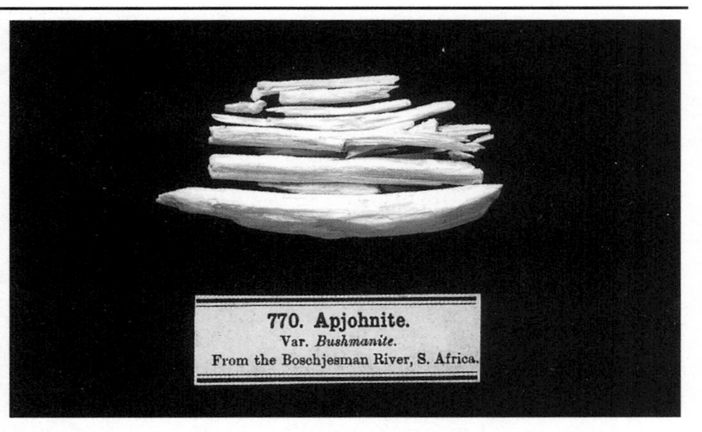

Large crystals of the manganese alum named after Apjohn (courtesy of the Trustees of the National Museums of Scotland)

From the late 1830s Apjohn's interests developed strongly towards mineralogical chemistry. In 1838 he analysed and described a new mineral from Grahamstown, South Africa, which turned out to be a manganese alum. This was subsequently named Apjohnite. Three years later he described a new lead mineral from Kilbricken, Co. Clare, and in 1852 a new yellow-green garnet mineral (Jellettite), which had been found near Zermatt in Switzerland by J. H. Jellett, later Provost of Trinity College. Many other analyses of rocks were performed – on a meteorite that had fallen at Adare, Co. Limerick, in 1813, on jade, on hyalite (a glassy silicate with an anomalous polarising effect from Mexico), and on zeolites. Additionally, he investigated some inorganic compounds, especially complexes including potassium iodide. Monographs produced by Apjohn were a descriptive catalogue of the mineral collection at Trinity College (1850), a pamphlet describing the chemical constitution of well water in the College grounds and an inorganic chemistry textbook, *Manual of the Metalloids,* first published in 1864, which went into a second edition in 1865.

Apjohn's medical background was not often directly called into use, though he did contribute several articles to the *Cyclopaedia of Practical Medicine* of 1833–35. (It is said that Charles Dickens founded his account of the death of Krook in *Bleak House* on Apjohn's description of spontaneous combustion.) He also carefully checked the chemical processes described in the first edition of the *British Pharmacopoeia* of 1864. After his retirement from the Chair of Chemistry at Trinity in 1874 he gradually gave up his other professional interests. He died at the advanced age of 90.

Further reading:

Obituary Notices in *Journal of the Chemical Society,* **51**, 469–70, 1887, and *Chemical News* **53**, 296, 1886.

E.M. Philbin: Chemistry, in T. O Raifeartaigh (ed.), *The Royal Irish Academy: A Bicentennial History, 1785–1985,* 275–300, Dublin, 1985.

R.G.W. Anderson, British Museum, London.

ROBERT JAMES GRAVES Physician

Born: Dublin 1796.
Died: Dublin 1853.

Robert James Graves – sculpture in marble by Albert Bruce Joy (courtesy of the Royal College of Physicians of Ireland – photograph by David Davison)

Family:
Son of Richard Graves, scholar and divine, Archbishop King's Professor of Divinity, Professor of Laws and Regius Professor of Greek and Divinity at Trinity College and Dean of Ardagh.

Address:
4 Merrion Square South, Dublin

Distinctions:
King's Professor, Trinity College Dublin 1827; President of the Royal College of Physicians of Ireland 1843; Fellow of the Royal Society 1849; Honorary Member of the Medical Societies of Berlin, Vienna, Hamburg, Tübingen, Bruges and Montreal.

The 'Dublin School of Medicine' is a title which has been given to a group of physicians who made remarkable contributions to medicine in the nineteenth century. Its international reputation had its foundations in Robert Graves. Having spent some time at Edinburgh, Graves graduated from Trinity College in 1818, and set off to study at Berlin, Göttingen, Vienna, Copenhagen, Paris and Italy.

A facility for foreign languages landed him in an Austrian prison for ten days on the suspicion of being a German spy. While travelling through the Mount Cenis pass in the Alps in the autumn of 1819 he met a young artist and the pair travelled together for some time, neither seeking the other's name. The artist was James Mallard William Turner.

Graves was appointed physician to the Meath Hospital in 1821. His opening lecture did little to endear him to his seniors. He claimed that many fatalities resulted from indifferent treatment, and he deplored the attitude of medical students who walked the wards in pursuit of entertainment rather than medical knowledge. Graves had been much impressed by the method of bedside clinical teaching on the continent, especially in Germany. He praised the gentleness and humanity of the German physicians who, unlike their Irish and English colleagues, did not have 'one language for the rich, and one for the poor', and whose practice it was to put unpleasant diagnoses into Latin, rather than upset their unfortunate patients.

Graves was joined by his young colleague William Stokes in 1826, and together they began to reform clinical practice in the Meath Hospital. Their revolutionary methods caused some resentment, but it is to the credit of the Meath Hospital that it permitted its young physicians to effect their reforms.

A particularly intriguing aspect of this partnership, which was destined to influence medicine far beyond the shores of Ireland, was the fact that Graves anticipated such a happening. In 1834, he lamented the failure of Ireland to find a place on the international stage: 'It is not unusual to find the publications of France, Germany, Italy and England, simultaneously announcing the same discovery, and each zealously claiming for their respective countrymen an honour which belongs equally to all. I am sorry to say that, with some splendid exceptions, this interesting and innocent controversy has been carried on by other countries, while Ireland has put no claim for a share of the literary honours awarded to the efforts of industry and genius.'

More intriguingly, he identified two young colleagues who would make this audacious ambition become reality; William Stokes and Dominic Corrigan. Of Stokes, he said: 'His labours have placed him in a position, as far elevated above the necessity of praise, as above the fear of censure.' Of Corrigan: 'I may assert that his paper [on sounds and motions of the heart] is written in the true spirit of philosophical enquiry, and that he deserves opponents of a far higher grade than those who have endeavoured to refute his arguments in the English periodicals.'

It was not long before Graves and Stokes had earned an international reputation that was later acknowledged by the great William Osler, who said: 'I owe my start in the profession to James Bovell, kinsman and devoted pupil of Graves, while my teacher in Montreal, Palmer Howard, lived, moved and had his being in his old masters, Graves and Stokes'.

In 1843, Graves published his famous *Clinical Lectures on the Practice of Medicine,* which was subsequently translated into French, German and Italian. In this book we find evidence of the gift that was common to these Victorian masters of clinical expression – the ability to describe their observations in clear and lively prose. It was the French physician Trousseau who proposed that the illness exophthalmic goitre, described in the *Lectures,* be named 'Graves' disease'.

Graves was revolutionary in his treatment of patients with fever: he advocated supportive therapy rather than starvation, bleeding and blistering. The story goes that one day on his rounds he was struck by the healthy appearance of a patient recently recovered from severe typhus fever and said to his students: 'This is the effect of our good feeding, and gentlemen, lest when I am gone you may be at a loss for an epitaph for me, let me give you one in three words: "He fed fevers"'.

During his professional career he received many honours. He died from cancer of the liver in 1853 at the age of 57.

Further reading:

E. O'Brien, A. Crookshank and G. Wolstenholme: *A Portrait of Irish Medicine – an Illustrated History of Medicine in Ireland*, Dublin, 1984.

W. Stokes (Editor): The Life and Labours of Graves, in *Studies in Physiology and Medicine by the Late Robert James Graves*, London, 9–83, 1863.

J.F. Duncan: The Life and Labours of Robert James Graves, *Dublin Journal of Medical Science*, **65**, 1–12, 1878.

S. Taylor: *Robert Graves. The Golden Years of Irish Medicine*, London, 1989.

D. Coakley: *Irish Masters of Medicine*, Town House, Dublin, 1992.

Eoin O'Brien, Royal College of Surgeons in Ireland, Dublin 2.

38 EDWARD COOPER Astronomer

Born: Dublin, May 1798.
Died: Markree Castle, Co. Sligo, 23 April 1863.

Family:
Married: 1. Sophia L'Estrange of Moystown in County Offaly – but she died the year after the marriage; 2. Sarah Frances Wynne (died 1862) of Haslewood, County Sligo.
Children: There were no children of his first, brief marriage, but by his second marriage he had five daughters (Laura, Charlotte, Emma, Selina, and Cecily).

Distinctions:
Member of the Royal Irish Academy 1832; Fellow of the Royal Society 1853; Cunningham Medal of the Royal Irish Academy 1858; Honorary LLD University of Dublin 1863.

Edward Cooper was born at a house in St Stephen's Green, Dublin, but in 1800 he was taken to County Sligo when his father assumed responsibility for his family's extensive estates centred upon Markree Castle near Collooney, some twelve kilometres to the south of Sligo itself. His mother is said to have given him his earliest interest in astronomy, but that interest was further developed when, as a schoolboy in Armagh, he paid visits to the Armagh Observatory. His education continued at Eton College and at Christ Church, Oxford, but he left Oxford without taking a degree. Travel now became his principal activity. In 1820–1821 he toured in Persia and Turkey, and in Egypt he went up the Nile as far as Wadi Halfa; in 1824–1825 he travelled in Denmark and Sweden, and in Norway he went as far to the north as North Cape. During all these travels he made a practice of observing his latitude and longitude and in this manner he rapidly became a proficient astronomical observer.

In 1830 he took over the Markree Castle estates upon the death of his father and he immediately resolved to establish there a finely equipped astronomical observatory. He purchased instruments from various manufacturers, including Thomas Grubb of Dublin and in 1851 his establishment was described as 'undoubtedly the most richly furnished of private observatories'. A multitude of novel and important observations were made, much of the work being carried out by Cooper's efficient assistant Andrew Graham between his appointment in March 1842 and his resignation in June 1860. On 25 April 1848 the new minor planet Metis was discovered by the observers at Markree, but the most important work carried out was the measurement of the approximate positions of 60,066 stars, only 8,965 of which were previously known. These observations appeared in a *Catalogue of Stars near the Ecliptic, Observed at Markree* published by the government in four volumes between 1851 and 1856, the work having been recommended to the government by the Royal Society.

Cooper took an active part in public life and he was a Member of Parliament for County Sligo from 1830 to 1841 and from 1857 to 1859. At his death he was succeeded by his nephew Lieutenant Colonel Edward Henry Cooper of the Grenadier Guards, but this gentleman took no interest in astronomy and the observatory was left to go to rack and ruin. When the estates passed into new ownership in 1874 some revival of astronomical work was attempted, but little was accomplished and all astronomical work at Markree ceased in 1902.

Markree Castle, Co. Sligo

Further reading:

Dictionary of National Biography
Henry Boylan: *A Dictionary of Irish Biography*, Dublin, 1978.
S.M.P. McKenna: Astronomy in Ireland from 1780, *Vistas in Astronomy*, **9**, 1968, 283–296.

Gordon L. Herries Davies, Department of Geography, Trinity College, Dublin.

Born: 1798 (or, possibly, late 1797).
Died: Dublin 31 October 1881 (aged 83).

Courtesy of the National Botanic Gardens, Glasnevin, Dublin

Family:
Married: c1816 Jane Goodshaw, daughter of T. Goodshaw, Collumswell, Leixlip. Nine children including Revd Richard Turner born 1817, died 7 February 1849, and William Turner born 1827, died 9 June 1888.

Address:
Ballsbridge, Dublin

Richard Turner is remembered principally for his glasshouses, especially the magnificent, 'ornamental, light, useful and ... everlasting' Curvilinear Range in the National Botanic Gardens, Glasnevin, and for the wings of the equally remarkable glasshouse in the Belfast Botanic Gardens. He also built the Great Palm House at the Royal Botanic Gardens, Kew, perhaps the most famous glasshouse in the world, a building that exemplifies the vigour, wealth and excitement of the Victorian era.

Richard Turner's antecedents were ironmongers. Nothing has yet been discovered about his childhood. Leixlip seems to have been his early home. He took over the family ironmongery on St Stephen's Green, Dublin, from his uncle Richard. He engaged in building speculation in Dublin during the 1820s and 1830s and, with the capital gained, was able in 1834 to acquire land at Ballsbridge on which he established the Hammersmith Iron Works: the foundry site is now occupied by the Veterinary School. There Turner forged his spectacular glasshouses and many other more mundane iron artefacts – demesne gates (for example, at Donadea Castle, County Kildare), roofs for railway stations (Broadstone Station, Dublin, had one), and railings for Trinity College, Dublin (those along Pearse Street bear Turner's mark).

As for his glasshouses, the earliest known commission was built at Colebrooke, County Fermanagh. The story of the Glasnevin Curvilinear Range, constructed in stages between 1843 and 1869, is too complicated to repeat here, but Turner certainly influenced the design and erected all of it apart from the earliest, south-facing half of the east wing. The history of the Palm House at Kew is also complicated, but Turner was intimately involved in the design, collaborating with Decimus Burton, and he was entirely responsible for the construction of the building. Kew's Palm House is loftier and more extensive in plan than the Glasnevin Range, yet the buildings are 'sisters'. Turner was working on both during the late 1840s while the Great Potato Famine was at its peak. In fact the Kew glasshouse was pre-fabricated at the Hammersmith Iron Works, Ballsbridge.

Several contemporary pen-portraits of Richard Turner provide insights into this remarkable innovator. In 1848 John Lyons of Ladiston, Mullingar, described 'Dick Turner [as] ... an ingenious, tasty, clever fellow, without a depth of science'. More than thirty years later, Thomas Drew characterised Turner as 'vigorous, although advanced in years and retired'. Drew continued: 'Many middle-aged architects will remember Mr. Turner in his vigorous days when he was ubiquitous, with a stock of daring and original projects always on hand, remarkable for his rough-and-ready powers of illustration of them, and sanguine belief in them, and his eloquent, plausible, and humourous advocacy of them'. Turner's stock of daring schemes included designs for 'crystal palaces', including the Great Exhibition in London (1851) and the Dublin Exhibitions (1853) – he competed unsuccessfully in competitions for these.

What made Turner so remarkable? His buildings are light, bright and brilliantly constructed. The thin glazing bars and pierced pilasters let in as much light as possible. Turner was able to use malleable iron in novel ways to create free-standing, robust structures of hitherto unimaginable size. His innovations included a glazing bar of wrought iron, just one and a half inches deep and half an inch wide – more than that, he could curve these bars to create airy spaces.

As Thomas Drew correctly observed, Turner's 'brilliant, innovative and constructive genius did not bring the financial successes he deserved', but he is honoured as the builder of several of the World's finest buildings, including the sumptuous Curvilinear Range at Glasnevin which was restored to its original state for the bicentenary of the National Botanic Gardens in 1995.

Further reading:

E.J. Diestelkamp: Richard Turner (c.1798–1881) and his Glasshouses, *Glasra*, **5**, 51–53, 1981.
E.J. Diestelkamp: The Design and Building of the Palm House, Royal Botanic Gardens, Kew, *Journal of Garden History*, **2**, 233–272, 1982.
E.J. Diestelkamp: The Curvilinear Range at the National Botanic Gardens, *Moorea*, **9**, 6–34, 1990.
E.J. Diestelkamp and E.C. Nelson: Richard Turner's Legacy – the Glasnevin Curvilinear Glasshouse, *Taisce Journal*, **3**(1), 4–5, 1979.
E.C. Nelson: Richard Turner, An Introductory Portrait, *Moorea*, **9**, 2–5, 1990.
E.C. Nelson and E.M. McCracken: *The Brightest Jewel: A History of the National Botanic Gardens, Glasnevin, Dublin*, Boethius Press, Kilkenny, 1987.

Glasshouses by Richard Turner:

Colebrooke, Fermanagh (c1830); Belview, Fermanagh (c1835); Marlfield, Tipperary (c1835); Peach House, Vice-Regal Lodge (now Áras an Uchtharáin), Phoenix Park, Dublin (1836); Palm House (wings), Botanic Gardens, Belfast (1839); Chief Secretary's Lodge (now residence of the American Ambassador), Phoenix Park, Dublin (1842); Killakee, Co. Dublin (1843); Curvilinear Range, National Botanic Gardens, Glasnevin, Dublin (1844); Haddo House, Aberdeenshire, Scotland (1844); Palm House, Royal Botanic Gardens, Kew (1844); Winter Garden, Royal Botanical Society's Garden, Regent's Park, London (1845); Rath House, Laois (c1847); Ballynagall, Westmeath (c1850); Middleton Park, Westmeath (c1850); Ballyfin, Laois (c1850); Edermine, Wexford (c1850); Roxborough Castle, Tyrone (c1850); Victoria Regina House, Royal Botanic Gardens, Kew (1852); Bessborough, Cork (c1855); Woodstock, Kilkenny (c1860); Longueville, Cork (1866); Knapton, Laois.

E. Charles Nelson, Outwell, Wisbech, England.

Born: Dromiskin, near Ardee, Co. Louth, 20 December 1799.
Died: Maynooth, Co. Kildare, 10 January 1864.

Distinctions:
Awarded the Degree of STD, University of Rome 1826; Appointed Professor of Natural Philosophy at Maynooth 1826.

Callan's most notable contribution to electrical science was the Induction Coil, the forerunner of the modern step-up Voltage Transformer.

Influenced by his former professor, Dr Denvir, Callan early acquired a great interest in electrical phenomena in general. He constructed several electromagnets one of which had a lifting power of two tons. Influenced also by the work of Joseph Henry of Princeton, who independently discovered the phenomenon of self-induction, Callan constructed a coil for giving electric shocks. Working on this, he first separated the primary coil from the secondary and, more importantly, wound it round an iron core. This apparatus gave shocks of great intensity. Later he increased the number of turns of wire in the secondary coil and obtain sparks from the free ends. In his 'Giant' induction Coil there are three secondary coils connected in series. Each coil contains about ten miles of very fine wire insulated with a mixture of beeswax and gutta-percha. In 1837 this apparatus gave sparks fifteen inches long. Callan devised a 'point and plate' type of 'valve' which rectified the secondary current. This was used later in X-ray apparatus. Also, he was the first to note that the 'intensity' (voltage) of the secondary current depended, among other things, on the rapidity of interruption of the primary current.

Callan sent a smaller replica of this induction coil to his friend Sturgeon in London who demonstrated it to members of the Electrical Society of London. It evoked great interest amongst the scientific coterie at the time. It was copied by many of them.

The invention of the induction coil made it possible to produce X-rays. It also provided the means of studying electric discharges in rarefied gases. This later contributed to the elucidation of atomic structure.

Callan constructed electric motors and even drew up plans for a battery-driven train to ply between Dublin and Kingstown (now Dun Laoghaire). The plan fell through owing to practical snags.

To supply electric current for his researches Callan experimented with various types of battery. Eventually he arrived at a cast iron/zinc cell. A cast-iron trough acted as positive plate; it contained nitric acid. The negative plate was of zinc standing in a porous pot containing a mixture of sulphuric and nitric acids. This cell gave quite heavy current for a considerable period. It was subsequently manufactured commercially by the firm of E.M. Clarke at the Adelaide Galleries, 428 The Strand,

Callan's giant induction coil gave 15 inch sparks in 1837

London, who named it the 'Maynooth Battery'.

Callan found that the nitric acid rendered the cast-iron highly resistant to corrosion and obtained a patent for this process of protecting exposed iron-work from rusting.

The Museum at Maynooth houses Callan's induction coils, electric motors, his 'Maynooth Battery' and other interesting 'Callaniana'.

Callan's 'Maynooth Battery'

Further reading:

P. J. McLaughlin: *Nicholas Callan, Priest-Scientist 1799–1864*, Dublin, Clonmore & Reynolds; London, Burns & Oates, 1965.

Michael T. Casey: Nicholas Callan; inventor of the Induction Coil, *The School Science Review,* No. 160, June 1965.

Michael T. Casey: Nichoas Callan, Priest, Professor and Scientist, *Physics Education,* **17**,1982.

Charles Mollan and John Upton: *The Scientific Apparatus of Nicholas Callan and Other Historic Instruments*, St Patrick's College, Maynooth & Samton Limited, Dublin, 1994.

Reverend Michael T. Casey, OP, 1902–1997.

Born: Dublin, 16 April 1800.
Died: Dublin, 17 January 1881.

Courtesy Royal Society, London

Family:
Married: Dorothea Bulwer in July 1840.
No children.

Addresses:
No. 35 Trinity College and Provost's house.
17 Fitzwilliam Square South, Dublin 2.
Killcroney Abbey, Enniskerry, Co. Wicklow.
Victoria (now Ayesha) Castle, Killiney, Co. Dublin.

Distinctions:
Fellow of the Royal Society 1836; DD University of Dublin 1840; President of the Royal Irish Academy 1846–1851; DCL University of Oxford 1855; President of the British Association 1857; Cunningham Medal of the Royal Irish Academy 1862; German order 'Pour le Merite' 1874; He was a Fellow of the Royal Societies of London and Edinburgh and an Honorary Member of the Philosophical Societies of Cambridge and Manchester as well as of many other learned societies of Europe and America; Lloyd had a brilliant academic career in the University of Dublin of which the following were the main landmarks; Erasmus Smith Professor of Natural and Experimental Philosophy 1831; Senior Fellow in Trinity College 1843; Vice-Provost 1862; Provost 1867. He was an outstanding university teacher and played an important role in the foundation of the School of Engineering in Trinity College (1841).

Humphrey Lloyd worked mainly in physical optics and terrestrial magnetism. In the former he had two important successes. The first (1832) was the discovery of a new kind of refraction which takes place when light passes through biaxial crystals, that is crystals with two axes in which no double refraction occurs. There are two species of conical refraction depending on whether it takes place on entering or leaving the crystal. Lloyd made this discovery through very delicate and precise experiments following a brilliant mathematical prediction by his colleague William Rowan Hamilton.

The second success in optics was the design of 'Lloyd's Mirror Experiment' for which he is remembered by physicists. In 1834, while examining Fresnel's classic experiment in which an interference pattern is produced by light from a single source reflected by two mirrors, he discovered that the same effect can be produced using a single mirror when the reflected light is made to interfere with the direct light from the source, the angle of incidence being almost 90°. The experiment proved that a change of phase takes place when light is reflected by a mirror.

In the early nineteenth century there was great interest in terrestrial magnetism; programmes for gathering data concerning the earth's field were developed and there was extensive international co-operation. In Britain and Ireland magnetic surveys were undertaken under the auspices of the British Association (founded 1831). Together with the Dublin-born Edward Sabine FRS (1788–1883), Lloyd designed a new instrument for measuring dip and intensity at the same time.

Lloyd's magnetical observatory at Trinity College

In 1832 the German mathematician Carl Friedrich Gauss introduced a method for measuring the intensity of the earth's field absolutely, that is referred to standard units of length, mass and time. This marked the beginning of a new era. The realisation of Gauss' ideas required, however, the ingenuity of a generation of younger physicists who developed a series of new instruments. Humphrey Lloyd was the outstanding figure in the British Isles involved in this work. He designed several new instruments on Gaussian principles for his magnetical observatory at Trinity College.

The instruments of the Dublin Observatory – built mostly in Rathmines by Thomas Grubb and Howard Grubb – were copied and installed in observatories in the British colonies and at many continental observatories. The instruments designed by Lloyd provided great accuracy. For example, the instruments for measuring the changes in the two components of the intensity of the earth's field could read to 1/50,000th of the whole. The magnetical observatory, which stood in the Provost's garden from the time of its erection in 1838, was removed in 1974 but now stands on the campus of University College at Belfield.

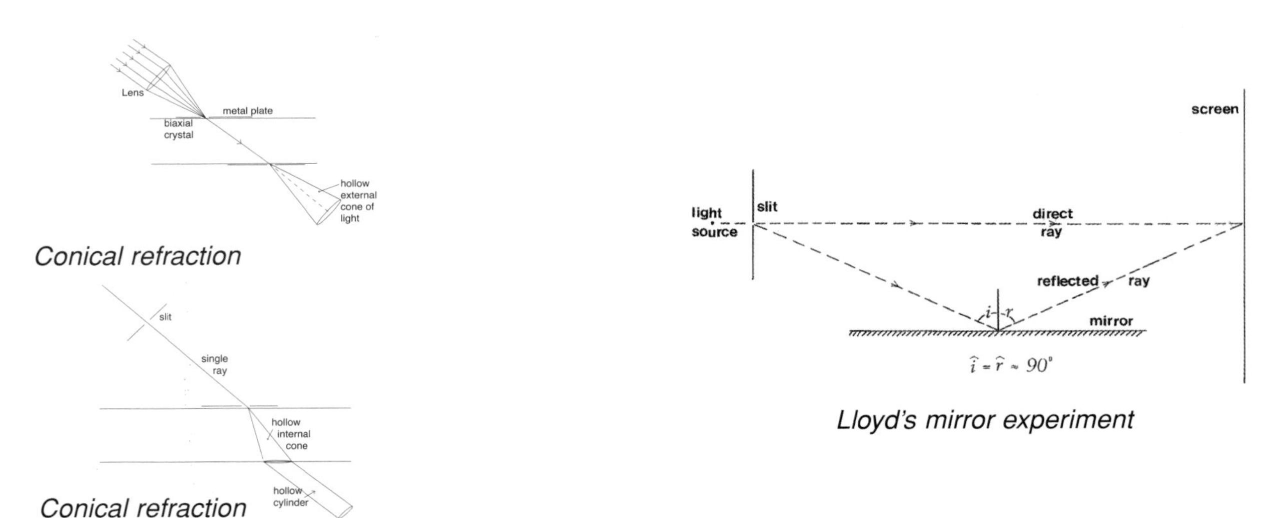

Conical refraction

Conical refraction

Lloyd's mirror experiment

Further reading:

J.G. O'Hara: The Prediction and Discovery of Conical Refraction by William Rowan Hamilton and Humphrey Lloyd (1832–3), *Proceedings of the Royal Irish Academy,* **82A**, 231–257, 1982.

J.G. O'Hara: Gauss and the Royal Society: the Reception of his Ideas on Magnetism in Britain (1832–1842), *Notes and Records of the Royal Society,* **38** (1), 17–78, 1983.

James G. O'Hara, Leibniz Archiv, Niedersächsische, Landesbibliothek, Hannover, Germany.

42 WILLIAM PARSONS Third Earl of Rosse, Astronomer and Amateur Engineer

Born: York, England, 17 June 1800.
Died: Monkstown, Co. Dublin, 31 October 1867.

Family:
Eldest son of Sir Laurence Parsons (second Earl of Rosse 1807) and Alice Lloyd.
Married: Mary Field of Heaton, Yorkshire (see Mary, Countess of Rosse).
Children: Alice, Laurence, William, John, Randal, Clere, Charles.

Addresses:
Birr Castle, County Offaly
13 Connaught Place, London

Distinctions:
Lord Oxmantown 1807; Member of the Royal Irish Academy 1822; MP for King's County 1821–1835; Fellow of the Royal Society 1831; Lord Lieutenant of King's County 1831; Representative Peer for Ireland 1845; Knight of the Order of Saint Patrick 1845; President of the Royal Society 1848–54; Member of the Imperial Academy of Science St Petersburg 1852; Knight of the Legion of Honour 1855; Chancellor of the University of Dublin 1862; Honorary Degree from Dublin 1863, and Cambridge 1842.

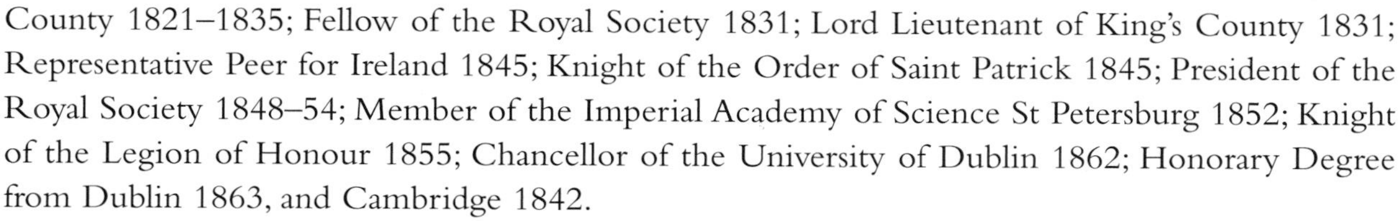

William Parsons was educated at home in Birr Castle, and entered Trinity College Dublin in 1818, transferring to Magdalen College Oxford from where he graduated first class in mathematics in 1822. In 1824 he joined the Astronomical Society and he made a decision to follow in the footsteps of William Herschel by increasing the light grasp of telescopes so that he could study faint objects like nebulae.

In his workshop at Birr he began to experiment with the construction of mirrors for reflecting telescopes using speculum metal, a bronze alloy which is very brittle and difficult to handle. At first he tried to make large mirrors from small pieces of bronze soldered to a brass base. He also made a careful study of shaping the surface of the mirrors to a parabola. To do this he built a special machine which allowed him to systematically improve the process of grinding and polishing. It was driven by a small steam engine constructed by him at Birr in 1827. In 1830 he published an account of his researches leading to the mounting of a 24 inch diameter mirror on a wooden altazimuthal stand (i.e. one which could rotate in both horizontal and vertical directions).

After his marriage to a wealthy heiress in 1836 he resumed his programme of work, refining his methods of casting, fabricating and shaping mirrors. Progress was rapid and was crowned with success. His paper in 1840 explained the secret, which was to construct the mould for castings with a base of close packed steel strips through which gases could escape. He had made and mounted a perfect casting for a mirror of three feet diameter. He quickly proceeded to cast, in one piece, a giant six foot mirror which weighed 3.5 tons. With its mounting and the 54 foot long telescope tube, this was so massive that stone walls were required to support and orient the telescope in an altazimuthal arrangement of restricted movement. All of this work was carried out with a team of local craftsmen trained by him at Birr at a cost reputed to lie between £20,000 and £30,000. Almost immediately the huge light grasp revealed that one nebula, M51 in Canes Venatici, has a spiral structure. A pencil drawing made by him was shown to a meeting of the British Association at Cambridge in 1845 where it created a sensation.

His father died in 1841 and he succeeded as third Earl of Rosse. His mastery of engineering skills was recognised in 1849 by election as an Honorary Member of the Institution of Civil Engineers. In 1851 his scientific work was honoured by the award of the Royal Society's Gold Medal.

The 'Leviathan' of Birr remained the largest telescope in the world for seventy years, but eventually its supremacy as an instrument for discerning faint objects was lost as the technique for making photographs with long exposures was perfected. Lord Rosse had tried to use the new invention of photography with the telescope, but the materials available to him were too insensitive and the telescope could not be driven accurately enough to follow the movements of celestial objects.

The outbreak of the Great Famine in the autumn of 1845 brought a devastation to Ireland which Lord Rosse had foreseen. With Mary, Countess Rosse, he now worked hard to ease the hardship round about. He published his thoughts, which were very critical, on the behaviour of the Government in London. When the crisis began to ease in 1848, he was elected President of the Royal Society and, because of this and pressure of parliamentary business, there were long absences from Birr, so he funded a succession of able astronomical assistants to continue the work of the observatory.

William Parsons' interests lay not just in science but in engineering and manufacturing, which took him to many workshops and shipyards. As president of the Royal Society he argued for continued funding of Charles Babbage's mechanical 'difference engine'. He also urged the Admiralty to make more use of technically educated personnel and put forward a proposal for an iron-clad vessel to be used in the Crimean war. In later life he purchased a yacht in which he brought his wife and children cruising, around Britain and to Spain and France.

Lord and Lady Rosse were very close to their children, who were educated by tutors at Birr and were encouraged to use the workshops for themselves. Of their eleven children only four boys survived childhood. The eldest, Laurence, continued an interest in astronomy. The two younger sons graduated as engineers, Clere from TCD, and Charles, the inventor of the steam turbine, from Cambridge. William Parsons died after an operation to remove a tumour from his knee.

Further Reading:

C.A. Parsons (Editor): *The Scientific Papers of William Parsons*, London, 1926.
P. Moore: *The Astronomy of Birr Castle*, Birr, 1971.
W.G. Scaife: *From Galaxies to Turbines*, Institute of Physics, Bristol, 1999.

W. Garrett Scaife, Fellow Emeritus, Trinity College Dublin.

43 THOMAS (1800–1878) and HOWARD (1844–1931) GRUBB

Born: Thomas: Waterford, 4 August 1800;
Howard: Rathmines, Dublin, 28 July, 1844.

Died: Thomas: Dublin 19 September 1878;
Howard: Monkstown, Co Dublin,
16 September, 1931.

Thomas Grubb (1800–1878) from Manville G.E., Two Fathers and Two Sons, *Reyrolle Parsons Group, Newcastle Upon Tyne, 1971*

Family:
Thomas Grubb's father was William Grubb (?–1831) and his mother was Eleanor Fayle. He married Sarah Palmer and they had ten children.
Howard married Mary Hester Walker, a lady of Irish descent from New Orleans, in 1871, and they had six children.

Addresses:

Thomas:

1835	Parnell Place, Parnell Road, Harold's Cross, Dublin
1878	141 Leinster Road, Rathmines, Dublin

Howard:

1871	17 Leinster Square, Rathmines, 1871
Before 1888	51 Kenilworth Square, Rathgar
1888	141 Leinster Road, Rathgar
After 1888	Rockdale and later Aberfoyle, Orwell Road, Rathgar
1925	1, de Vesci Terrace, Kingstown (now Dun Laoghaire)
1931	13, Longford Terrace, Monkstown

Distinctions:
Thomas: Member of the Royal Irish Academy 1839; Fellow of the Royal Society 1864.
Howard: Hon. Master in Engineering Trinity College Dublin 1876; Fellow of the Royal Society 1883; Knighted 1887; Boyle Medal of the Royal Dublin Society 1912.

Thomas Grubb's name appears around 1832 as the owner of a foundry near Charlemont Bridge in Ranelagh, Dublin, specialising in cast-iron billiard tables. Probably through friendship with T.R. Robinson of Armagh Observatory, he became an amateur astronomer and constructor of telescopes. Amongst these were the largest refractor (lens telescope) of the time (1834), a 13.3-inch instrument for E.J. Cooper of Markree, with lens by R.A. Cauchoix of Paris. In 1835 he provided Armagh with a 15-inch Cassegrain (reflecting telescope) using optics of his own manufacture and containing the first of the now widely-used 'equilibrated lever' mirror support systems, which he invented. This approach was adopted by William Parsons, third Earl of Rosse, for his largest telescope, the 'Leviathan of Parsonstown'. Other well-known examples of his workmanship are the 12-inch South refractor at Dunsink Observatory, Co. Dublin, still in use, and his largest instrument, the 48-inch 'Great Melbourne Telescope' of 1869, constructed in partnership with his son, Howard.

In the field of optics, Grubb designed grinding and polishing machinery of a type still in widespread use. He was probably the first person to use the technique of mathematical 'ray-tracing' to compute the behaviour of lenses, now, in the age of computers, the main tool for their design.

He also was the first to understand the field properties of camera lenses. He held an important patent on an achromatic meniscus lens and manufactured many examples. He was active in photographic circles during the 1850s. Thomas Grubb earned his living for most of his life as 'Engineer to the Bank of Ireland', for which he designed some unique printing machinery that automated the production of banknotes from copper-plate engravings.

Howard Grubb, (courtesy of the Mary Lea Shane Archives, Lick observatory, University of California, Santa Cruz)

Howard Grubb, youngest son of Thomas, at first studied engineering at TCD but was withdrawn in 1866 to partner his father in the construction of the Great Melbourne Telescope. His entry to telescope manufacturing coincided with the rise of astrophysics and he secured some notable contracts during the 1870s. Amongst these was the 'Great Vienna Refractor' of 27 inches, completed in 1878. He built up his business at the 'Optical and Mechanical Works' located in Rathmines, Dublin. He supplied six 'Astrographic' telescopes for the large-field photographic survey project known as the *Carte du Ciel*. His largest refractor (28 inches) was made for the Royal Greenwich Observatory and his largest reflector (40 inches) for Simeis in Crimea. He improved his father's designs and invented many sophisticated mechanisms for the accurate control of telescopes. He also supplied astronomical measuring equipment and surveying instruments.

Around 1900, he turned his attention to military optics and became the major supplier of periscopes to the Royal Navy, though the telescope business continued. Because of security problems, the plant was moved from Dublin to St Albans in England at the end of the World War I (1918). Beset with economic problems following the return to peace, the firm limped on until 1925, when it failed. It was re-constituted as Sir Howard Grubb Parsons by Sir Charles Parsons, the entrepreneurial youngest son of the third Earl of Rosse. The Grubb interest ended in 1929 when R.R. Grubb, the youngest surviving son of Howard, resigned.

Thomas and Howard Grubb were active participants in the Dublin scientific scene and were active members of the Royal Irish Academy and the Royal Dublin Society. Unlike many opticians of their time, they freely described their methods and published many papers in scientific journals.

Further reading:

I.S. Glass: *Victorian Telescope Makers: the Lives and Letters of Thomas and Howard Grubb*, Institute of Physics Publishing, Bristol and Philadelphia, 1997.

Ian S. Glass, South African Astronomical Observatory, PO Box 9, Observatory 7935, South Africa.

Born: Thomas Street, Dublin 1801.
Died: Merrion Square, Dublin 1880.

Dominic Corrigan – sculpture in marble by John Foley (courtesy of the Royal College of Physicians of Ireland – photograph by David Davison)

Family:
Son of John Corrigan, merchant, farmer, shopkeeper, chapman and collier-maker, and Celia O'Connor.

Addresses:
4 Merrion Square West, Dublin
'Inniscorrig', Coliemore Road, Dalkey

Distinctions:
MD Edinburgh University 1825; Physician to the Charitable Infirmary, Jervis Street 1830; Physician to House of Industry Hospitals 1840; Physician-in-Ordinary to Queen Victoria in Ireland 1847; Senate of the Queen's University 1850; President of King's and Queen's College of Physicians 1859–63; Baronet of the Empire 1866; Member of Parliament for the City of Dublin 1870; Vice-Chancellor of the Queen's University 1871; Corresponding Member of Academie de Medicine de Paris.

Dominic Corrigan was the first of the Catholic middle-class to rise to fame in Dublin medicine. Appointed to the staff of the Charitable Infirmary in the year after Catholic emancipation, he quickly fulfilled the promise he had shown in his famous paper – 'Permanent Patency of the Aortic Valve' – a condition still called 'Corrigan's disease'.

Corrigan, like Robert Graves and William Stokes, had the ability, firstly, to take the time to observe illness and its effects, clinically at the bedside, and later in the laboratory or autopsy room. They were then able to relate cause to effect but, most importantly, they had the literary ability to depict their observations in a style that was unique and lasting. Here is an example: 'Let me suppose you now at the bedside of a fever case; stand there quietly, don't disturb the patient, don't at once proceed to examine the pulse, or chest, or abdomen, or to put questions. If you do, you may be greatly deceived, for under a sharp or abrupt question a patient may suddenly rouse himself in reply, answer your question correctly, and yet die within three hours.'

Corrigan was an expert on fevers and was in charge of the Hardwicke Fever Hospital in the Richmond. In the days before the discovery of bacteria, the diagnosis of fever was very much a clinical art. 'In the child in fever, there is another sign revealed to you from merely looking at the countenance, and always to be dreaded – it is frowning, however slight. A frown is not natural to a child, and it is often the first sign of commencing mischief in the child's brain.'

Like Graves, Corrigan recognised that, between them, they had created a unique medical movement in Dublin: 'In short the true practical physician adapts his practice to his patient, not his patient to his theory. This constitutes the true practice of medicine. It is on its steady adherence to these principles that the high character of the Dublin school of medicine has been raised and which I am sure it will maintain. It is known throughout Europe and America as essentially the "eclectic school of Europe". Having no theory or hypothesis to support, it accepts information, and is ready to test alleged improvements, come from where they may. It tests them cautiously and carefully in its hospitals, adopts them if worthy of being adopted, or rejects them if found erroneous.'

As a leading member of the Central Board of Health during the Great Famine, Corrigan incurred the wrath of his professional colleagues, most notably Robert Graves, for making what was considered to be a derisory five-shilling-a-day award to doctors working in the famine areas. This had a profoundly adverse effect on his career when, in 1847, he was black-balled ignominiously for the Honorary Fellowship of the College of Physicians.

Government, however, saw matters differently and rewarded his commitment to the famine cause by making him Physician-in-Ordinary to Queen Victoria in Ireland, and later a Baronet of the Empire.

In 1855, Corrigan sat the licentiate examination of the Royal College of Surgeons and Physicians, after which he could no longer be debarred from the fellowship and, two years later, his friend and colleague, William Stokes, proposed him for the highest office of the College – the Presidency – a position he held for an unprecedented term of five years, and during which time he built the College Hall in Kildare Street.

He was a member of the Senate of the Queen's University and a Commissioner for National Education, and in these roles he advocated much needed reform in education.

He was elected a Liberal MP for the City of Dublin in 1870. At Westminster he supported the temperance cause, but his greatest efforts were directed to the issue of university education in Ireland. With remarkable courage and foresight he advocated non-denominational national university education believing that race and creed should not be considerations in third-level education. This stance brought him into bitter conflict with the hierarchy of his church. Corrigan refused to compromise his liberal principles for what he saw to be doctrinaire Catholicism. He feared that the Protestant religious bigotry to which Ireland had been subjected for so long might, with emancipation and the disestablishment of the Church, be replaced by an equally pernicious form of religious intolerance. This conflict with the Catholic Church left him disillusioned and he did not seek re-election to parliament.

He died on 1 February 1880 and was interred in the family vault in St Andrew's Church in Westland Row.

Further reading:

E. O'Brien: *Conscience and Conflict – A Biography of Sir Dominic Corrigan, 1802–1880*, Dublin, 1983.
E. O'Brien, A. Crookshank and G. Wolstenholme: *A Portrait of Irish Medicine – an Illustrated History of Medicine in Ireland*, Dublin, 1984.
D. Coakley: *Irish Masters of Medicine*, Town House, Dublin, 1992.

Eoin O'Brien, Royal College of Surgeons in Ireland, Dublin.

WILLIAM STOKES Medical Author, Researcher and Educationalist

Born: Dublin, July 1804.
Died: Dublin, 7 January 1878.

Pencil portrait of William Stokes by Frederick Burton (courtesy of the National Gallery of Ireland)

Family:
William Stokes was the son of Dr Whitley Stokes, Professor of Medicine to the Royal College of Surgeons Ireland.
Married: Mary Black of Glasgow 1828.
Children: Nine children, of whom Whitley became a distinguished Celtic scholar, Margaret an authority on archaeology and William President of the Royal College of Surgeons of Ireland.

Addresses:
16 Harcourt Street, Dublin
York Street, Dublin
5 Merrion Square North, Dublin
Carraig Breac, Howth, Co. Dublin

The second quarter of the nineteenth century is called the 'golden age' of Irish Medicine. Notabilities in Dublin during that period included doctors whose names are remembered today: Abraham Colles ('Colles' fracture'), Sir Dominic Corrigan ('Corrigan's pulse'), Sir William Wilde ('Wilde's incision'), Robert Graves ('Graves' disease') and William Stokes ('Cheyne-Stokes breathing'). Dr Stokes is unique in this group in being not merely a distinguished individual but a member of an intellectual dynasty.

William Stokes was educated privately before studying medicine in Dublin and Edinburgh; he graduated in Edinburgh University in 1825. While still a student, he published a book on the use of the stethoscope, the first in English on that recent invention.

At the Meath Hospital, Dublin, where Stokes was appointed physician in 1826, he and Robert Graves established a famous teaching centre. Hitherto, medical students had learned from books and lectures with little direct contact with patients. Under the new system favoured by Stokes and his senior colleague, the students were given responsibility for the examination and treatment, under supervision, of patients in the wards and thus gained clinical experience that added a new dimension to their reading.

In addition to this important innovation, William Stokes made contributions to medicine by his descriptions of what are now called 'Cheyne-Stokes breathing' and the 'Stokes-Adams syndrome'. The former, a waxing and waning of breathing, is a sign of ill omen in acute illness and with the elderly. The Stokes-Adams syndrome, characterised by a slow pulse and episodes of unconsciousness, is now known to be due to heart block and can be successfully treated. His *Diagnosis and Treatment of Disease of the Chest (1837)* and *Diseases of the Heart and Aorta* (1854) were influential textbooks with a wide circulation.

The Meath Hospital, founded 1753

The closing phase of Dr William Stokes' life was shadowed by ill health. A widower, he lived in retirement, cared for by his daughter at Carraig Breac, Howth, where, after some months of helplessness resulting from a stroke, he died on 7 January 1878.

Further reading:

F.O.C. Meenan: The Georgian Squares of Dublin and Their Doctors, *Irish Journal of Medical Science,* 149–54, 1966.
M. Neuberger: The Famous Irish Triad – Graves, Stokes, Corrigan, *Irish Journal of Medical Science,* 35–40, 1948.
Sir William Stokes: *William Stokes,* London, 1898.

J.B. Lyons, Department of the History of Medicine, Royal College of Surgeons in Ireland, Dublin.

Born: Dublin, Midnight 3–4 August 1805.
Died: Dunsink Observatory, 2 September 1865.

Sir William Rowan Hamilton and one of his sons (circa 1845) (courtesy of Trinity College Dublin)

Family:
Married: Helen Bayly, 1833 (she died in 1869).
Children: Two sons and one daughter.

Addresses:

1805–1808	Dominick Street, Dublin
1808–1823	Trim, Co. Meath
1823–1827	South Cumberland Street, Dublin
1827–1865	Dunsink Observatory

Distinctions:
Andrews Professor of Astronomy, Trinity College, Dublin, and Royal Astronomer of Ireland, Dunsink Observatory (these appointments were made while Hamilton was still an undergraduate at Trinity College) 1827; Member of the Royal Irish Academy 1832; Cunningham Medal of the Royal Irish Academy 1834, 1848; Knighted 1835; Royal Medal of the Royal Society for his work in optics 1836; President of the Royal Irish Academy 1837–1846; Elected Correspondent (Geometry) of the Académie Des Sciences, Paris 1844; First Foreign Associate of the American National Academy of Sciences 1864.

A line with a direction in space is called a vector. A scalar, on the other hand, is a magnitude without direction. It was Hamilton who introduced these terms. He is best known for his method of quaternions, which was a solution to the problem of multiplying vectors in three-dimensional space. His method, which employs the imaginary number $\sqrt{-1}$, involves three unit imaginaries i, j, k (with $i^2 = j^2 = k^2 = -1$) such that each of them is perpendicular to the others. The product of any two of them corresponds to a 90° rotation of one about the other in a direction depending on the order in which they are multiplied. Thus:

$$ij=k,\ jk=i,\ ki=j,$$
$$\text{but } ji = -k,\ kj = -i,\ ik = -j,$$

The product of two vectors consists therefore of a real scalar part plus three terms in i, j, and k. Thus it has four terms in all, hence the name *quaternion* (from the Latin *quaternio*, a set of four).

The revolutionary aspect of Hamilton's discovery, which deeply influenced later developments in algebra, is that it does not correspond with traditional multiplication. If we multiply a real number a by another real number b, the product is ab. If we multiply b by a, the answer is the same; that is ab = ba. However, in the case of Hamilton's imaginaries i, j, k, we find that ij = -ji, jk=-kj, and ki = -ik (see above).

Quaternions played a seminal role in the invention of vector analysis, and have found applications in physics. Hamilton's major treatises on the subject are *Lectures on Quaternions* (Dublin, 1853) and *Elements of Quaternions* (London, 1866).

Among Hamilton's other important works are his early optical researches, in which he sought to make optics a mathematical science based on general principles. This work enabled him to predict (1832) that under certain conditions a light ray undergoes an unusual kind of refraction, called 'conical refraction', on passing through biaxial crystals (see Humphrey Lloyd). Hamilton's search for general principles extended to dynamics. His reformulation of the equations of motion of Joseph-Louis Lagrange (1736–1813) became – and remains – a powerful tool in classical mechanics and in modern wave mechanics.

Dunsink Observatory, Hamilton's home from 1827 to 1865 (courtesy of P.A. Wayman)

Hamilton had quite extraordinary linguistic gifts. Under his uncle's special tuition at Trim, he could read Greek, Latin, and Hebrew at the age of 3 or 4, and it seems he had acquaintance with some 15 languages by the time he was 10. When at Trinity College he was unbeaten in every examination in both Classics and Science in which he entered, achieving the highest grade in every case. He could also perform prodigious feats of mental calculation.

He was a habitual scribbler: he would scribble his ideas on literally anything – scraps of paper (he always carried a notebook with him), his finger nails, even the shell of his morning egg! His most famous 'scribble' came on 16 October 1843, while walking to the Royal Irish Academy with his wife along the Royal Canal. As he was passing Brougham Bridge, the idea of quaternions suddenly came to him. He stopped, took out his penknife, and on the stone parapet of the bridge scratched the fundamental formulae of his quaternion algebra.

His personal life was burdened with anguish: the real, unattained love of Hamilton's life was Catherine Disney, whom he first met on 17 August 1824, but who married a Rev. William Barlow in May 1825.

Further readling:

Thomas L. Hankins: *Sir William Rowan Hamilton,* Johns Hopkins University Press, Baltimore and London, 1980.

Alan Gabbey, Department of Philosophy, Barnard College, Columbia University, New York.

47 WILLIAM THOMPSON Naturalist

Born: Belfast, 2 November 1805.
Died: London, 17 February 1852.

Family:
William Thompson was the eldest son of William and Elizabeth Thompson of Wolfhill, Belfast, who owned a prosperous linen business.
He was unmarried.

Addresses:

1805–c1820	Wolfhill, in north Belfast, where the family linen export business was situated
c1820–1852	1 Donegall Square West, Belfast, (now the Scottish Provident Insurance Building) the city centre business office Holywood House, Holywood, Co. Down was the family's country residence

Distinctions:

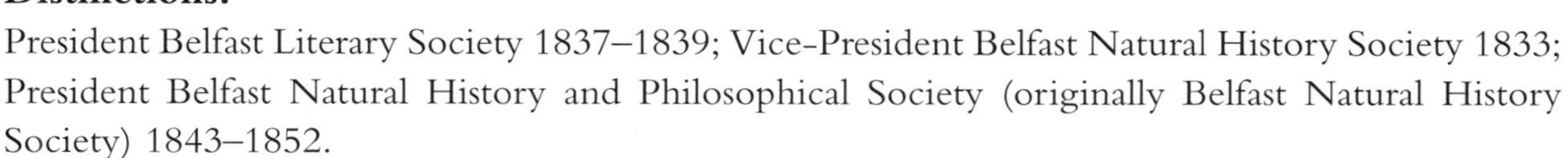

President Belfast Literary Society 1837–1839; Vice-President Belfast Natural History Society 1833; President Belfast Natural History and Philosophical Society (originally Belfast Natural History Society) 1843–1852.

The composition of the Irish fauna and flora has been determined by many different influences – the distance of Ireland from the mainlands of Britain and Europe, its size, geology and climate, probably being the most important. These factors combine to give Ireland its unique assemblage of plants and animals which differs considerably from that of Britain and Western Europe. The differences had been apparent to casual observers for a very long time. However, until William Thompson's work, no formal attempt had been made to quantify and list the Irish species. This is an essential first step to further work on the formulation of theories as to the origins of Ireland's fauna and flora – the study of biogeography – one of today's most topical avenues of investigation.

Although without formal scientific education, William Thompson left the family business after a few years, and became a naturalist of wide interests. From about 1827, he started investigating, noting and listing details of Ireland's fauna and flora which, until then, had attracted less study than those of most other European countries. Though he specialised in ornithology and molluscs, he also collected other vertebrates and invertebrates and formed a fine herbarium. His appetite for natural history had been stimulated by a close family friend, George Crawford Hyndman, his senior by nearly ten years, who brought him into early contact with Belfast Natural History Society.

Through careful encouragement and acknowledgement of their assistance, he built up a large network of reliable correspondents and acquaintances throughout Ireland, and they sent him their observations and records. Thompson realised the importance of publication of his observations to encourage further work on the Irish fauna. Between 1827 and 1852, he published some 80 scientific papers on Irish natural history in journals such as *Magazine of Botany and Zoology* and *Proceedings of the Zoological Society of London* to which he was a frequent contributor. His ability brought him to the notice of other leading British naturalists such as P.J. Selby, George Johnston and Professor Edward Forbes, and Thompson became an active member of the British Association for the Advancement of Science. He prepared for them a series of Reports on the vertebrate and invertebrate fauna of Ireland. These were to form the basis of a multi-volume book on the natural history of Ireland which Johnston, considering Thompson ideally fitted for the task, had encouraged him to undertake. The first three volumes of *The Natural History of Ireland*, dealing with Irish birds, were published between 1849 and 1851. Then, unfortunately, his health declined and he died at the age of forty-seven without completing the work. A fourth volume was posthumously published in 1856, having been compiled and edited from Thompson's copious notes by his naturalist friends, Robert Patterson, Robert Ball and George Dickie. Thompson's landmark book remains an indispensable reference work on early nineteenth century Irish zoology as it contains detailed notes and references to many Irish collectors, collections and first records of species in Ireland.

Selection of birds from the Thompson collection

Belfast Museum, about 1840

Altogether Thompson added about 1000 species to Irish fauna lists, among which were a number of species he described as new to science. In 1852, as a tribute to his achievements as one of the foremost naturalists that Ireland has produced, Belfast Museum opened the 'Thompson Room', to display his life's work.

Further reading:

The Natural History of Ireland, Volume 4, contains a detailed memoir of Thompson.

Helena Chesney, 35 Colenso Parade, Belfast BT 9 5AN.

Born: Mallow, 1806.
Died: London, 12 March, 1843.

A copy of a painting of Robert Murphy (courtesy of The Master and Fellows of Gonville and Caius College, Cambridge)

Family:
His father was a local shoe-maker. He was the seventh in a family of ten. Nothing is known of his siblings. He never married. He is buried in an unmarked grave in London.

Addresses:
1825–1833 Cambridge University
1834–1843 Various lodgings, London

Distinctions:
Fellow of Gonville and Caius College Cambridge 1829; Fellow of the Royal Society 1834; Honorary Member of the Royal Cork Institution 1838 (Institution ceased in 1861).

At the age of eleven, Murphy broke his thigh-bone and was confined to bed where he read almanacs, Euclid and also a book on algebra. He started to solve mathematical puzzles which were printed in a Cork newspaper, attracting the attention of the author, a Mulcahy man, and a local miller, John Dillon Croker, who subsequently became his patron. Free schooling was provided by a Mr Hopley and, when he finished, a public subscription, lead by Croker, enabled him to go to Cambridge. His entry to Cambridge was assisted by a Mr McCarthy from Cork, himself a Cambridge graduate. Murphy published a 20 page pamphlet refuting a claim made by a Co. Tipperary priest and a graduate in logic from Maynooth College on how to make a cube a double of a cube, it being one of the topical problems of the time.

Murphy obtained his BA degree in 1829, qualifying as third wrangler in mathematics. He took deacon orders in 1831 and a MA in 1832. He published numerous mathematical papers written at a 'very high level of mathematical sophistication...far beyond those of his contemporaries'. He wrote the first analytical textbook on electrical phenomena.

He left Cambridge at the end of 1833, with personal debts accumulated due to his 'dissipative habits'. He subsequently worked as examiner in mathematics and natural philosophy for the University of London. He wrote a book on algebraic equations for The Society for the Diffusion of Useful Knowledge. George Boole mentions Murphy's work on analytical operations. William Thomson (later Lord Kelvin) also cites Murphy's work.

Further Reading:
G.C. Smith: *History of Mathematics: Robert Murphy,* Monash University, Australia, 1984.
N. Barry: Mallow's Prodigy – Robert Murphy, *Mallow Field Club Journal,* **16,** 157–175, 1998.

Noel Barry, Department of Electrical and Electronics Engineering, Cork Institute of Technology.

49 WILLIAM WILLOUGHBY COLE (Third Earl of Enniskillen) Geologist

Born: Florence Court, 25 January 1807.
Died: County Fermanagh, 12 November 1886.

The third Earl of Enniskillen

Family:
The Earls of Enniskillen were prominent landowners settled in County Fermanagh since the Plantation of Ulster in 1610. The third Earl succeeded to the title in 1840.
Married: 1. Jane Casamaijor 1844 (died in 1855).
2. Hon. Mary Emma Broderick 1865, daughter of the sixth Viscount Midleton.
Children: There were seven children from his first marriage.

Distinctions:
Fellow of the Geological Society 1828; Fellow of the Royal Society 1829; MP for Fermanagh 1831–40; afterwards sat in the House of Lords; Member of the Royal Irish Academy 1846; First President of the Royal Geological Society of Ireland 1865; Honorary Doctor of Civil Law Oxford University; Honorary LLD Universities of Dublin and Durham.

Little is known of Viscount Cole's boyhood days except that he was educated at Harrow School, but he must have developed an early interest in geology, for, on going to Christ Church, Oxford, in 1826 at the age of 19, he enrolled in the classes of William Buckland, first Professor of Geology at the University. Buckland, an eccentric but talented teacher, inspired Cole with an enduring passion for the study of fossils. At college, Cole met another young aristocrat and fellow student, Sir Philip Egerton, and the pair became life-long friends and later collaborators in collecting fossil fish, for which both became famous.

Cole and Egerton made their first foreign journey in the summer of 1830, when they travelled through Europe on a geological Grand Tour. Thus they met, in Munich Museum, Louis Agassiz, who was to influence profoundly the future course of their geological researches. Agassiz, a brilliant naturalist, was trying to make a complete classification of all the fishes, both fossil and living, and suggested that they could help by collecting all the different kinds of fossil fish in the world. The pair, realising that this was an entirely new field of research, henceforth made the collecting of fossil fish their main scientific pursuit. They returned home later that year, laden with some of the finest fossil fish that could be found in Europe.

Lord Enniskillen was a convivial man who enjoyed good company. In 1828, at the age of 21, he had joined the Geological Society of London, where he met the leading geologists of his day, playing host to them at his London rooms. They knew of his ambition to form a comprehensive collection of fossil fish and frequently helped to obtain for him rare or newly discovered examples. For his part, Lord Enniskillen never let slip any opportunity to secure a new specimen, and he used all his charm and persistence to obtain what he wanted, by purchase, exchange or unashamed begging from friends. By the late 1830s he had many hundreds of these fossils arranged in his private museum at Florence Court – in a pavilion in one of the flanking wings to the main house. He inserted a large lantern window in the roof for illumination and lined the walls with cases to

Florence Court, 1870, the Pavilion, left, housed the third Earl of Enniskillen's private museum (photograph courtesy Earl of Erne)

display his treasures. They eventually numbered nearly 10,000 individual objects, most of them fossil fish. They were the best obtainable, and many were important to science. It remains one of the largest and scientifically most important collections of fossil fish ever assembled.

As well as his parliamentary duties, Lord Enniskillen was prominent in other public affairs, especially in the Orange Order, of which he was an enthusiastic member for over fifty years. This brought him controversy and danger, but his loyalty never wavered and he became first imperial Grand Master of the Order world-wide in 1866.

Lord Enniskillen transferred his collection to the British Museum in London in 1883 (for £3500) where it was arranged in the newly opened Natural History building, alongside the collection of his old friend Sir Philip Egerton, who had died a few years earlier. Lord Enniskillen himself died shortly after at the age of 79 and was buried in the family vault in St Macartan's Cathedral, Enniskillen. Lord Enniskillen published nothing on his scientific work. His contribution to science was his collection and his generosity in allowing others to use it. His specimens figured – and continue to figure – in the standard works on fossil fish.

Today Florence Court is owned by the National Trust, but there is nothing left there to remind the visitor of the third Earl of Enniskillen's scientific life. He deserves to be more widely known as a gracious Irishman, who lived life to the full and whose legacy to geology is safely preserved in the Natural History Museum, London.

Further reading:

K.W. James: *Damned Nonsense! – The Geological Career of the Third Earl of Enniskillen,* Belfast, 1986.

Kenneth W. James, Geology Department, Ulster Museum, Belfast.

Born: Dundee, Forfarshire, Scotland, 23 April 1808.

Died: Glasnevin, Dublin, 9 June 1879.

Courtesy of the National Botanic Gardens, Glasnevin, Dublin

Family:

Married three times:

1. 7 April 1836, Hannah Bridgford (died December 1840), daughter of Thomas and Sarah Bridgford; two children.
2. 1843, Isabella Morgan (died 7 November 1847); two children.
3. 7 December 1854, Margaret Baker (died 14 April 1917, aged 88), daughter of Thomas Baker; five children including Frederick William Moore, born 3 September 1857, died 23 August 1949.

Address:

Botanic Gardens, Glasnevin, Dublin

Distinctions:

Honorary Doctorate, University of Zurich 1863; Member of the Royal Irish Academy.

David Moore was a Scot and it is said that he retained his Scottish accent until his death. He came to Dublin in 1828 as assistant to James Townsend Mackay, Curator of the College Botanic Garden, Ballsbridge, Dublin. Moore had trained as a gardener in his native Scotland, and under Mackay, who was also a Scot, he honed his skills and improved his knowledge of exotic plants. Undoubtedly he also gained a good knowledge of Irish native plants from Mackay. In 1834, having been unsuccessful in the election for the vacant curatorship at the Royal Dublin Society's Botanic Gardens, Glasnevin, Moore joined the Irish Ordnance Survey and worked for almost four years in the north of Ireland, collecting and cataloguing the native flora for an ambitious survey of Ireland's natural resources that was never completed. In 1838, when the Glasnevin post became vacant again, on the resignation of Ninian Niven, Moore was successful in obtaining this post and he remained at Glasnevin until his death in 1879, by which time he was styled Director. Until 1847, Professor Samuel Litton was the Royal Dublin Society's resident Professor of Botany (and nominal director), but his successor William Henry Harvey (1811–1866) did not live in the Botanic Gardens and took little part in the management.

David Moore was both an expert horticulturist and an excellent field botanist. He was respected by his contemporaries, and included among his numerous correspondents were Charles Darwin, the explorer Richard Spruce, and Sir William Hooker (Director of the Royal Botanic Gardens, Kew). Moore's contributions to Irish botany ranged from the botany section of the Ordnance Survey's famous *Templemore Memoir* (first issued in 1835, revised and reissued in 1837), to a later (1873) catalogue of native mosses, and, jointly with Alexander Goodman More, the remarkable species-by-species account of the distribution of native flowering plants, conifers and ferns entitled *Contributions Towards a Cybele Hibernica* (1866). *Cybele Hibernica* and a companion paper included a pioneering example of maps portraying plant distribution patterns.

Under Moore's direction, the Glasnevin Botanic Gardens were greatly improved and expanded. The plant collections were built up and numerous projects successfully accomplished, including the construction of three separate glasshouse ranges including Richard Turner's curvilinear range. With these facilities, Moore was able to accomplish two remarkable advances in horticulture. The first was the raising of orchids from seeds. No-one else had succeeded in germinating the minute seeds and keeping the seedings alive until they were mature, blooming plants, yet this was achieved at Glasnevin between 1845 and 1849. The second advance was less difficult but no less remarkable. Several species of North American insectivorous pitcher-plants, *Sarracenia*, were artificially cross-pollinated and hybrid seedlings were raised. They created a sensation when exhibited at the International Botanical and Horticultural Congress in Florence in 1872, and Moore was awarded a gold medal.

David Moore must also be given credit for the melancholy observation of blight on potato plants growing at the Glasnevin Botanic Gardens on 20 August 1845, the first signs of the impending catastrophe, the Great Famine. Moore continued to observe the progress of the murrain (as the blight was called) and in the following year accepted that its cause was a microscopic fungus. He tried to find a treatment to cure or prevent the disease but failed. In the wake of famine came disease. David Moore's second wife apparently succumbed and died in November 1847.

The University of Zurich awarded Moore an honorary doctorate in 1863 yet, while he was elected a Member of the Royal Irish Academy, he received no other honours in Ireland. However David Moore is honoured by the names of a number of native and exotic plants. The French botanist Charles Lemaire named the genus comprising the gigantic pampas grasses *Moorea* to mark the fact that the genus was first grown at Glasnevin, but these plants are now called *Cortaderia*. Among Irish plants named after him are several hybrids; a marshwort, *Apium* x *moorei*, a horsetail *Equisetum* x *moorei*, and a remarkable willow *Salix* x *grahamii* nothovar. *moorei* that Moore found on Muckish Mountain, County Donegal, in 1866 (it has not been found in the wild again). *Sarracenia* x *moorei*, the first hybrid pitcher-plant also bears his name, as does the splendid *Crinum moorei* from Natal and an enigmatic American passionflower *Passiflora mooreana*.

Dr David Moore was succeeded at Glasnevin by his son Frederick (1857–1949). Father and son were in charge of the Botanic Gardens for a continuous period of 84 years, 1838 to 1922.

Further reading:

E.C. Nelson and M.R.D. Seaward: Charles Darwin's Correspondence with David Moore of Glasnevin on Insectivorous Plants and Potatoes, *Biological Journal of the Linnean Society*, **15**, 157–164, 1981.

E.C. Nelson: In Honour of David and Frederick Moore, *Moorea*, **1**, 14, 1982.

E.C. Nelson and E.M. McCracken: *The Brightest Jewel: A History of the National Botanic Gardens, Glasnevin, Dublin*, Boethius Press, Kilkenny, 1987.

E.C. Nelson: Mapping Plant Distribution Patterns: Two Pioneering Examples from Ireland published in the 1860s, *Archives of Natural History*, **20**, 391–403, 1993.

E.C. Nelson: *The Cause of the Calamity. Potato Blight in Ireland, 1845–1847, and the Role of the National Botanic Gardens, Glasnevin*, The Stationery Office, Dublin. 1995.

E. Charles Nelson, Outwell, Wisbech, England.

ROBERT KANE Chemist and Educationalist

Born: Dublin, 24 September 1809.
Died: Dublin, 16 February 1890.

Family:
Son of John Kane, a member of the United Irishmen, who studied chemistry while a refugee in Paris and on his return to Dublin started a chemical works, making lime, sulphuric acid and bleaching powder.
Married: Katherine Baily of Newbury, Berkshire 1890.
Children: Four sons, Robert Romney, Henry Coey, Francis Bailey, Roderick, and a daughter, Valentine.

Addresses:

1809–1832	48 Henry Street, Dublin
1832–1845	23 Gloucester Street, North (now lower Sean MacDermott Street)
1846–1856	Gracefield, Booterstown
1857–1873	Wickham, Dundrum
1874–1878	21 Raglan Road
1879–1889	Fortlands, Killiney
1889–1890	2 Wellington Road

Distinctions:
Professor of Chemistry, Apothecaries' Hall 1831–45; Member of the Royal Irish Academy 1832; Lecturer in Natural Philosophy and (later) Professor of Chemistry at the Royal Dublin Society 1834–47; Director Museum of Irish Industry 1845; Knighted 1846; First President Queen's College Cork 1846–73; Fellow of the Royal Society 1849; Honorary LLD University of Dublin 1868.

Even if it were allowed nowadays, to take seven years to obtain a pass BA degree would not be considered an auspicious start to a career in science and education. This is how long Robert John Kane took to obtain his degree in Trinity College, where he became a student in 1828. However, to have published several learned papers on chemistry, to have qualified to practise medicine and won a gold medal in doing so, to have successfully analysed a mineral as the arsenide of manganese and have it called after you (in this case Kaneite), to have become a Professor of Chemistry, to have been elected a Member of the Royal Irish Academy, to have published a textbook on pharmacy, and to have founded a scientific journal that still continues in print today, over the same period of seven years, may perhaps excuse Kane's tardiness in completing his studies as an undergraduate at Trinity College.

Despite his brilliant medical beginnings, Kane's main scientific interests were firmly attached to chemistry. His contributions to chemical knowledge were many. While working in Liebig's laboratory at Giessen in the German province of Hesse in 1836 he isolated acetone, $(CH_3)_2CO$, from wood spirit. On returning to Ireland and while working at the laboratory of the Royal Dublin Society, he changed this simple aliphatic compound into an aromatic hydrocarbon, which he called mesitylene, C_9H_{12}. His method was to distil acetone carefully with sulphuric acid. Mesitylene is the symmetrical tri-methyl benzene and it was the first cyclic hydrocarbon to be synthesised from a straight-chain compound. Kane was an active chemical experimenter and a bold theorist. He was also the first to suggest that alcohol, ether and some esters contained a 'radical', C_2H_5. His researches on the combination of ammonia with metallic salts acquired for him a European reputation. Both the Royal Irish Academy and the Royal Society acknowledged his work on the chemistry of the natural dyestuffs, achil and litmus, which can be extracted from lichens, by presenting him with medals. Kane's role as an active chemical researcher ended after 1841 when he published a textbook on chemistry, which was to become a best-seller. In 1844 he published his brilliant and detailed assessment of Ireland's natural resources and sources of industrial power.

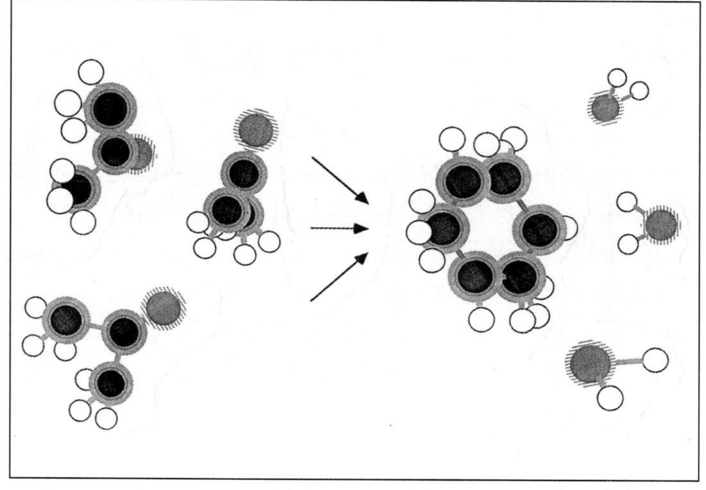

Computer-generated simulation of Kane's synthesis of mesitylene from acetone

Concurrently with all these activities Sir Robert (he was knighted in 1846) was a very busy administrator. Not only was he advising the government about education, at the national level as well as at university level, but during the Famine years he was giving advice, as one of the eight Irish relief commissioners, on food distribution and later, as a member of the Committee of Health, on how to deal with the outbreaks of typhus fever which were occurring all over the country as a result of the terrible conditions. As director of the Museum of Irish Industry, an institution that was what we now would call a research association, he oversaw a number of investigations, including one into the possibility of growing beet as a source of sugar. The Museum also involved itself in the schemes for the provision of teaching in applied sciences, and some of its staff were incorporated into the Royal College of Science when it was founded in 1867.

After Kane was appointed the first president of the newly formed Queen's College Cork, he continued as director of the Museum of Industry in Dublin for many years. His non-residence in Cork over the 28 years for which he held the latter appointment caused a great deal of controversy. A man of strong opinions, his influence on the moulding of science and education in nineteenth-century Ireland was immense.

Further reading:

T.S. Wheeler and others: *The Natural Resources of Ireland – A Series of Discourses Delivered Before the Royal Dublin Society* Dublin, 1944.
R. Kane: *The Industrial Resources of Ireland* Dublin, 1844.
D. Reilly: *Sir Robert Kane*, Cork, 1942.
D. Reilly: Robert John Kane (1809–90): Irish Chemist and Educator, *Journal of Chemical Education,* **32,** 404, 1955.
B. Kelham: Robert Kane, *Studies,* **56,** 297, 1967.

William J. Davis, University Chemical Laboratory, Trinity College Dublin.

Born: Dublin, 15 October 1809.
Died: Donnybrook, 29 October 1899.

Family:
Married: Mary Elizabeth (Eliza) who died on 5 January 1894, but whose maiden name is unknown.
Children: We know of only one son, Bernard Patrick, who died at the age of 20, in 1874.

Addresses:
In 1856 he gives an address at 78 Capel Street, Dublin, premises which were then an hotel. 1872–1894 his address was 29 Bath Avenue, Sandymount Co. Dublin. No. 29 was demolished about thirty years ago, but many similar, neighbouring houses survive. He died, in 1899, at 52 Donnybrook Main Street, which was owned by members of the Ganly family until the house was demolished for road widening in about 1930.

Patrick Ganly's family came from Co. Roscommon to Dublin where, he tells us, his father practised as an architect. Ganly himself joined Richard Griffith's staff in the Boundary Survey in 1827; he worked in the Ordnance Survey from 1830 to 1833; and then he worked, under Griffith, in the Valuation Office until he was made redundant in 1860.

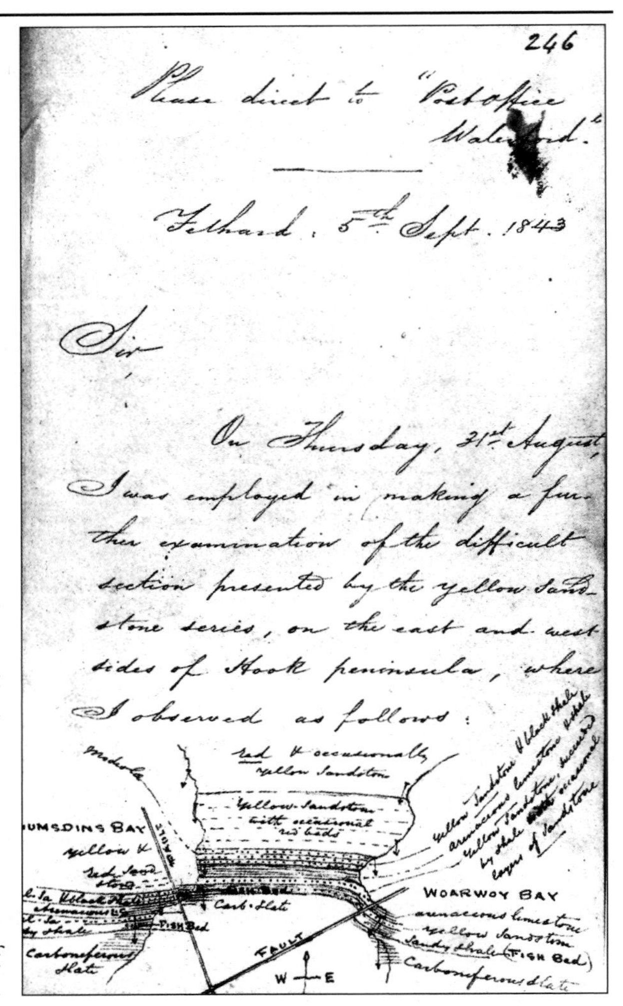

246

Please direct to "Post office Waterford."

Fethard, 5th Sept. 1843

Sir

On Thursday, 21st August I was employed in making a further examination of the difficult section presented by the yellow sandstone series, on the east and west sides of Hook peninsula, where I observed as follows:

The first page of a typical letter from Ganly to Griffith. Each of his letters describes his geological work during the previous few days and contains maps and diagrams illustrating his most recent geological discoveries.

We would know very little indeed about Ganly's geological work had not three volumes of his letters been found in the Valuation Office during the 1940s. A fourth volume has since been discovered in a private library. The letters – over 600 survive – were written by Ganly to Griffith between the years 1837 and 1847 and they show that from 1835 until 1847 Ganly worked for long periods on the construction of Griffith's great geological map of Ireland. The letters prove that it is largely thanks to Ganly's hard work and ability that the later versions of this map are so remarkably accurate.

It was while working on Griffith's map that Ganly made the discovery for which he is now internationally known. Near Carndonagh in Co. Donegal, in the mid-1830s, he observed that the structure most commonly found in sandstones known to geologists as 'cross-stratification', represents the cross-sections of ancient sand ripples. It occurred to him that the characteristic shape of the ancient ripple-profiles, as revealed in cross-stratification, could be used to distinguish rock layers that are right way-up from those that have been turned upside-down by intense folding. He tested his hypothesis and proved it to be correct on the shores of the Dingle Peninsula, at Fahan, in June 1838. Ganly did not make his discovery known until 1856 at a meeting of the Dublin

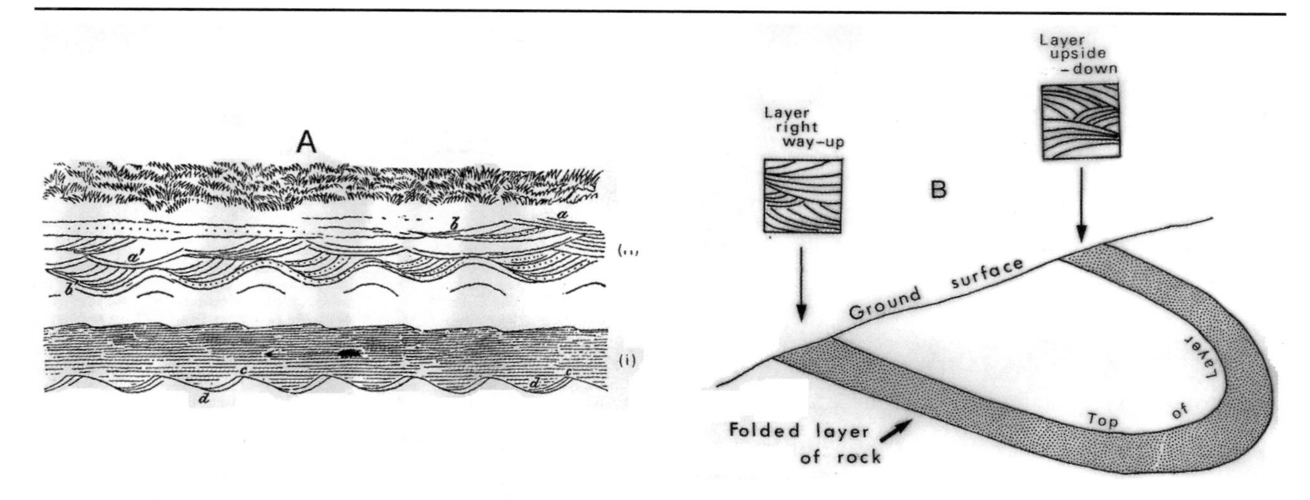

Ganly's findings on cross-stratification. (A) An illustratlon from his 1856 paper showing:
(i) below, cross stratification forming on the downstream side of sand ripples on the bed of a stream flowing from right to left; (ii) above, cross section showing the cross stratification formed by several generations of ripples built one on top of another.
(B) A cross-section of a folded layer of rock which only appears at the ground surface below the arrows. The insets show the appearance of the cross stratification where the layer is upside-down and where it is right-way-up. Note that the fold can be traced, using Ganly's method, even though it is underground. Folds of this type are widespread in parts of southwestern Ireland.

Geological Society, which published his findings in its journal. Unfortunately his contemporaries did not appreciate the significance of his discovery and it was forgotten. Nearly seventy years elapsed before the use of cross-stratification as evidence of way-upness was independently re-discovered by American geologists. Ever since, it has been in day-to-day use by geologists the world over and is described in most basic geological textbooks.

Ganly was a hard-working and talented man who clearly derived great satisfaction from his geological work, but it was insufficient to satisfy his intellectual curiosity. In the early 1840s he enrolled as an external student in Trinity College – the restrictions on Roman Catholic attendance had been lifted some time earlier – studying Greek, Latin, logic, astronomy, mathematics and physics. He received his BA degree in 1849. Oddly, when he received his degree, Griffith was in the congregation because, at the very same ceremony, Griffith received an honorary LLD Ganly obtained no public honours in his lifetime and his enormous contribution to Griffith's map remained a closely guarded secret. Griffith had no authority to employ a valuator as a geologist and he had no wish to share credit for his map with anyone. Ganly married long before the keeping of centralised records of births, deaths and marriages started in 1863. As a result we know almost nothing about his family.

Further reading:

New Dictionary of National Biography.
J.B. Archer: Patrick Ganly – Geologist, *The Irish Naturalists' Journal,* **20**, 142–148, 1980.
G.L. Herries Davies: *Sheets of Many Colours, the Mapping of Ireland's Rocks 1750–1890,* Royal Dublin Society Historical Studies in Irish Science and Technology No. **4**, Dublin, 1983.
R.C. Simington and A. Farrington: A Forgotten Pioneer: Patrick Ganly, geologist, surveyor and civil engineer (1809–1899), *Journal of the Department of Agriculture,* **46**, 36–50, 1949.

Jean Archer, Nenagh, Co. Tipperary.

53 JAMES MacCULLAGH Mathematician and Mathematical Physicist

Born: Landahussy, County Tyrone 1809.
Died: Dublin, 24 October 1847.

Marble bust by Christopher Moore RHA (courtesy of Trinity College Dublin)

Family:
Son of James (1777–1857) and Margaret (1784–1839) MacCullagh and the eldest of the eight of their twelve children who survived beyond childhood. The family circumstances were modest but two of his brothers were to follow him to Trinity College. One of these, John (1811–91), was subsequently called to the Bar and appointed a resident magistrate. James MacCullagh never married.

Addresses:
1809–1824 Landahussy, Upper Badoney, and Strabane, both in County Tyrone
1826–1847 Trinity College Dublin; he lived at various times in houses 14, 7 and 40

Distinctions:
Fellow of Trinity College Dublin 1832; Member of the Royal Irish Academy 1833; Erasmus Smith's Professor of Mathematics, University of Dublin 1835–43; Cunningham Medal of the Royal Irish Academy 1838; Copley Medal of the Royal Society 1842; Secretary of the Royal Irish Academy 1842–46; Erasmus Smith's Professor of Natural and Experimental Philosophy, Trinity College Dublin 1843–47; Fellow of the Royal Society 1843.

James MacCullagh was only fifteen when he was admitted to Trinity College Dublin in 1824. It was a good time to study mathematics in Trinity. A major reform of the syllabus had just been implemented, chiefly owing to the efforts of Bartholomew Lloyd, who held the chairs of mathematics and natural philosophy in succession before being elected Provost in 1831. There were several able mathematicians among the Fellows of the College and the lectures and prescribed texts reflected the new developments in mathematics taking place on the Continent, especially in France. Among his near contemporaries as undergraduates, both a few years senior to MacCullagh and both to become his colleagues as professors in the University of Dublin, were Humphrey Lloyd, the son of Bartholomew, and William Rowan Hamilton, who was to achieve international renown as a mathematician.

MacCullagh's real mathematical talent lay in geometry. He was especially interested in the ellipsoid and other, so-called, surfaces of the second order. As Professor of Mathematics he lectured on geometry, and exercised a profound influence on a generation of students some of whom, such as George Salmon, were to make their own major contributions in the subject. MacCullagh is, however, better known for his work in mathematical physics, particularly in the development of mathematical models for the aether. One of the major scientific quests throughout the nineteenth century was to identify and to describe in mathematical and mechanical terms this medium, to which the name aether had been given, whose existence was supposed to be essential in order for

Sketches of James MacCullagh made after his death by F.W. Burton (courtesy of the National Gallery of Ireland)

The Cross of Cong, MacCullagh's gift to the nation (courtesy of the National Museum of

light waves to propagate. It was only in the opening years of this century, following the experimental observations of Michelson and Morley and the radical re-interpretation of basic physical concepts by Einstein in his Special Theory of Relativity, that the aether hypothesis was put aside. MacCullagh's models combined mathematical subtlety, which used his skill as a geometer to the full, with a profound appreciation of the essential physics of light propagation. After his death, when it came to be realised that light was an aspect of the wider and newly discovered phenomenon of electro-magnetic radiation, G.F. Fitzgerald showed that McCullagh's aether could equally well describe the more general phenomenon.

MacCullagh played a key role in building up the Academy's collection of Irish antiquities. Although not a wealthy man, he himself bought the Cross of Cong for 100 guineas to present to the Academy and contributed towards the purchase of other treasures. This was the start of the collection now in the National Museum. In August 1847 he stood, unsuccessfully, as parliamentary candidate for one of the Dublin University seats. Later that year, in one of his periodic bouts of depression, he took his own life. An obituary notice in *The Nation* described him as 'a warm and ardent Nationalist'. This is misleading; he was not a Nationalist, but he was a patriotic Irishman devoted to his country and profoundly critical of the lack of national self-respect which he saw about him.

Further reading:

Dictionary of National Biography and *Dictionary of Scientific Biography* and *Obituary Notices* of the Royal Irish Academy and the Royal Society.

T.D. Spearman: James MacCullagh 1809–1847, in J. Nudds, N. McMillan, D. Weaire and S. McKenna Lawlor (eds), *Science in Ireland 1800–1930*, Dublin, 1988.

B.K.P. Scaife: James MacCullagh, MRIA, FRS, 1809–1847, *Proceedings of the Royal Irish Academy*, Section C, **90**(3), 67-106, 1990.

David Spearman, Trinity College, Dublin.

Born: Dublin, 3 June 1810.
Died: London, 5 November 1881.

Courtesy of the Royal Society

Family:
Married: Cordelia Watson in 1831, by whom he had three sons and three daughters. His son, John W. Mallet FRS, became Professor of Chemistry, successively at the Universities of Alabama, Louisiana and Virginia. Following the death of his first wife in 1854, Robert married Mary Daniel in London in 1861.

Addresses:

1810–1836	7, 8, 9 Ryders Row, Dublin
1836–1858	Delville, Glasnevin
1858	1, Grosvenor Terrace, Monkstown
After 1860	11 Bride Street, Westminster

Distinctions:
Fellow of the Royal Society 1854; Cunningham Medal of the Royal Irish Academy 1862; Honorary MAI (Master of Engineering) 1862 and Honorary LLD 1864, University of Dublin; President, Institution of Civil Engineers of Ireland 1866; President, Geological Society of Dublin 1846.

Mallet entered his father's iron founding business at the age of twenty-one and built the firm into one of the most important engineering works in Ireland, undertaking large contracts, including much of the ironwork required by the major railway companies.

In the Victoria Foundry at Ryder's Row, he researched extensively into the properties and strengths of materials, corrosion, and problems associated with the cooling of large iron castings – being one of the few early nineteenth century iron founders to attempt a scientific explanation of fracture in terms of the 'molecular structure' of the metal. In 1852 he patented the 'buckled plate', used particularly to strengthen flooring, combining the maximum of strength with the minimum of weight.

Mallet held a lifelong conviction that the most effective approach to the study of the earth sciences was through measurement, quantification, and experimentation. He is regarded as one of the founders of modern seismology (the study of earthquakes) and measured the speed of travel of shock waves through the earth. Together with his son, John W. Mallet, he compiled an earthquake catalogue and seismic map of the world (1850–1858), and was the first to determine an earthquake's focus or epicentre (Naples,1858). This, and associated later work on volcanoes, represented attempts to understand the behaviour of the materials in the earth's crust in terms of their underlying fundamental structures.

Further reading:
R.C. Cox (ed.): *Robert Mallet, F.R.S., 1810–1881,* Proceedings of Centenary Seminar, Institution of Engineers of Ireland and Royal Irish Academy, Dublin, 1982.

Ronald Cox, Centre for Civil Engineering Heritage, Trinity College, Dublin 2.

Born: Summerville House, Limerick, 5 February 1811.
Died: Torquay, England, 15 May 1866.

Pencil sketch c.1850 by F.W. Burton (courtesy of the National Gallery of Ireland)

Family:
Son of Joseph Massey and Rebecca (née Mark) Harvey. Married: Elizabeth Lecky Phelps in Limerick on 2 April 1861. There were no children.

Distinctions:
Fellow of the Royal Society 1858; Fellow of the Linnean Society of London, 1857; Member of the Royal Irish Academy; MD (hon. causa), University of Dublin, 1844.

As a child Willliam Harvey often spent his summer holidays at Milltown Malbay on the coast of County Clare where he became fascinated by the sea, by the flotsam it deposited on the beach and by the shells and sea-weeds. Sea-weeds (marine algae) became his lifelong passion.

The Harveys were Quakers, and William, who was the youngest of eleven children, went to the great Quaker schools at Newtown, Co. Waterford, and Ballitore, Co. Kildare, where his love of natural history was encouraged. After he left school William took an interest in non-flowering plants – mosses and liverworts, for example – and this led to his first contacts with Professor William Hooker, who was to become a close friend.

Harvey's mother died in 1831 and, after the death of his father in 1834, William sought a government position overseas. Through the good offices of Cecil Spring Rice, MP for Limerick, his brother Joseph was appointed colonial treasurer at Cape Town – this was an error, for William should have been appointed. However, William decided to accompany his brother, hoping to spend his time collecting and studying the flora of the Cape of Good Hope, but Joseph Harvey became ill soon after reaching Cape Town and he died two weeks after beginning a return voyage to Europe. William was then appointed in his brother's stead and returned to Cape Town in 1836. He resumed his interrupted botanical studies, joking that he spent more time collecting plants than collecting taxes and thus should be called 'her majesty's pleasurer-general not treasurer-general'.

William Harvey too became ill during his second tour of duty and resigned his position in 1841. However, he had done valuable work on South African botany and encouraged others to collect plants for him: all this was to lead to a series of important publications, including the first three volumes of *Flora Capensis.*

Harvey returned to Ireland and, in 1844, after the death of Dr Thomas Coulter, he was appointed Curator of the Herbarium in Trinity College Dublin; he had no teaching duties, only the care of the collection of dried and labelled plants that comprised the herbarium. He spent many hours sorting, identifying and naming the specimens and publishing the results of his studies. Thousands of specimens from overseas collectors were sent to Trinity College for Harvey to catalogue and study, especially specimens from the Cape of Good Hope, Natal and Transvaal, and also seaweeds.

In 1848 Harvey was elected Professor of Botany to the Royal Dublin Society; this position obliged him to give public lectures throughout Ireland. Eight years later he was appointed Professor of Botany in the University of Dublin (Trinity College), and he held both of these professorships until his death. He visited the United States in 1849–50 and undertook a tour to Australia, New Zealand and the South Pacific in 1854–56 to collect seaweeds and other plants.

William Henry Harvey published three important works on the botany of southern Africa, and he also produced three monographs on seaweeds, one on Irish and British species *(Phycologia Britannica)* and the others on species from North America and Australia. He illustrated some of these with his own lithographic drawings for he was a skilful, self-taught draughtsman.

During his career Harvey classified and named hundreds of new species of flowering and non-flowering plants – for example the beautiful Californian tree-poppy, *Romneya coulteri,* after his predecessor Thomas Coulter and the Armagh astronomer Thomas Romney Robinson. He was himself commemorated by a South African genus *Harveya,* a group of parasites related to the European foxgloves. This amused William Harvey, whose voluminous correspondence displays his twinkling sense of humour – frequently he made fun of his pompous fellow botanists.

Principal publications:

1836 *The Genera of South African plants*
1844 *The Seaside Book*
1846–51 *Phycologia Britannica*
1847 *Nereis Australis*
1851–53 *Nereis Boreali-Americana*
1859–63 *Thesaurus Capensis*
1859–65 (with Otto Sonder) *Flora Capensis* (not completed)

Sources and further reading:

S.C. Ducker (ed.): *The Contented Botanist, Letters of W.H. Harvey about Australia and the Pacific,* Melbourne, 1988.
[Lydia Fisher]: *Memoir of W.H. Harvey,* London, 1869.
R.L. Praeger, William Henry Harvey, in F.W. Oliver (eds): *Makers of British Botany,* 20–24, Cambridge, 1916.
D.A. Webb: W.H. Harvey and the Tradition of Systematic Botany at TCD, *Hermathena,* **103**, 32–45, 1966.
E.C. Nelson and E.M. McCracken: *The Brightest Jewel: A History of the National Botanic Gardens, Glasnevin, Dublin,* Kilkenny, 1987.
E.C. Nelson: William Henry Harvey as Colonial Treasurer at the Cape of Good Hope: A Case of Depression and Bowdlerized History, *Archives of Natural History*, **19**, 171–180, 1992.
E.C. Nelson: The Juvenile Correspondence of William Henry Harvey, *Occasional Papers National Botanic Gardens, Glasnevin*, **8**, 55–61, 1996.
E.C. Nelson: William Henry Harvey. A Portrait of the Artist as a Young Man, *Curtis's Botanical Magazine* **13**, 36–41, 1996.

E. Charles Nelson, Outwell, Wisbech, England.

56 JOSEPH BEETE JUKES Geologist 1811–1869

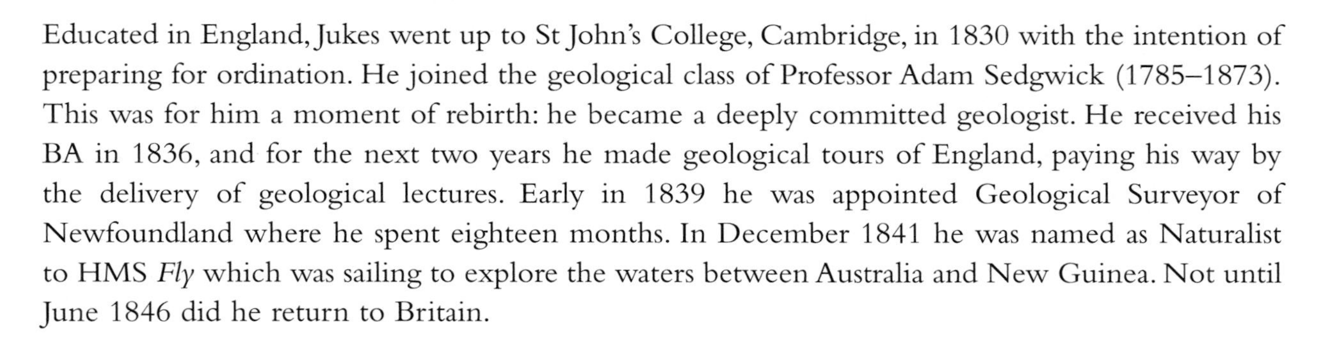

Born: Summerhill, near Birmingham, 10 October 1811.
Died: Hampstead House, Glasnevin, Dublin, 29 July 1869.

Family:
Son of John Jukes, a Birmingham button manufacturer, and his wife, Sophia.
Married: Georgina Augusta Meredith in 1849. No children.

Addresses:
1855–1869 72 Upper Leeson Street, Dublin

Distinctions:
Fellow of the Geological Society of London 1836; Member of the Royal Irish Academy 1852; Fellow of the Royal Society 1853; President of the Geological Society of Dublin 1853–1855; President of the Geological Section of the British Association for the Advancement of Science 1862.

Educated in England, Jukes went up to St John's College, Cambridge, in 1830 with the intention of preparing for ordination. He joined the geological class of Professor Adam Sedgwick (1785–1873). This was for him a moment of rebirth: he became a deeply committed geologist. He received his BA in 1836, and for the next two years he made geological tours of England, paying his way by the delivery of geological lectures. Early in 1839 he was appointed Geological Surveyor of Newfoundland where he spent eighteen months. In December 1841 he was named as Naturalist to HMS *Fly* which was sailing to explore the waters between Australia and New Guinea. Not until June 1846 did he return to Britain.

The following October he was appointed to the staff of the Geological Survey of the United Kingdom, his own work lying in North Wales and the English midlands. He soon became one of the finest field-geologists of his generation. In 1850 he was promoted to be the Local Director (Director in 1867) of the Geological Survey of Ireland in succession to Thomas Oldham. From 1854 Jukes was also Professor of Geology in the Museum of Irish Industry, an institution which in 1867 became the Royal College of Science for Ireland. He was a splendid chief for the Irish Survey. Under his supervision there were published 120 one-inch geological sheets covering more than one half of Ireland. It was a remarkable achievement. His personal most significant research contribution was a paper on the river pattern of the south of Ireland presented first to the Geological Society of Dublin in May 1862. That paper holds an honoured place in the international history of geomorphology. His *Student's Manual of Geology* (first edition 1857) was a widely used textbook.

Further reading :
J.B. Jukes: On the Mode of Formation of Some of the River-Valleys in the South of Ireland, *Quarterly Journal of the Geological Society of London*, **18**, 378–403, 1862.
G. L. Herries Davies: *North from the Hook: 150 Years of the Geological Survey of Ireland*, Geological Survey of Ireland, Dublin, 1995.

Gordon L. Herries Davies, Department of Geography, Trinity College, Dublin.

57 MARY BALL Naturalist

Born: Cobh, Co. Cork, 15 February 1812.
Died: Dublin, 17 July 1898.

Family:
Sister Anne Elizabeth (1808–1872) studied marine flora and fauna.
Brother Robert (1802–1857), a civil servant with wide interests in natural history, became Director of Trinity College Museum.

Addresses:

1812–1815	Cobh, Co. Cork
1815–1837	Youghal, Co. Cork
1837–1862	28 Eccles Street, Dublin
1862–1898	Belmont Avenue, Donnybrook, Dublin

The father of Robert, Anne Elizabeth and Mary, was Robert (Bob) Stawell Ball, a Customs officer at Cobh, Co. Cork. He was a natural history enthusiast who furnished his children with knowledge and curiosity about local plants and animals. The Ball children grew up in an age when the rising fashion for natural history study was sweeping away many social and sexual barriers. For women, the range of suitable leisure pursuits had greatly increased with the rise in popularity of natural history study, and conchology, entomology and botany became fashionable studies for them as they conformed to contemporary ideas of femininity and self-expression. Mary and Anne Elizabeth had an added advantage through their brother's involvement in natural history, and the contacts it brought with other naturalists must have stimulated and greatly encouraged their researches. From about the age of twenty-one, Mary started to specialise in collecting and studying butterflies and moths and 'the ardent papiliologist', her brother's description, became a dedicated student of insects. Robert, a founder member of the Royal Zoological Society of Ireland in 1830, himself organised scientific and literary 'conversaziones', often attended by up to 150 leading intellectuals and scientists, which would have enabled his sisters to build up a wide scientific circle. Mary came to know personally Alexander Haliday of Holywood, near Belfast, one of the foremost nineteenth century entomologists, with whom she would have discussed her observations and records when he visited their home. Likewise, she was held in high regard by many others including Baron de Selys-Longchamps, the Belgian authority on dragonflies who, like Haliday, included her work in their publications.

One of her observations has an interesting history. As recently as 1982, the eminent zoologist Professor Evelyn Hutchinson of Yale University recognised that about 1840 Mary had been the first person to discover the habit called stridulation (the production of a harsh grating noise) in corixid water bugs. Her observations of this behaviour had been obscured as they had been published by her brother Robert, for, even as late as the 1850s, women did not usually publish under their own names in scientific journals.

By 1843, her collections of insects and shells were among the 30 notable Irish collections listed by William Thompson for the British Association for the Advancement of Science for their meeting in Cork that year. Thompson was another close friend of the Ball family, and details of Mary's zoological collections and observations were also included in his book *The Natural History of Ireland*. His high esteem of her abilities led him to pen a tribute to her in naming a shell *Rissoa balliae* (=

Chrysallida indistincta) as 'a very trivial compliment to that Lady's acquirements in different departments of the invertebrata of Ireland'.

Apart from Mary's entomological collection, some of which still remains in Trinity College Museum, and her shell collection, now dispersed, she also collected and studied a wide range of marine invertebrates such as sea urchins, starfish, jellyfish, hydroids and foraminifera, especially when still living at Youghal, Co. Cork.

With the death of William Thompson, followed in 1855 by that of her brother a short time later, her interest in such studies started to wane, although she continued to exhibit 'green fingers' in her garden and fernery. She survived her sister Anne by some twenty-four years and died at the venerable age of 86.

Mary Ball is one of a number of Irish women who made substantial contributions to the biological sciences, whose interest may have been encouraged by being part of the small close-knit community of naturalists in Ireland at the time.

Further reading:

J. Hanley and P. Deevy: *Stepping Stones in Science.*

G. E. Hutchinson: The Harp that Once.....A Note on the Discovery of Stridulation of the Corixid Water Bugs, *Irish Naturalists' Journal*, **20**, 457–466, 1982.

Helena Chesney, 35 Colenso Parade, Belfast BT9 5AN.

MARY, COUNTESS OF ROSSE Pioneer photographer

Born: Heaton near Bradford, England 23 July 1813.
Died: London 1885.

Examining a steroscopic transparency

Family:
Elder daughter of John Wilmer Field, a wealthy Yorkshire landowner, and his wife Anne, née Wharton-Myddelton.
Married: William Parsons, third Earl of Rosse.
Children: Eleven – four sons (Laurence, Randal, Clere and Charles) reached adulthood.

Distinctions:
Silver medal of the Photographic Society of Ireland 1859; Contributed to the Dublin International Exhibition 1865.

Following the death of his wife whilst Mary and her sister Delia were very young, Squire Field entrusted their upbringing to Susan Lawson, who recognised Mary's artistic talent and stimulated its development. Evidence of this artistic ability is to be seen in the furniture she designed for Birr Castle and the extensive redesigning of the demesne which she undertook in conjunction with her uncle, Richard Wharton Myddelton. Two of her most notable achievements are the entrance gates and a drawbridge tower known as the 'keep gate', for both of which she made beautiful scale models. Her interest in heraldry was expressed in the embellishment of her wrought iron gates installed in the 'keep gate'. By the close of 1853, work on the remodelling of the demesne was nearing completion: she had extended the castle to house her large family though sadly this proved to be unnecessary, as only four of her children survived their teens.

At this time, presumably seeking a new challenge, she turned to photography. It would appear that she first experimented with the recently introduced 'wet collodion' process and with the 'daguerreotype' apparatus purchased by her husband in 1842. These first photographs, which were of the great telescope, were sent by her husband to W.H. Fox Talbot, who responded in most complimentary terms requesting permission to have them framed and exhibited at what was to be the first exhibition of the London Photographic Society, later to become the Royal Photographic Society.

The Countess created a fine darkroom in the castle and gradually acquired a wide range of cameras, including an early single lens stereo camera by Knight of London which was used to make many of the stereoscopic views in the collection. She also used a superb 12" x 15" 'folding' double box camera, with which she took pictures by both the collodion and the 'waxed paper' processes.

During the following years, Mary Rosse exhibited her work in Ireland and England and became a recognised exponent of the 'waxed paper' process. Her pictures received favourable comment in the photographic journals, and the Photographic Society of Ireland awarded her a silver medal. In 1865, the jury of the Dublin International Exhibition made special mention of her photographic exhibit. She became well known as an active member of the Amateur Photography Association, through which organisation her work, including general views and detail pictures of the six-foot reflecting telescope, was widely distributed.

Richard Wharton Myddelton, Mary Rosse's uncle. He was deeply involved in the design of the improvements to the castle and Demesne undertaken by the Countess. This is one of her most powerful portraits.

The six-foot telescope from the south-west. A group of family members is composed within this picture. The man wearing a top hat is the third Earl, whilst Charles Parsons is seated on the steps. Circa *1857–58.*

Although the Countess mastered the complex chemistry of several photographic processes and perfected a form of the waxed paper process, her scientific and technical accomplishment was merely a means to an end. Her greatest achievement was artistic. Her superb compositions, whether individual portraits, groups or landscapes, are of the highest order. Her ability to infuse an atmosphere of spontaneity whilst making exposures of around twenty seconds is quite remarkable. It is unfortunate that only one of her paper negatives is known, for this process allowed great flexibility to the photographer as the material could be prepared in advance, used and processed when convenient, being suited thereby to use when out of reach of a darkroom. It is tempting to speculate on the content and quality of these negatives, especially as records show that she took pictures in Spain whilst on a cruise in her husband's yacht Titania.

It would seem that she took no further photographs after the death of her husband in 1867, and, as her relationship with the new Countess became increasingly strained, she moved out of the castle and, following a brief sojourn in Dublin, she moved in 1870 to the house in Connaught Place, London, where she and her husband had entertained most of the leading scientists of the era, and hosted the functions at which Talbot's 'photoglyphic engravings' and the world's first news photograph were first shown. She lived at this address until her death in 1885. She was buried in Birr beside her husband. Her four sons commissioned Robert Kemp to design a memorial window which is still to be seen in the parish church at Heaton.

Further reading:

David H. Davison: *Impressions of an Irish Countess,* Birr, 1989.

David H. Davison, Dundrum, Dublin.

59 THOMAS ANDREWS Physical Chemist, Physician & University Administrator

Born: Belfast, 19 December 1813.
Died: Belfast, 26 November 1885.

Family:
Thomas Andrews was the son of a Belfast linen merchant.
Married: Jane Hardie Walker 1842.
Children: Three daughters and two sons.

Addresses:

1813–1849	3 Donegall Square South, Belfast
1849–1879	Queen's College, Belfast.
1879–1885	5 Fort William Park, Belfast.

Distinctions:
First Professor of Chemistry at the Queen's College Belfast 1845; Member of the Royal Irish Academy 1849; Fellow of the Royal Society 1849; Honorary Fellow of the Royal Society of Edinburgh 1870; President, British Association for the Advancement of Science 1867.

Described as 'a modest, silent boy with a great capacity for general knowledge', Andrews was introduced to chemistry at the University of Glasgow in 1828 at the age of fifteen. After two years there, and following a period in Paris studying with J.B. Dumas and L.J. Thenard, he returned to Dublin and Trinity College where he took a four year course in which he distinguished himself in classics as well as in science. He completed medical studies in Edinburgh.

In 1836, still only twenty-three and with his long, exacting, formal education behind him, he set up as a physician in Belfast as well as accepting the post of Professor of Chemistry in the Belfast Academical Institute ('Inst'). His ambition had always been to 'study chemistry profoundly as a great branch of human knowledge' and, while building up a successful medical practice, he found time to carry out the chemical researches which made him one of the foremost physical chemists of his day.

When the Queen's College (now The Queen's University) was opened in 1845 he became not only its first Professor of Chemistry but also its first Vice-President. His ability and success as a university administrator were renowned. At the same time he began his most famous work on the properties of gases. First, he proved that ozone was an allotrope (that is a different form) of oxygen. Then, by using heavy glass capillary tubes to contain gases, he was able to subject them to extreme pressures and observe the changes of state which took place.

In a series of experiments remarkable for both their simpliciry and delicacy, he showed that Boyle's Law does not describe the behaviour of gases under extreme conditions and that the so-called 'permanent' gases could be liquefied as long as they were below a characteristic temperature called the 'critical-point'. This discovery led to the liquefaction of all gases, as well as adding considerably to our knowledge of the nature of gases and liquids. His scientific advice was sought by many of Europe's leading scientists and his discoveries opened the way to many modern developments.

Andrews' original critical point apparatus
(courtesy of the Science Museum, London)

When you open a fridge or watch a space launch be reminded of Andrews, for it was his work that laid the foundations for their operation.

His apparatus is still to be seen in the Chemistry Department of The Queen's University of Belfast.

Further reading:

P.G. Tait and A. Crum Brown: *The Scientific Papers of Thomas Andrews with a Memoir,* Macmillan, London, 1889.
H. Riddell: *Proceedings of the Belfast Natural History and Philosophical Society,* 107–138, 1920/21.
W.J. Davis: In Praise of Irish Chemists. Some Notable Nineteenth Century Chemists, *Proceedings of the Royal Irish Academy,* **77**, 309–316, 1977.

William J. Davis, University Chemical Laboratory, Trinity College, Dublin.

MAXWELL SIMPSON Chemist

Born: Beech Hill, Co. Armagh, 15 March 1815.
Died: London, 26 February 1902.

Family:
The youngest of nine children of Thomas Simpson.
Married: Mary Martin of Loughorne, Co. Down, a sister of John Martin, a Young Irelander.
Children: Two daughters

Addresses:

1815–1845	Beech Hill, Co. Armagh
1845–1857	11 Wellington Road, Dublin
1860–1870	33 Wellington Road, Dublin
1870–1892	11 Dyke Parade, Cork
1893–1902	7 Darnley Road, Holland Park, London

Distinctions:
Fellow of the Chemical Society London 1857, Vice-President 1872–74; Fellow of the Royal Society 1862; Honorary Fellow of the King's and Queen's College of Physicians of Ireland 1865; Professor of Chemistry, Queen's College Cork 1872–92; President, Chemical Section of the British Association 1878; Honorary LLD University of Dublin 1878; Honorary DSc Queen's University 1882.

Maxwell Simpson's long life spanned a major portion of the period in which the first developments of modern chemistry were being made. In the year of his birth Faraday was 24 and was working on the chemistry of simple hydrocarbons; his electrical researches were still about ten years away. In that same year Humphry Davy invented the miner's safety lamp, Dalton was 49, Berzelius was 36, Liebig was a boy of 12, and Kekulé was not yet born. When Simpson died in 1902, Planck's quantum theory was just two years old and the scene was set for the great leap in the understanding of physical science. Simpson received his early education in Newry at the same school attended by John Martin and John Mitchel, both of whom were to make a considerable impact in Irish politics as 'Young Irelanders'. In 1832 Simpson went on to study medicine at Trinity College where Martin and Mitchel were also students. Simpson did not, at that time, complete his medical course, but left Trinity in 1837 with the degree of Bachelor of Arts. On a visit to Paris a few years later he attended a lecture on chemistry given by Jean Baptiste André Dumas. This seems to have made such an impression on Simpson that he decided to take up the study of chemistry. For two years he attended the lectures at London University given by Thomas Graham, best known for his famous law on the diffusion of gases. He also worked in Graham's laboratory. In 1845 Simpson came back to his friends of student days in Dublin, and married the sister of 'Honest John' Martin. He seems to have maintained his friendship with the leaders of the Young Irelanders throughout his scientific career in Dublin and later in Cork: shortly before he was transported to Van Diemen's land (now Tasmania) Martin records a visit from Simpson.

After his marriage Simpson resumed his medical studies so that he could become a lecturer in chemistry in the Park Street Medical School in Dublin. He qualified in 1847. In 1849 he moved to a similar post in the Peter Street or 'Original' School of Medicine. While in this position he was

Justin Von Liebig's famous laboratory at Giessen 1842

granted three years' leave of absence, during which he studied in Germany under Adolph Kolbe in Marburg and Robert Bunsen in Heidelberg.

Simpson then returned to Dublin and continued to lecture till 1857, when he resigned and went to Paris. Here he assisted Wurtz, who had discovered methylamine and ethylamine in 1849. These men with whom he collaborated were some of the foremost organic chemists in Europe. After two years with Wurtz he returned to Dublin, where he set up his own laboratory in a back kitchen of his house in Wellington Road. It was here that Simpson carried out many of the researches that have made him a significant figure in the history of Irish chemistry. During the eight years which Simpson spent at work in his home-based laboratory not a season passed without his publishing one or two papers of first-class scientific importance. He was expert both in organic synthesis and in the elemental analysis which was needed to get information about the mechanisms of reactions. He was the first chemist to synthesise succinic acid, $C_2H_4(COOH)_2$, until then only available from natural sources, from ethylene cyanide. He was also first to point out that the acidity of organic acids arises from the presence in the molecule of what we would now call the functional carboxylic group, -COOH, but which he called the 'semi-molecule of oxatyl'. As well as working on di- and tri-basic acids he also synthesised many new halogen-containing organic compounds. His publications run to over twenty-five papers. For a short time Simpson returned to Paris to do further work with Wurtz, and later he spent a few years in London, where he was examiner to many public services, such as the Indian Civil Service, and to the Queen's University in Ireland. In 1872, at the age of 57, he went to Queen's College Cork as Professor of Chemistry. During his twenty years in Cork, Simpson devoted himself almost entirely to his teaching duties. He published only two further research papers.

Further reading:

Royal Society Year Book, 252–8, London, 1903.
Obituary Notice in the *Journal of the Chemical Society: Transactions* **81,** 631–5, 1902.

William J. Davis, University Chemical Laboratory, Trinity College Dublin.

GEORGE BOOLE Mathematician and Logician

Born: Lincoln, 2 November 1815.
Died: Cork, 8 December 1864.

Family:
Married: Mary Everest (1832–1916). Her uncle George was the surveyor after whom Mount Everest is called.
Children: Five daughters including Alicia (1860–1940) a notable mathematician; Lucy (1862–1905) the first woman professor of chemistry in England; Ethel (1864–1960) a novelist whose book *The Gadfly* has sold more copies than any other book written by an Irish-born author. Boole's grandson, Geoffrey Taylor FRS, was a noted applied mathematician who worked on the development of the atomic bomb. His great-grandson, Howard Hinton FRS, was one of the world's foremost entomologists.

Addresses:
1849–1855 5 Grenville Place, Cork
1855–1857 Sunday's Well, Cork
1857–1862 Castle Road, Blackrock, Cork
1862–1864 Lichfield Cottage, Ballintemple, Cork

Distinctions:
Awarded the first ever gold medal for mathematics by the Royal Society 1844; Awarded honorary LLD by Trinity College, Dublin 1851; Elected president of the Cuvierian Society 1855; Elected Fellow of the Royal Society 1857; Awarded the Keith prize by the Royal Society of Edinburgh 1857; Elected member of the Cambridge Philosophical Society 1858; Honorary degree of DCL from Oxford University 1859.

Boole was a pioneer in mathematics whom Bertrand Russell described as the 'founder of pure mathematics'. He invented a new branch of mathematics – invariant theory – and made important contributions to operator theory, differential equations and probability. However, his most significant discoveries were in mathematical logic. Boole was a deeply religious man, a Unitarian, whose ambition was to understand the workings of the human mind and to express the 'laws of thought' in mathematical form. He invented a new type of algebra, called boolean algebra, which today's engineers and scientists have found to be ideal for the design and operation of electronic computers. Perhaps in some uncanny way Boole foresaw that the human brain behaves like a very complicated computer. Boolean algebra is also essential for the design and operation of the electronic hardware responsible for today's technology. Much of the 'new mathematics' now taught in schools can be traced back to Boole's work – for example, set theory, binary numbers and probability.

Boole was the eldest son of a struggling Lincoln shoemaker who was more interested in building telescopes than mending shoes. When his father's business failed, George left school at fourteen and became a junior teacher to support his family. Later he opened a school in Lincoln, helped by his sister and brothers. In his spare time he taught himself Latin, Greek, French, Italian and German. Later he studied optics and astronomy and finally he turned to mathematics. In addition, he read and wrote poetry and supported movements for adult education and social reform.

George Boole's family – Mary Boole (seated), their five daughters (standing) and some grandchildren

George Boole's family

The Boole Library, University College, Cork

In 1849 Boole was appointed first Professor of Mathematics at Queen's College (now University College) Cork, despite being almost entirely self-taught and having neither secondary schooling nor a university degree. While in Cork he produced his greatest work *The Laws of Thought* which earned him the title 'Father of Symbolic Logic'. This book contains the mathematics of today's computer technology. Boole was an excellent and devoted teacher and met his death after walking in the pouring rain to give a lecture. He is buried beside St Michael's Church of Ireland in Blackrock near Cork City.

Further reading:

George Boole: *The Laws of Thought,* Dover, 1958 (Reprint of 1854 publication by Walton and Maberley, London).

Desmond MacHale: Boolean Algebra, in *The Handbook of Applicable Mathematics,* Ledermann Wiley (ed.), Wiley, 1980.

Desmond MacHale: *George Boole – his Lfe and Work,* Boole Press, Dublin, 1985. Illustrations are taken, with kind permission, from this publication.

Desmond MacHale, Department of Mathematics, University College, Cork.

THOMAS OLDHAM Geologist 1816–1878

Born: Dublin, 4 May 1816.
Died: Rugby, 17 July 1878.

Family:
Eldest son of Thomas Oldham, a broker with the Grand Canal Company, and his wife Margaret (née Boyd).
Married: Miss L.M. Dixon of Liverpool, 1850.
Children: five sons and one daughter – one of his sons, Richard Oldham, also joined the Geological Survey of India.

Addresses:

c.1846	28 Trinity College Dublin
c.1850	18 Pembroke Road, Dublin
c.1851–1876	1 Hastings Street, Calcutta
After 1876	Eldon Place, Rugby, England

Distinctions:
Fellow of the Geological Society of London 1843; Member of the Royal Irish Academy 1844; Fellow of the Royal Society 1848; President of the Geological Society of Dublin 1848–1850; Member of the Royal Asiatic Society of Bengal 1857; Gold Medal from the Emperor of Austria 1873; Royal Society Royal Medal 1875.

Entering Trinity College Dublin in 1830, Thomas Oldham proceeded to a BA in 1836 and an MA in 1846. After receiving his BA, he moved to Edinburgh where he attended the lectures of Professor Robert Jameson (1774–1854). In 1838 he returned to Ireland to take a geological post with the Ordnance Survey. He worked upon the famed geological survey of County Londonderry under Joseph Ellison Portlock (1794–1864) before returning to Dublin to become Curator to the Geological Society of Dublin (1843–1845) and Assistant to Professor Sir John Benjamin MacNeill (1793?–1880), the first Professor of Civil Engineering in Trinity College Dublin. He became the College's Professor of Geology in 1845. Between July 1846 and November 1850 he was the Local Director of the Geological Survey of Ireland (founded 1845), and under his aegis there were published the earliest official geological maps for any part of southern Ireland. Early in 1851 he left Ireland to become the first Superintendent of the new Geological Survey of India, in which capacity he proved highly successful. He retired from India in 1876.

His renowned Irish discovery was that fossils were present in the ancient rocks of Bray Head in County Wicklow. At that time those were the oldest fossils known anywhere in the world. Oldham made his discovery during 1840, but he did not make public his find until 1844. In 1848, the naturalist Edward Forbes (1815–1854) suggested that the fossils should be named *Oldhamia* in honour of their discoverer. As such they have ever since been universally termed.

Further reading :

G.L. Herries Davies: *North from the Hook: 150 Years of the Geological Survey of Ireland*, Geological Survey of Ireland, Dublin, 1995.

Gordon L. Herries Davies, Department of Geography, Trinity College, Dublin.

63 GEORGE VICTOR DU NOYER Geologist, Archaeologist and Artist

Born: Dublin, 1817.
Died: Antrim, 3 January 1869, from scarlet fever.

Family:
Descendants of Huguenot refugees from Provence, who settled in Ireland about three hundred years ago. Son of Louis Victor Du Noyer, teacher.
Married: 4 January 1858, Frances Adélaide Du Bedat (1833–1914).
Children: Five – his eldest daughter pre-deceased him by one day.

Addresses:

1845	Seafort Avenue, Sandymount
1848	53 Charlemont Street, Dublin
1862–1863	Dunloe, Frankfort Avenue, Rathgar
1863–1867	Albert Ville, Sydney Avenue, Blackrock

Distinctions:
Honorary Life Member of the Royal Irish Academy 1863.

George Victor Du Noyer used his pencil, pen and pallet to record natural features and archaeological sites of the Irish countryside, much as the modern field worker uses a camera. He also prepared many important illustrations of fossils and carried out detailed geological surveys of large areas of Munster, Leinster and Ulster, as well as publishing numerous scientific and archaeological papers.

Educated at Mr Jones's school in Dublin's Great Denmark Street, Du Noyer studied art under George Petrie (1790–1866), the romantic landscape painter and pioneer of serious archaeological studies in Ireland. Petrie probably arranged for the young Du Noyer to join the archaeologists, folklorists and natural historians employed by the Ordnance Survey during the 1830s in a regrettably short-lived venture to document the minutiae of the Irish countryside. From 1835 until 1842 Du Noyer was engaged in illustrating botanical, geological and zoological specimens for the Ordnance Survey's field parties, as well as in preparing many topographic pen and ink drawings. The four fine plates of grasses included in the Ordnance Survey's so-called *Templemore Memoir* were selected for publication from the dozens of superb botanical drawings and watercolours by Du Noyer that are now preserved in the National Botanical Gardens at Glasnevin. The hundreds of fossil illustrations, gathered into thirty-eight plates in the *Report on the Geology of Londonderry and Parts of Tyrone and Fermanagh* by Joseph Ellison Portlock (1794–1864) are all by Du Noyer and include scientifically very important records of the type specimens of individual species.

Made redundant by the Ordnance Survey, Du Noyer taught art at St Columba's College, then at Stackallen Co. Meath, before joining the staff of the Geological Survey of Ireland in 1846, the year after its foundation. Du Noyer had been trained by Portlock in making geological sections, and his first assignment in the Geological Survey was to record the geological structure exposed in the cuttings then being excavated for construction of the railways northwards and westwards from

Dublin. As records of temporary exposures of the solid rock, Du Noyer's railway sections have proved to be of lasting value, particularly in areas of thick glacial deposits where there are few natural surface showings of the solid rock.

Watercolour illustration by G.V. Du Noyer of the rocks at Waterford Harbour showing flat-lying layers of Old Red Sandstone resting, uncomformably, on steeply dipping, Lower Palaeozoic slates

The railway sections completed, Du Noyer joined the ranks of the geologists who were systematically mapping-out Ireland's geological structure. He surveyed all, or parts of, forty-two of the 205 one-inch to the mile (1:63,360) geological map sheets that cover Ireland, and he contributed to seventeen of the accompanying 'explanatory memoirs', as well as constructing eight of the longitudinal geological sections published by the Geological Survey during his lifetime. Our present knowledge of the geological structure of large areas of counties Kerry, Cork, Tipperary (South Riding), Waterford, Wexford, Carlow, Wicklow, Dublin, Meath, Louth, Cavan, Longford, Westmeath and Antrim owes much to Du Noyer's work.

As well as official publications, Du Noyer published numerous papers in learned journals. Of particular interest are his contributions to the contemporary debate on the origin of the so-called 'drift' deposits, because the papers include a fine account of many important glacial features discovered by him among the mountains of south Munster. A compulsive artist, he decorated his geological worksheets with sketches and watercolours. Many of the woodcuts that illustrate the Geological Survey's 'explanatory memoirs' or feature in contemporary textbooks are based on Du Noyer's work, but the woodcuts scarcely do justice to the beauty of the originals, many of which are preserved in the archives of the Geological Survey of Ireland.

Besides his geological legacy, Du Noyer left behind a wealth of archaeological material, including detailed drawings of sites which have since been destroyed. Eleven volumes of his archaeological drawings were presented by him to the Royal Irish Academy and a further dozen volumes were acquired, after his death, by the Royal Society of Antiquaries in Ireland.

Further reading:

P. Coffey: George Victor Du Noyer (1817–1869), Artist, Geologist and Antiquary, *Journal of the Royal Society of Antiquaries of Ireland*, **123**, 102–119, 1993 (1995).

Fionnuala Croke: *George Victor Du Noyer 1817–1869 Hidden Landscapes*, National Gallery of Ireland, Dublin, 1995.

Gordon Herries Davies, *Sheets of Many Colours, the Mapping of Ireland's Rocks 1750–1890*, Royal Dublin Society Historical Studies in Irish Science and Technology No. **4**, Dublin, 1983.

Jean Archer, Nenagh, Co. Tipperary.

Born: Skreen, Co. Sligo, 13 August 1819.
Died: Cambridge, 1 February 1903.

Family:
Married: Mary Susannah Robinson, 1859, daughter of Thomas Romney Robinson.
Children: One son and two daughters, one of whom died in childhood.

Addresses:

1819–1834	Skreen Rectory, Co. Sligo
1837–1859	Pembroke College, Cambridge
1859–1903	Lensfield Cottage, Cambridge

Distinctions:
Fellow of the Royal Society 1851 (President 1885–1890); Member of Parliament for Cambridge University 1887–1892; Baronetcy (Sir George Gabriel Stokes) given by Queen Victoria in 1889; Copley Medal of the Royal Society in 1893; Professorial Jubilee (50 years as Lucasian Professor) by the University of Cambridge 1899.

Sir George Gabriel Stokes (courtesy of The Royal Society)

The Stokes family is better known in Ireland for its distinguished contributions to medicine (Cheyne-Stokes respiration and the Stokes-Adams syndrome) rather than to mathematics. Both sprang from a common ancestor, Gabriel Stokes, born in 1682, an instrument maker residing in Essex Street, Dublin, who became Deputy Surveyor-General of Ireland. In 1798, his direct descendant, another Gabriel Stokes, Rector of Skreen, married Elizabeth Haughton: George Gabriel was the youngest of their seven children. A memorial opposite the Church marks his birthplace. Stokes was greatly influenced by his upbringing in the West of Ireland, and he returned regularly for the long summer vacation. After the death of his parents he continued to visit his brother John, a clergyman in Tyrone, and his sister Elizabeth in Malahide, where there is a plaque to him in St Andrew's Church.

His first teacher was the Clerk of Skreen Parish, who recorded Stokes as 'working out for himself new ways of doing sums, better than the book'. Later he attended Dr Wall's School in Hume Street, Dublin, where he attracted attention by his elegant solution of geometrical problems. In 1834 Stokes moved to Bristol College, where he was taught by Francis Newman, brother of the Cardinal. Newman wrote that Stokes 'did many of the propositions of Euclid as problems, without looking at the book'.

Stokes entered Pembroke College, Cambridge, in 1837. He was Senior Wrangler (that is, placed first in mathematics in the whole university) in 1841 and elected to a Fellowship at Pembroke. His early research was in the area of hydrodynamics, both experimental and theoretical, during which he put forward the concept of 'internal friction' of an incompressible fluid. Stokes' methods could also be applied to other continuous media such as elastic solids. He then turned his attention to oscillatory waves in water, producing the conjecture on the wave of greatest height, which now bears his name, as do the fundamental Navier-Stokes equations of fluid motion. But perhaps his major advance was in the wave theory of light, examining mathematically the properties of the aether which he treated as an incompressible elastic medium. This enabled him to obtain major results on the mathematical theory of diffraction, which he confirmed by experiment, and on fluorescence, which led him into the field of spectrum analysis. His last major paper on light was the dynamical theory of double refraction.

Such was Stokes' reputation as a promising young man, familiar with the latest Continental literature, that in 1849 he was appointed to the Lucasian Chair of Mathematics. Although appointed Professor for his outstanding research, Stokes showed a concern in advance of his time for the welfare of his students, stating that he was 'prepared privately to be consulted by and to assist any of the mathematical students of the university'. Stokes' manuscript lecture notes still exist in the University Library in Cambridge, although his writing was so bad that he eventually became one of the first people in these islands to make regular use of a typewriter. In 1859 Stokes married Mary Robinson of Armagh. The couple reputedly met at Birr Castle, while Stokes was advising the Earl of Rosse on his telescope mirror. Stokes, who was a tireless writer of letters, carried on an extensive (one letter runs to 55 pages) and frank correspondence with his fiancée. In one letter, the theme of which will be familiar to all spouses of research scientists, he states 'When the cat's away the mice may play. I have been doing what I guess you won't let me do when we are married, sitting up till 3 o'clock in the morning fighting hard against a mathematical difficulty'. They lived in Lensfield Cottage, Cambridge, where Stokes had a 'simple study' and conducted experiments 'in a narrow passage behind the pantry, with simple and homely apparatus'.

The mathematical results of Stokes arose mainly from the needs of physical and industrial applications. Besides his links with the School of Mines, he acted for many years as consultant to the telescope maker Howard Grubb. His major work on the asymptotic expansion of integrals and solutions of differential equations arose from the optical research of the astronomer G.B. Airy. The well-known theorem in vector calculus which bears his name is sadly not due to Stokes, but was communicated to him in a letter by Lord Kelvin. There is justice in this, however, as Stokes was undoubtedly generous in sharing his own unpublished ideas with others. In its leader of 3 February 1903, following his death two days earlier, *The Times* wrote that 'Sir G. Stokes was remarkable ... for his freedom from all personal ambitions and petty jealousies'.

Further Reading:

Joseph Larmor: *Memoir and Scientific Correspondence of the late Sir George Gabriel Stokes*, Cambridge University Press, 1907.

G.G. Stokes: *Mathematical and Physical Papers*, Cambridge University Press, 1880–1905.

David B. Wilson: *The Correspondence between Stokes and Kelvin*, Cambridge University Press, 1990.

Alastair Wood, School of Mathematical Sciences, Dublin City University, Dublin 9.

Born: Cork, 25 September 1819.
Died: Dublin, 22 January 1904.

Courtesy of Trinity College Dublin

Family:
Son of Michael Salmon, a Cork linen merchant, and his wife Helen (née Weekes). He had three sisters.
Married: Frances Anne Salvador.
Children: Four sons and two daughters.

Addresses:
1848–1880 4 Heytesbury Terrace (81 Wellington Road)
1888–1904 Provost's House, Trinity College.

Distinctions:
Fellow of Trinity College 1841; Member of the Royal Irish Academy 1843; Cunningham Medal of the Royal Irish Academy 1858; Fellow of the Royal Society 1863; Regius Professor of Divinity, University of Dublin 1866; Royal Medal of the Royal Society 1868; Fellow of the Accademia dei Lincei, Rome 1885; Provost of Trinity College 1888–1904; Copley Medal of the Royal Society 1889; Fellow of the British Academy 1902; Honorary degrees from Oxford, Cambridge, Edinburgh, Christiania; Honorary member of the Berlin, Gottingen and Copenhagen Academies.

George Salmon entered Trinity College Dublin at about the same time as James MacCullagh was appointed to the chair of mathematics, so it is not surprising that his interests were led towards geometry where most of his significant mathematical contributions were to be made. Unlike MacCullagh and William Rowan Hamilton, Salmon had no interest in physics; he pursued mathematics for its own intrinsic interest. Salmon also differed from MacCullagh in his readiness to undertake formidable calculations: in this he resembled Hamilton, who seemed almost to enjoy heavy calculation for its own sake, whereas MacCullagh would prefer to devote long hours to searching for an elegant and simple way to reach a result. Hamilton tried but failed to interest him in quaternions.

Salmon's researches did extend into algebra, where he collaborated with the English mathematicians Cayley and Sylvester in pioneering work on matrices and the theory of groups. In fact this work in algebra, which was particularly concerned with the determination of invariants, was very relevant to geometry. The theory of invariants was subsequently used by Felix Klein in his so-called Erlangen Programme as a basis for a systematic description of geometry. Klein, who learned about invariants from a German translation of Salmon's book, was to meet his mentor in 1892 when he represented his university at the tercentenary celebrations in Trinity College at which Salmon presided as Provost. To Klein's disappointment the Provost no longer wished to talk about mathematics.

Although his original research contributions were substantial, Salmon's fame in mathematics stems primarily from his four textbooks: *Conic Sections, Higher Plane Curves, Lessons Introductory to the*

The Provost's House, Trinity College, Dublin
(courtesy of Trinity College Dublin)

Modern Higher Algebra and *Geometry of Three Dimensions*. Written with elegance and clarity, these books enjoyed great success – they appeared in many editions and were translated into several languages. *Conic Sections* remained in use as a textbook until comparatively recent times and it would not be altogether surprising to see a copy of one of the others on the desk of a research mathematician today.

As was normally required of a fellow of Trinity College at that time, George Salmon was ordained a priest in the Church of Ireland. In 1866 he was appointed Regius Professor of Divinity and his main interest switched from mathematics to theology. He was a respected New Testament scholar, but the best known of his theological writings is *The Infallibility of the Church*, a trenchant criticism of papal infallibility, which reflects its author's deep distrust of any authority over the individual conscience. In the reorganisation of the Church of Ireland which followed its disestablishment in 1870 Salmon played a central and constructive part. In 1888 George Salmon was appointed Provost of Trinity College and remained head of the college until his death in 1904. He was highly respected at home and abroad and he presided with much dignity over the college's tercentenary celebrations in 1892.

Further reading:

Dictionary of National Biography.
Dictionary of Scientific Biography.
Obituary notices: *Proceedings of the British Academy*, **1**, xxii, 1903–4; *Proceedings of the London Mathematical Society*, **2nd ser.**, **1**, xxii, 1903–4; *Proceedings of the Royal Society* **75,** 347, 1905; *Nature* **69,** 324, 1903–4.
T.D. Spearman: Mathematics and Theoretical Physics, in T. Ó Raifeartaigh (ed.), *The Royal Irish Academy, a Bicentennial History, 1785–1985,* Dublin,1985.

David Spearman, Trinity College Dublin.

JOHN TYNDALL Physicist and Populariser of Science

Born: Leighlinbridge, Co. Carlow, 2 August 1820.
Died: Hindhead, Surrey, England, 4 December 1893.

John Tyndall lecturing at the Royal Institution in 1870 (from The Illustrated London News)

Family:
Father: John Tyndall, member of the Royal Irish Constabulary.
Married: In 1876, Louisa Charlotte Hamilton, daughter of Lord Claud Hamilton.

Addresses:
London
Bel Alp (in the Alps)
Hindhead (Surrey)

Distinctions:
Fellow of the Royal Society 1852; President of British Association for the Advancement of Science 1874.

John Tyndall was one of the great public figures of nineteenth-century British science. As a researcher, an educator, a lecturer and a controversialist, he played a major role in both the professionalisation and popularisation of science.

Born into a relatively poor family in Co. Carlow, Tyndall received his early education in Carlow before going to work for the Ordnance Survey, first in Ireland and then in England. While he worked, he attended lectures at the local Mechanics Institute, where members of the working class could receive basic instruction in the sciences. Tyndall was dismissed from the survey in 1843 for protesting the working conditions of the Irish labourers, but he soon found work as a surveyor for the booming railroad industry. In 1847 he became a teacher at Edmondson School, Queenwood College, in Hampshire, whose facilities included one of the first teaching laboratories in Britain. There, he first became interested in the teaching of practical science and engineering, which he would pursue for the rest of his life.

In 1848, Tyndall left England to get his PhD at Marburg University in Germany. The PhD degree was a recent innovation, and the German universities were then world leaders in scientific research training. He studied chemistry under Robert Bunsen (of Bunsen burner fame) as well as mathematics and physics. After two years of study, he completed his degree and went on to do research in Berlin, where he mingled with many of the great German scientists of the age. When he returned to Britain in 1851, however, he was unable to find a university position. His unconventional education and working class background had little appeal for those in the British educational establishment. He returned to Queenwood College and competed unsuccessfully for professorships in Toronto, Sydney, Cork and Galway. Finally, in 1853, after a brilliant lecturing performance, he was offered the chair in Natural Philosophy at the Royal Institution in London. Unlike the universities, the Royal Institution was a professional, urban institution whose activities included frequent public lectures. He spent the rest of his career there, succeeding Michael Faraday as its Director in 1867.

At the Institution, Tyndall engaged in a number of lines of research. In Germany he had become interested in the behaviour of crystals in a magnetic field. This eventually led him to study the compression of crystal substances, and he soon took an interest in glaciers. He even became a pioneer in the sport of mountaineering, and later spent much of his free time at his house in the Alps. He also studied solar and heat radiation. He was particularly interested in the interaction of heat, light and atmospheric gases, and he made a study of the scattering of light by particles in the atmosphere. In fact, it was Tyndall who first explained that the sky is blue because the different wavelengths of sunlight are scattered to different degrees by the atmosphere. He also made the first pollution measurements of the London atmosphere and became involved in early work on bacteriology. He was an early advocate of Pasteur's germ theory of disease and publicly defended it against its opponents. His pioneering work is represented by the number of scientific phenomena named after him, including the Tyndall effect, the Tyndall cone, Tyndall scattering, Tyndallisation and the Tyndallo-meter.

Tyndall was one of the first people to adopt the term 'physicist' to differentiate himself from the traditional 'natural philosopher'. The typical natural philosopher had been an Oxford or Cambridge don and often a clergyman. Tyndall, by contrast, was unconnected with any university and earned a living giving public lectures and writing textbooks. He also wrote for newspapers and magazines and helped to found the now famous scientific journal *Nature* in 1869. With the support of his politically influential and scientifically-minded friends in the X-Club, he worked for increased government support for scientific research and the reform of scientific education. He also gave evidence to government commissions on education, copyright, and accidents in coalmines and, on Faraday's death, Tyndall took over his role as science advisor to the Brethren of Trinity House (the administrators of British lighthouse system). In this way, he pioneered the now common role of the scientist as expert witness and government advisor.

Tyndall grew famous for both his theatrical style of lecturing and his public battles with famous figures. Most famous was his 1874 presidential address to the British Association for the Advancement of Science in Belfast. Condemning the attitudes of the Catholic hierarchy in Ireland to science, he proposed that science and reason rather than faith are the only acceptable guides to truth. Like T.H. Huxley, one of the major proponents of evolution and the originator of the term 'agnostic', Tyndall saw the separation of science from religion as a key element in its professionalisation. His highly publicised comments on religion, however, led to his demonisation as an advocate of materialism and atheism. Politically Tyndall was a liberal, but he broke with the liberal party over Gladstone's Home Rule bill.

Further Reading:

W.H. Brock, N.D. McMillan and R.C. Mollan (eds.): *John Tyndall: Essays on a Natural Philosopher*, Royal Dublin Society, Dublin, 1981.
Dictionary of Scientific Biography.
Dictionary of National Biography.

David Attis, Program in History of Science, Department of History, Princeton University, Princeton, NJ, USA.

Born: Carlow, 21 December, 1821.
Died: Dublin, 31 October 1897.

Portrait by Sarah Purser (courtesy of Trinity College Dublin)

Family:
The Haughtons were a Quaker family, although Samuel's parents were apparently not practising members of the Society of Friends and he had no difficulty in being ordained into the ministry of the Church of Ireland. Samuel Haughton and his wife Louisa had four sons.

Addresses:

1847–1874	17 Heytesbury Terrace (afterwards 51 Wellington Road), Dublin
1875–1886	31 Baggot Street Upper, Dublin
1887–1897	12 Northbrook Road East, Dublin

Distinctions:
Fellow of Trinity College 1844; Member of the Royal Irish Academy 1845; Cunningham Medal of the Royal Irish Academy 1848; Professor of Geology in the University of Dublin 1851–81; Fellow of the Royal Society 1858; President of the Royal Zoological Society of Ireland 1885–90; President of the Royal Irish Academy 1886–91; Honorary degrees from Oxford, Cambridge, Edinburgh, Bologna.

Samuel Haughton has been appropriately described as a Victorian polymath. It was more common in Victorian times than it is today for a scholar or scientist to possess a broadly-based expertise but, even by the standards of his time, the range of Haughton's knowledge and the variety of fields to which he made original contributions were quite unusual. Mathematics was his first interest; James MacCullagh's lectures, which he attended as an undergraduate at Trinity College Dublin, made a big impact. Later after MacCullagh's death, Haughton, with J.H. Jellett, published the collected papers of their former professor.

The problems which interested Haughton were usually of a practical type, preferably amenable to a quantitative description. One continuing field of interest was elasticity, what would today be called continuum mechanics. In 1848, four years after his election to fellowship in Trinity, he was awarded the Cunningham Medal of the Royal Irish Academy for his memoir *On the Equilibrium and Motion of Solid and Fluid Bodies*. Haughton also shared MacCullagh's interest in the refraction of polarised light by crystalline media, which led him naturally to mineralogy. It was probably this, set against the background of an enthusiam for natural history that he had developed as a schoolboy in Carlow, that caused geology to take over from mathematics as his primary interest and, in 1851, he was appointed to the chair of geology. Haughton's geological work was wide-ranging. He did pioneering work on the chemical composition of rocks and also applied his mathematics to such topics as tidal motion and planetary equilibrium. He compiled comprehensive tide tables, which he applied to such varied topics as the incidence of shipwrecks, the sequence of events at the Battle of Clontarf, and the evidence which had been presented at a murder trial some years previously.

Sir Patrick Dun's Hospital when completed in 1816

This last study was one of several investigations of a forensic nature which attracted his curisoity. Haughton was a pioneer in the geological study of climatic change and, in this context, he studied solar radiation and examined the effect of ocean currents on climate.

It was apparently his study of fossils which gave rise to an interest in anatomy and in 1859 – the year in which Darwin published *The Origin of Species* – Haughton, by now a Fellow of the Royal Society, entered the Trinity medical school where he studied for three years, while still retaining his geology chair, before graduating MB, MD in 1862. He carried out research into a variety of physiological problems, still favouring those which could be treated mathematically. He investigated muscular action and the mechanism of joints, publishing his results in 1873 as a book *The principle of Animal Mechanics.*

Following his election to Fellowship Haughton took holy orders in the Church of Ireland. His unwavering opposition to Darwin's evolutionary theory was probably due both to religious and scientific conviction. Apart from his research and teaching, he played an active administrative role – in Trinity where he was Registrar of the medical school, in the Academy where he served on Council for many years and was President from 1886 to 1891, on the board of Sir Patrick Dun's Hospital, which he served for 35 years, and as Honorary Secretary of the Royal Zoological Society of Ireland. The restaurant building at the zoo, still known as the Haughton House, was built in his memory by public subscription. One other facet of this remarkable man, linked to his pedagogical interests, was the publication in joint authorship with his colleague J. A. Galbraith of a popular series of scientific manuals covering a wide range of topics.

Further reading:

W. J. E. Jessop: Samuel Haughton: a Victorian polymath, *Hermathena,* **116**, 5, 1973.

David Spearman, Trinity College Dublin.

Born: Dripsey, Co. Cork, 1822.
Died: Cork, 12 May 1890.

Family:
His father introduced the first Irish mechanical paper-mill at Dripsey. It was burnt down by 'Luddite' workers. He married Frances Hennessy, a sister of Henry G. Hennessy in 1849. A grandson, Thomas Dillon, became Professor of Chemistry in University College Galway in 1919.

Addresses:

1849–1873	Various Dublin Addresses
1873–1890	President's Lodge, Queen's College, Cork

Distinctions:
Member of the Royal Irish Academy 1857 (Vice-President 1866, Secretary 1867–1874); President of Queen's College, Cork 1873–1890; DSc The Royal University of Ireland 1882.

Like Robert Kane, Sullivan was introduced to chemical technology through his father's industrial enterprise, a factory making paper. Educated by the Christian Brothers in Cork, Sullivan's first lecturing post was at the Mechanics Institute in the city. In the early 1840s he studied chemistry at Liebig's famous laboratory in Giessen. His earliest paper, in the *Philosophical Magazine* in September 1845, is the first description of the iron test for phosphates which was to become a standard. He was appointed assistant to Robert Kane, the then Director of the Museum of Economic Geology of Ireland (subsequently the Museum of Irish Industry). He was to become Assistant Chemist in 1846 and later Chemist to the Museum.

His main work was in chemical analyses, and in mineralogical and industrial investigations. He was particularly concerned with the utilisation of two of Ireland's natural resources, peat and sugar-beet. His researches showed that beet of high sugar content could be grown in Ireland (presaging the setting-up of the Irish Sugar Company in 1935).

In 1854 the Museum was expanded to become a School of Science, and Sullivan became Professor of Chemistry. In 1856 he accepted the Professorship of Chemistry in Cardinal Newman's Catholic University of Ireland while retaining the Chair of Theoretical Chemistry in the Museum. When Kane retired from the presidency of Queen's College Cork in 1873, Sullivan was appointed. Under his presidency the College prospered. The student numbers grew – from under 200 in 1873 to 402 in 1882.

Sullivan had many interests outside science. He was an active member of the Young Ireland movement. His three-volume work on pre-Christian Irish culture (1873) was an outstanding achievement.

Further reading:
T.S. Wheeler: Life and Work of William K. Sullivan, *Studies*, **34**, 21–36, 1945.

William J. Davis, University Chemical Laboratory, Trinity College Dublin.

Born: Belfast, 16 February 1822.
Died: Glasgow, 8 May 1892.

Courtesy of The Queen's University of Belfast

Family:
Eldest son of Professor James Thomson (1786–1849) and brother of William (Lord Kelvin). Married: Elizabeth, daughter of William John Hancock of Lurgan, County Armagh, in 1853. They had one son, James, and two daughters.

Distinctions:
Fellow of the Royal Society 1877; LLD University of Glasgow 1870; DSc Queen's University of Ireland 1875; LLD University of Dublin 1878.

Educated by his father, Thomson attended the University of Glasgow from the age of twelve, matriculating in 1834 and gaining a MA in 1840 in mathematics and natural philosophy. He was set on pursuing an active civil engineering career, but suffered much ill-health, and returned to Belfast in 1851, where he became resident engineer for the water commissioners. In 1857 he was appointed Professor of Civil Engineering at Queen's College in Belfast and held the post until 1872, when he was elected to succeed William John Macquorn Rankine at Glasgow.

From an early age, Thomson exhibited an inventive bent and in 1850 he patented the inward flow (or vortex) turbine. He investigated the properties of whirling fluids, which led him to devise improvements in the action of blowing fans, to the invention of a centrifugal pump, and to important improvements in the design of turbines. His work on the measurement of flow by notches (especially the v-notch) was fundamental. In 1848 Thomson began his many contributions to scientific literature. In a paper to the Royal Society of Edinburgh in 1849 on 'The effect of pressure in lowering the freezing-point of water' he expounded the principles which, in 1857, he used as the foundation of his explanation of the plasticity of ice.

He extended our knowledge of the continuity of gaseous and liquid states of matter, the circulation of currents in the atmosphere, and investigated the nature of the rock structures of the Giant's Causeway in County Antrim.

Further Reading:
Proceedings of the Royal Society, **53**, 1893, i–x.
Michael H. Gould: *150 Years of Civil Engineering at Queen's 1849–1999,* School of Civil Engineering, Belfast, 1999.
Dictionary of National Biography.

Ronald Cox, Centre for Civil Engineering Heritage, Trinity College, Dublin 2.

Born: Dublin 1823.
Died: Melbourne, Australia, 13 May 1899.

M'Coy in later life wearing his KCMG insignia (courtesy of The Queen's University of Belfast)

Family:
Second son of Bridget (1799–1876) and Dublin-based medic Simon M'Coy (1795–1875), who taught anatomy and published on cholera, and later became the first Professor of Materia Medica in Queen's College, Galway (1849).
Married: Anna Maria Harrison (died 1886) in 1843 and was survived by a son Frederick Henry (born 1844): they also had a daughter Emily (1845–1861).

Addresses:

1829–1843	Digges Street and French Street (off Mercer Street), Dublin
1843–1845	Ranelagh and Rathmines, Dublin
1867	'Maritima', 45 South Road Brighton, Melbourne, Australia

Distinctions:
Fellow of the Royal Society 1880; Murchison Medal, Geological Society, London; DSc Cambridge; DSc University of Dublin; MA Edinburgh; Honorary Member, Cambridge Philosophical Society; Knight Commander of the Order of St Michael and St George 1891; Chevalier of the Order of the Crown, Italy; Emperor of Austria Great Gold Medal for Arts and Sciences.

Described as having a 'sturdy frame, reddish hair, florid complexion and a warm heart', Frederick M'Coy initially rose to fame as the author of two volumes (1844, 1846) describing the fossils found during the geological mapping of Ireland by a team under Sir Richard Griffith. He established over 500 new species of extinct animals, which are now in the National Museum of Ireland. He was the first Professor of Natural Science in Queen's College, Belfast, and also worked with the geologist Adam Sedgwick in Cambridge. M'Coy emigrated to Australia in 1854 to take up the post of Professor of Natural Science at Melbourne University for the princely sum of £1,000 a year and a house. He worked as a state geologist and mineralogist and soon established a Botanic Garden and the National Museum of Victoria, which he built into one of the most significant natural history museums in the British Empire.

Further reading:
Obituaries in: *Nature*, **60**, 83, May 1899; *Geological Magazine*, **6**, 283–287, June 1899; *Quarterly Journal of the Geological Society, London*, **56**, lix–lx, 1900.
G.L.H. Davies: *Sheets of Many Colours – the Mapping of Ireland's Rocks 1750–1890*, Royal Dublin Society, Dublin, 1983.
P.N. Wyse Jackson and N. T. Monaghan: Frederick M'Coy: An Eminent Victorian Palaeontologist and his Synopses of Irish Palaeontology of 1844 and 1846, *Geology Today*, **10**, 231–234, 1994.

Nigel T. Monaghan, National Museum of Ireland, Collins Barracks, Dublin 7.

Born: Belfast, 26 June 1824.
Died: Netherhall, Largs, Scotland, 17 December 1907.

Courtesy of the Royal Society

Family:
Father: James Thomson (1786–1849), Professor of Mathematics at the Belfast Academical Institution and later at Glasgow University. Brother: James Thomson (1822–1892), Professor of Engineering at Queen's College Belfast and later at Glasgow University.
Married: (1) Margaret Crum (1852 – she died in 1870); (2) Frances Blandy (1874). No children.

Addresses:
Belfast; Glasgow; Eaton Place, London; Netherhall, Largs, Scotland.

Distinctions:
Colquhoun Silver Sculls 1844; Second Wrangler and First Smith's Prizeman, Cambridge University 1845; Fellow of the Royal Society 1851; Knighted 1866; President of the British Association for the Advancement of Science 1871; President of the Society of Telegraph Engineers 1874; Copley Medal of the Royal Society 1883; President of the Royal Society 1890–1895; Baron Kelvin of Largs 1892; Grand Cross of the Victorian Order 1896; Order of Merit 1902; Chancellor of the University of Glasgow 1904; innumerable honorary degrees and foreign honours.

One of the towering figures of nineteenth-century science, William Thomson not only made essential contributions to the fields of electromagnetism, mechanics and thermodynamics, he also championed the application of science to industry and made practical improvements to the operation of steam engines and the telegraph.

Thomson's father, a mathematics teacher at the Belfast Academical Institution, brought his family to Glasgow when William was eight. He was introduced to the study of natural philosophy at Glasgow University, and after taking his degree there he proceeded to Cambridge University, where he excelled in both mathematics and athletics. He became fascinated by the mathematical theories of the French engineers, and after graduation he spent time in France, studying Fourier's mathematical theory of heat and Carnot's theory of engines.

Thomson returned to Glasgow in 1846 as Professor of Natural Philosophy. Glasgow at that time was a major industrial centre, and Thomson believed that science should engage with the problems of industry. 'The life and soul of science,' he said, 'is its practical application.' To this end, he created the first teaching laboratory in Britain in which he trained students to make precise measurements, especially those useful for the growing telegraph industry. His involvement in electrical science led to one of his greatest achievements, the laying of the Atlantic Telegraph cable. Stretching 2,000 miles across the Atlantic from southwest Ireland to Newfoundland, the telegraph was one of the first major industrial enterprises to use theoretical science. Thomson's mathematical work provided a basic understanding of how signals were sent, his instruments helped to send and receive signals, and his laboratory tested the purity of the copper in the cable. The success of the cable in 1865 brought him a knighthood, and with the proceeds he was able to build a substantial seaside mansion and purchase a 126-ton yacht.

In addition to his practical work, Thomson played a central role in the creation of thermodynamics, the science of heat and energy. In 1848 he became interested in the measurement of temperature and created the absolute scale of temperature (the Kelvin scale) that now bears his name. He also befriended the German scientist Helmholtz, one of the first people to formulate the law of conservation of energy (the first law of thermodynamics). The concept of energy, now central to science, was then in its infancy, its importance growing as scientists struggled to make sense of the machines that increasingly surrounded them as industrialisation proceeded. Thomson helped to develop the second law of thermodynamics, which sets limits to the efficiency of engines. For him, the steam engine with its ability to transform energy from one form (heat produced by coal) to another (the movement of machines) became a metaphor for the entire universe. Just as a certain amount of heat is always lost in the operation of the steam engine, he argued, the entire universe will eventually run down as all of its energy is transformed into useless heat. Thomson also used his knowledge of thermodynamics to estimate the age of the earth. He calculated the amount of time it would have taken the earth to cool from a molten mass to its current temperature, claiming that his calculation invalidated the new theory of evolution, which required a longer history of the earth than physics allowed.

Thomson was famous for his mechanical view of physics. 'I never satisfy myself,' he explained, 'until I can make a mechanical model of something.' His love of mechanical models and his disdain for hypothetical entities, however, led him to oppose many innovative physical theories. Dissatisfied with Maxwell's theory of the electromagnetic field, he could only concede that it was 'curious and ingenious'. He also opposed modern ideas of nuclear energy and radiation. However, as J.J. Thomson said, 'Science never had a more enthusiastic, stimulating or indefatigable leader'. In the end, he published over six hundred papers, two dozen books and more than sixty patents.

Although Thomson left his native country at an early age, in later life called himself an Irishman, and he served in the House of Lords as an ardent supporter of Unionism. For his services to science he received many state honours and was buried in Westminster Abbey.

Further reading:

S.P. Thompson: *The Life of William Thomson, Baron Kelvin of Largs*, Two Volumes, London, 1910.
Crosbie Smith and M. Norton Wise: *Energy and Empire, A Biographical Study of Lord Kevin*, Cambridge University Press, 1989.
David B. Wilson: *The Correspondence between Sir George Gabriel Stokes and Sir William Thomson, Baron Kelvin of Largs*, Cambridge University Press, 1990.

Denis Weaire, Physics Department, Trinity College, Dublin 2.
David Attis, Program in the History of Science, Department of History, Princeton University, Princeton, New Jersey USA.

72 GEORGE JOHNSTONE STONEY Physicist and Educationalist

Born: Oakley Park, near Birr, King's County (Co. Offaly), 15 February 1826.
Died: London, 5 July 1911.

Family:
Married: His cousin, Margaret Stoney.
Children: Two sons and three daughters.
His eldest son, Gerald Stoney (1863–1942, FRS, Watt medallist of Institution of Electrical Engineers) was successively Manager of Parsons Turbine Works, Newcastle, and Professor at Manchester College of Technology (today University of Manchester Institute of Science and Technology, UMIST). His brother Bindon Blood Stoney (1828–1909, FRS) was Engineer to Dublin Port and Docks Board. Stoney's most distinguished relative was his nephew George Francis Fitzgerald.

Courtesy of the Royal Society

Addresses:

Born:	Oakley Park, King's County, where he spent his youth
1848–1852	Assistant at Parsonstown Observatory (Birr Castle)
1852–1857	Professor of Natural Philosophy at Queen's College (today University College) Galway
1857–1882	Secretary of The Queen's University in Dublin (predecessor of the Royal and National Universities), Dublin addresses: 1858 89 Waterloo Road; 1874 Weston House, Dundrum; 1878 3 Palmerstown Park, Rathmines
After 1893	Retirement in London.

Distinctions:
Stoney was a Fellow (1861) and a Vice-President (1898) of the Royal Society. In 1879 he was President of Section A of the British Association. He was a member of the Royal Irish Academy and was Honorary Secretary of the Royal Dublin Society for over 20 years and received the Society's first Boyle Medal. He was a foreign member of both the Academy of Sciences, Washington, and of the Philosophical Society of America.

Stoney's most important scientific work was the conception and calculation of the magnitude of the atom or particle of electricity for which he proposed the name 'electron'. In 1868 he estimated the number of molecules in a cubic millimetre of gas at room temperature and pressure from data obtained from the kinetic theory of gases. Similar determinations of this quantity, which is equivalent to Avogadro's number, were made independendy by J. Loschmidt of Vienna (1865) and by William Thomson (later Lord Kelvin) in 1870. In 1868 Stoney also proposed that light waves are produced by periodic 'orbital motions' within atoms or molecules. In his ultimate work on the atomic origin of spectra in 1891 *(Scientific Transactions of the Royal Dublin Society,* 1891, p. 583) Stoney proposed an 'electron' describing elliptical orbits in the molecule and used this idea to explain double and triple lines observed in gas spectra. This was the first time the word 'electron' was used in modern scientific writing. Joseph Larmor adopted Stoney's term and H.A. Lorentz, who developed the theory of electrons, acknowledged Stoney's contribution in his Nobel Lecture in 1902. In 1897 J.J. Thomson of Cambridge discovered cathode ray 'corpuscles', the date often taken for the discovery of the electron.

STONEY—*Cause of Double Lines in Spectra.* 583

1874, and printed in the *Scientific Proceedings* of the Royal Dublin Society of February, 1881, and in the *Philosophical Magazine* for May, 1881 (see pp. 385 and 386 of the latter). It is there shown that the amount of this very remarkable quantity of electricity is about the twentiethet (that is, $1/10^{20}$) of the usual electro-magnetic unit of electricity, *i.e.* the unit of the ohm series. This is the same as three-eleventhets ($3/10^{11}$) of the much smaller *C.G.S.* electrostatic unit of quantity. A charge of this amount is associated in the chemical atom with each bond. There may accordingly be several such charges in one chemical atom, and there appear to be at least two in each atom. These charges, which it will be convenient to call *electrons*, cannot be removed from the atom; but they become disguised when atoms chemically unite. If an electron be lodged at the point *P* of the molecule, which undergoes the motion described in the last chapter, the revolution of this charge will cause an electro-magnetic undulation in the surrounding æther. The

Copy of part of the page from the Transactions of the Royal Dublin Society, *Volume 4, 1891, showing the first use of the term 'electron'.*

As early as 1874 Stoney had calculated the magnitude of his electron from data obtained from the electrolysis of water and the kinetic theory of gases. The value obtained 10^{-20} Ampere (later called coulomb) was $^{1}/_{16}$th of the correct value of the charge of the electron. In this paper of 1874 (which was published in 1881) Stoney proposed the particle or atom of electricity as one of three fundamental units on which a whole system of physical units could be established. The other two fundamental units proposed were the constant of universal gravitation and the maximum velocity of light and other electromagnetic radiations. Several other nineteenth century scientists had contemplated the concept of an atom of electricity but no one had been bold enough to attempt a calculation of its magnitude with the crude data available.

Physicists like Larmor and Thomas Preston took their cue from Stoney's 1891 paper in investigating the Zeeman Effect – the splitting of spectral lines in a magnetic field. Stoney also partially anticipated Balmer's Law on the hydrogen spectral series of lines in 1871. He discovered a relationship between three of the four lines in the visible spectrum of hydrogen. The Swiss physicist J.J. Balmer found a formula to relate all four in 1885.

For most of his life Stoney was not a professional physicist and most of his practical investigations were done in his free time at the laboratory of the Royal Dublin Society.

Further reading:

J.G. O'Hara: George Johnstone Stoney FRS and the Concept of the Electron, *Notes and Records of the Royal Society,* **29** (2), 265–276, 1975.

Alex Keller: *The Infancy of Atomic Physics: Hercules in His Cradle,* Oxford University Press, New York, 1984.

J.G. O'Hara: George Johnstone Stoney and the Conceptual Discovery of the Electron, *Occasional Papers in Irish Science and Technology Number Eight: Stoney and the Electron*, Royal Dublin Society, 5–28, 1993.

James O'Hara, Leibniz-Archiv, Niedersächsische Landesbibliothek, Hanover, Germany.

73 HENRY HENNESSY Physicist and Mathematician 1826–1901

Born: Cork,19 March 1826.
Died: Bray, Co. Wicklow, 8 March 1901.

Courtesy of the Royal Society

Family:
His parents were John Hennessy of Ballyhennessy, Co. Kerry, and Elizabeth Casey of Cork.
A younger brother was John Pope-Hennessy MP.
Married: Rosa Corri.

Addresses:
1826–1855 Cork; 1855–1890 Dublin

Distinctions:
Fellow of the Royal Society 1858; Vice-President of the Royal Irish Academy, 1870–1873.

After schooling in Cork, Hennessy became a professional engineer, while maintaining an interest in physics and mathematics. His first paper, published in the *Philosophical Magazine* of London when he was only nineteen, was a proposal for the photographic recording of meteorological measurements. In the following year he published in the Royal Society's *Philosophical Transactions*.

In 1849 Hennessy became Librarian of the newly-founded Queen's College, Cork. He had hoped for a chair in natural philosophy at one of the new Colleges, and applied unsuccessfully for the vacant professorship of mineralogy at Cork in 1853. In 1855, however, Cardinal Newman invited him to fill the post of Professor of Physics at the new Catholic University in Dublin. He moved to the Royal College of Science in 1874 as Professor of Applied Mathematics, becoming Dean of the College in 1880. He retired in 1890 at the statutory age of 65.

Hennessy's scientific interests were broad, ranging through meteorology, climatology, geology, geophysics, practical mechanics and metrology, while a number of models of his inventions were preserved in the Museum of the Royal College. He took a strong interest in the promotion of institutions of higher education for Roman Catholics, particularly in science, and there is evidence that his ideas were not appreciated in the early days of Queen's College, Cork. The College was intended to be a secular foundation but, at the inauguration of the faculty of science at the Catholic University, Hennessy vigorously presented the case for institutions of scientific education 'guided, sanctioned and vivified by religion'.

Further reading:

H. Hennessy: *A Discourse on the Study of Science in its Relations to Individuals and to Society*, Dublin, 1858.
H. Hennessy: *On Freedom of Education*, Dublin, 1859.
Dictionary of National Biography.

Jim Bennett, Museum of the History of Science, University of Oxford.

74 HENRY JOHN STEPHEN SMITH Mathematician 1826–1883

Born: Dublin, 2 November 1826.
Died: Oxford, 9 February 1883.

Family:
Fourth child of John Smith (1792–1828), a Dublin barrister, and Mary Smith (née Murphy), who came from near Bantry Bay. Henry Smith was unmarried.

Addresses:

1826–1829	Near St Stephen's Green, Dublin
1830s	Ryde, Isle of Wight
1857–1874	64 St Giles', Oxford
1874–1883	Keeper's House, South Parks Road, Oxford

Distinctions:
Savilian Professor of Geometry, Oxford University 1860; Fellow of Royal Society 1861; Honorary LLD University of Dublin 1878.

Following the early death of his father, Smith's mother removed her family to England and eventually settled in Oxford in 1840. Smith was a student at Balliol College, where he gained his BA in 1849. He took up a mathematical lectureship at Balliol in 1850, and held this position until 1873. In the 1850s, he began his study of number theory. His researches were encapsulated in his *Report on the Theory of Numbers*, contributed in six parts to the annual meetings of the British Association between 1859 and 1865. The entire *Report*, occupying 325 pages of Smith's collected papers, still provides a valuable introduction to advanced number theory. In 1861, Smith published a paper on the integer-valued solutions to systems of linear equations with integer coefficients. This paper is the origin of the Smith normal form of an integer matrix, an important concept which has best perpetuated his name in mathematics.

In 1882, Smith submitted an entry for the Grand Prix des Sciences Mathématiques of the French Academy. The problem proposed for the prize, on the number of representations of a positive integer as the sum of the squares of five integers, had already been solved by Smith, and his solution, without detailed proofs, appeared in the *Proceedings of the Royal Society* in 1868. Smith died a few weeks before the announcement that he was to share the prize with the eighteen-year old Hermann Minkowski, who himself became a distinguished mathematician. Given Smith's priority of publication, the choice of prize problem and the competition's outcome did not do him justice.

Smith's work is a model of clarity and rigour, and is remarkable for its depth and modern spirit.

Further reading:
Dictionary of National Biography.
Dictionary of Scientific Biography.
J.W.L. Glaisher (ed.): *The Collected Mathematical Papers of Henry J. Smith* (two vols), Clarendon Press, Oxford, 1894.

Roderick Gow, Mathematics Department, University College, Dublin 4.

75 MARY WARD Microscopist, Astronomer, Naturalist, Artist

Born: Mary King, Ferbane, 27 April 1827.
Died: Birr, 31 August 1869.

Family:
Married: Henry William Crosbie Ward.
Children: Three sons and five daughters.

Addresses:
1827–1857 'Ballylin', Ferbane, King's County (now Co. Offaly)
1857–1861 Trimbleston, Booterstown, near Dublin
1861–1864 'Bellair', Moate, King's County
1864–1869 A number of addresses in or near Kingstown, Dublin

Courtesy National Trust of Northern Ireland

Mary King did not attend school or university but was educated at home in Co. Offaly by a governess. William Parsons, the third Earl of Rosse, was Mary's cousin and she was a frequent visitor to Birr Castle. She observed and chronicled the building of the giant telescope in the castle grounds. Through her famous cousin she met many of the most eminent men of science of the day.

Mary became well known as an artist, naturalist, astronomer and microscopist yet she never received any formal marks of distinction. It should be borne in mind that women could not become members of learned societies or institutions nor obtain degrees or diplomas during their lifetime. It was very difficult for them to become established or recognised in scientific or literary fields until well into the last quarter of the nineteenth century. Nevertheless Mary was the first woman to write and have published a book on the microscope in spite of the fact that it was very difficult to find publishers who would accept book manuscripts from women. When her first book on the microscope was published in London in 1858 Mary did not use her full name but was referred to as The Hon. Mrs W. She was to write three books on scientific subjects and numerous scientific articles while performing the duties of wife and mother of a rapidly growing family. Her book on the microscope was reprinted at least eight times between 1858 and 1880.

An exceptionally fine artist and painter, she illustrated all her own books and papers and also those of others. Sir David Brewster FRS came to visit her father's house and soon she was preparing microscopic specimens for him. These specimens she drew and painted, and the coloured illustrations may be seen in the *Transactions of the Royal Society of Edinburgh* in 1864. She also made the original drawings of Newton's and Lord Rosse's telescopes which can be seen in Brewster's *Life of Newton*. In 1864 Sir Richard Owen asked Mary to send him a copy of her painting of the natterjack toad for the collections of the British Museum. An article by Mary on 'Natterjack Toads in Ireland' had been published in a scientific journal and this paper was reprinted in full in *The Irish Times* in May 1864 with a very complimentary editorial comment. When eighteen years old her parents bought her a fine microscope which she continued to use and to demonstrate with enthusiasm until her death.

The natterjack toad
A painting by the Hon. Mrs Ward

Her first microscope book was produced privately by Sheilds of Parsonstown in 1857. It was called *Sketches with the Microscope* and only 250 copies were printed. It was published in 1858 by Groomsbridge of London as *The World of Wonders Revealed by the Microscope.* This was greatly expanded and became *Microscope Teachings* in 1864. *Telescope Teachings,* a companion volume to *Microscope Teachings,* was published in 1859.

On 31 August 1869, when she was 42, Mary, Henry and two of Lord Rosse's sons were travelling on a steam carriage invented by their father when it jolted and threw Mary to the ground where she was crushed by one of its heavy wheels and died instantly.

Plate 1 from 'Sketches with the Microscope'. 1. The Microscope. 2. Scales of Ghost Moth, magnified 80 diameters. 3. Scales on underside of Ghost Moth's wing, magnified 100 diameters. 4. Green Forester Moth. 5. Scales of Green Forester Moth, magnified 100 diameters. 6. Scale, magnified 300 diameters. 7. Six-spotted Burnet Moth. 8. Scale of Bumet Moth, magnified 420 diameters.

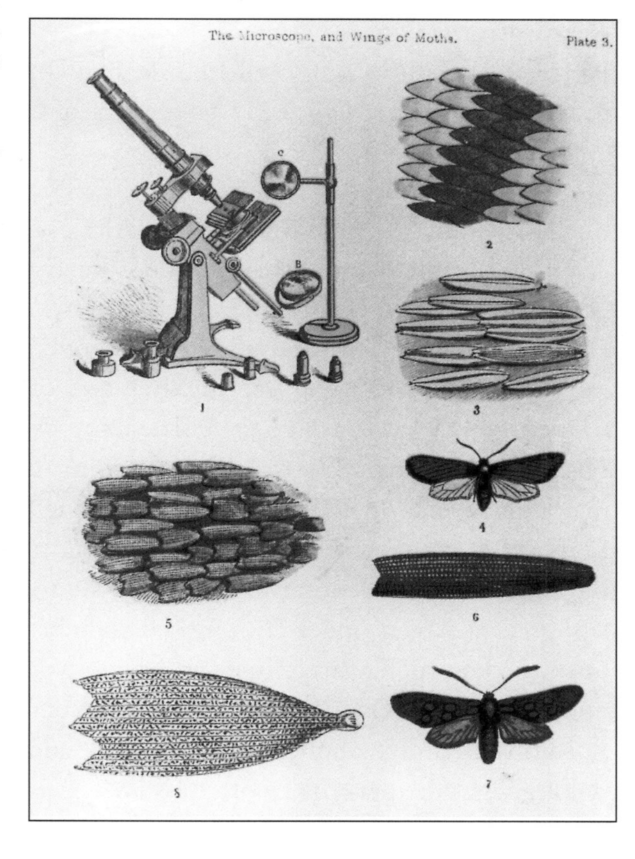

Further reading:

Transactions of the Royal Society of Edinburgh **23**, 1864.
Mary Ward: *Microscope Teachings,* Groomsbridge, London, 1864.
Mary Ward: *Telescope Teachings,* Groomsbridge, London, 1859.

Owen Harry, Department of Zoology, The Queen's University of Belfast.

76 BINDON BLOOD STONEY Civil Engineer 1828–1909

Born: County Offaly, 13 June 1828.
Died: Dublin, 5 May 1909.

Family:
His father, George Stoney, married Anne Blood of Cranagher, County Clare, in 1825. They had two sons and two daughters. The older son, George Johnstone Stoney was a distinguished physicist.
Married: Susannah Frances Walker of Grangemore, County Dublin, in 1879. They had a son, George Bindon (died 1909), and two daughters.

Addresses:

1828–1844	Oakley Park, King's County (now Co. Offaly)
1844–1864	89 Waterloo Road, Dublin
1864–1866	63 Wellington Road, Dublin
1866–1881	42 Wellington Road, Dublin
1881–1909	14 Elgin Road, Dublin

Distinctions:
President, Institution of Civil Engineers of Ireland 1871; Telford Medal and Premium, Institution of Civil Engineers 1874; LLD University of Dublin 1881.

Stoney was educated privately and at Trinity College Dublin, where he studied civil engineering under Sir John Macneill, graduating BAI in 1849. He was Resident Engineer on the Boyne Viaduct & Bridge under James Barton, where he carried out extensive full-scale tests on wrought-iron latticed girders. His comprehensive and innovative handbook on the theory of strains in structures and the strength of materials was first published in 1866–1869 in two volumes and filled an important gap in the engineering literature of the time. In 1856 Stoney was appointed as Assistant to George Halpin II at Dublin Port, taking over responsibility for all engineering matters in 1859. His Assistant from 1871 to 1898 was John (later Sir John) Purser Griffith (1848–1938). Stoney's major achievement at Dublin Port was the design and construction of new deep-water quays. Large monolithic concrete blocks, each weighing up to 350 tons, were constructed on shore and lifted into position by a floating shears crane, the foundation having first been excavated and levelled by labourers working under compressed air in a diving bell. This was the first time that concrete had been used on such a large scale for marine works. Also, under his direction, half the former shipping quays along the River Liffey were reconstructed as deep-water quays. His creative work in Dublin Port involved the application of sound engineering and scientific principles, coupled with a willingness to experiment with new materials and methods of construction.

Further Reading:
Ronald C. Cox: *Bindon Blood Stoney; Biography of a Port Engineer,* Institution of Engineers of Ireland, Dublin, 1991.

Ronald Cox, Centre for Civil Engineering Heritage, Trinity College, Dublin

Born: Antrim town, 21 May 1829.
Died: London, 18 October 1917.

Courtesy of the Royal Society

Family:
His father was Rev. J.D. Hull, a Church of Ireland curate in Antrim.

Addresses:

1829–1841	Antrim
1841–1843	Edgeworthstown, Co. Longford
1843–1846	Lucan, Co. Dublin
1846–1850	Dublin
1850–1867	England and Wales
1867–1868	Scotland
1868–1891	Dublin

Distinctions:
Fellow of the Royal Society 1867; Murchison Medal of the Geological Society, 1890.

Hull went to school in Edgeworthstown and Dublin. Deciding against a career in the church, he studied for the Diploma in Engineering at Trinity College Dublin, which he took in 1849, and graduated BA the following year. Geology was taught as part of the engineering course, and the lectures of Thomas Oldham were responsible for a shift in the direction of Hull's interests. Having failed to find employment as an engineer in Ireland, Hull spent seventeen years working for the Geological Survey of England and Wales, and from 1867 a further two years as a District Supervisor in the Geological Survey of Scotland. After a temporary appointment to the Geological Survey of Ireland to stand in for the ailing Jukes, Hull succeeded him as Director on his death in 1869. At the same time he was granted Jukes's other appointment, as Professor of Geology at the Royal College of Science.

One of Hull's early tasks as Director was to supervise the move of the Survey from St Stephen's Green to new premises in Hume Street. During his period of office, in 1883, he led an expedition charged with topographical and geological surveying in the Middle East on behalf of the Committee of the Palestine Exploration Fund, and he published two volumes on this work. Hull's tenure saw the completion of all the sheets of the one-inch Geological Map of Ireland in 1890, and his retirement the following year was part of the subsequent diminution in the staff of the Survey. He was a prolific writer of geological articles and books, an activity he continued into retirement, and he also published in areas outside his professional expertise.

Further reading:
E. Hull: *Reminiscences of a Strenuous Life*, London, 1910.
Obituary in *Proceedings of the Royal Society of London*, **90B**, xxviii–xxxi, 1919.
G.L. Herries Davies: *Sheets of Many Colours: the Mapping of Ireland's Rocks 1750–1890*, Royal Dublin Society, Dublin, 1983.

Jim Bennett, Museum of the History of Science, University of Oxford.

78 HENRY BENEDICT MEDLICOTT Geologist 1829–1905

Born: Loughrea, Co. Galway, 3 August 1829.
Died: Clifton, Bristol, 6 April 1905.

Family:
Second son of Rev. Samuel Medlicott and Charlotte (née Dolphin). Married: Louisa Maunsell of Balbriggan, Co. Dublin, 1857; four sons and two daughters.

Addresses:

1850–1853	Dublin
1853–1854	England
1854–1887	Roarkee, Bombay, India
1887–1905	Clifton, Bristol, England

Distinctions:
Fellow of the Royal Society 1877; President Asiatic Society of Bengal 1879–1881; Wollaston Medal Geological Society of London 1888. He rarely cited these distinctions, as he considered them to have been awarded for 'official' work.

Henry Medlicott was educated in France and Guernsey, and Trinity College, Dublin. He was one of three brothers employed by the fledgling Geological Survey of Ireland: Joseph (*c.*1825–1866) later worked for the Geological Survey of India (1851 to 1862), while Samuel (born 1832) was later ordained. Medlicott transferred to the Geological Survey in England in 1853 but was recruited after a few months by Thomas Oldham for the Geological Survey of India. However, he was immediately appointed Professor of Geology at Rukri College, and only rejoined the Survey in 1862 as Deputy-Superintendent. In 1857 he enlisted as a volunteer soldier during the Indian Mutiny, and narrowly escaped death. He is best known for coining the term 'Gondwana Series' in 1872 for Permian coal-bearing successions in India. It was later adopted, in the modified form 'Gondwanaland', as the name of the former southern super-continent, that included South America, India, Africa, Australia, and Antarctica. His other lasting legacy is the two volume *Manual of the Geology of India,* co-authored with W.T. Blanford, and published in 1879.

He succeeded Oldham as Superintendent (styled Director from 1885) in 1876, and retired in 1887 after 33 years service in India, during which time he only took 18 months leave. He was succeeded by another Irishman William King. Nearly twenty Irish geologists served with the Geological Survey of India, and for all but sixteen of the first seventy years the Director was Irish.

Medlicott retired to Bristol where he lived quietly and wrote philosophical pamphlets. In 1904 he became ill after straining himself while bicycling, and died seated in his study the following year.

Further reading:
Dictionary of National Biography.
Geological Survey of India News, **2**(3), 6–10, 1971.
W.T. Blanford, *Proceedings of the Royal Society, Series B,* **79**, 19-26, 1907.

Patrick N. Wyse Jackson, Department of Geology, Trinity College, Dublin.

Born: Dublin, 16 July 1830.
Died: Dublin, 27 February 1921.

Family:
Son of Captain Ewen Cameron of Locharber, Scotland. Married: Lucie Macnamara 1862; seven children.

Distinctions;
Knighted 1885; President of the Royal College of Surgeons in Ireland 1885–1886; President of the Society of Public Analysts 1893–1894; Companion of the Bath 1899.

Charles Alexander Cameron was educated and received his professional training in medicine and chemistry in Dublin, studying the latter with Dr Aldridge. One of his early posts was as Professor of Chemistry to the Dublin Chemical Society. This body was set up in 1852 to teach chemistry, and it lasted till 1862, by which time courses at the Royal College of Science made it unnecessary. He held numerous teaching posts concerned with chemistry and hygiene in the Dublin Medical Schools over the years.

The first Food and Drugs Act in 1860 established the post of Public Analyst. Dublin was one of the few authorities to use the Act and, in 1862, appointed Cameron to the post he held for 59 years. Following the 1870 Act he was chosen by 33 counties and boroughs and was humorously referred to as 'the Public Analyst for Ireland'. He was first Chairman of the Irish Analysts' Association set up in 1910 to deal with local problems and fees for professional work.

He published extensively original work on agricultural chemistry, analysis of food, diet and health and hygiene, in addition to text books on the same areas. He wrote a definitive *History of the Royal College of Surgeons in Ireland, and of the Irish Medical Schools* (1886, Second Edition 1916) and left personal records of his life and times *via Reminiscences* (1913) and an *Autobiography* (1921). In *Reminiscences* one can read of amusing anecdotes of court cases in which he had given evidence as an expert witness, details of formation of the Dublin Chemical Society and of the Corinthian Club, a club based on the ethos of the Savage Club, London, of which he was a member. He was clearly a character, lived a long active life and was well known in Irish Free Masonary. His obituary in the *Analyst* concludes 'that not the least of his achievements was in the retention of his full mental faculties and physical working powers beyond his 90th year'.

Further Reading:

Reminiscences of Sir Charles A. Cameron, CB, Hodges Figgis & Co., Dublin, 1913.
C.A. Cameron: *Autobiography*, Hodges Figgis & Co., Dublin, 1921.
B. Dyer: Obituary, *Analyst*, **46**, 175, 1921.

Duncan Thorburn Burns, Department of Analytical Chemistry, The Queen's University of Belfast.

Born: Drumcondra House, Dublin, 10 October 1832.
Died: Charlottesville, Virginia, USA, 7 November 1912.

Family:
Eldest son of Robert Mallet and his wife Cordelia (Watson). He had two brothers and three sisters. He married twice, firstly Mary Ormonde, daughter of an Alabama judge in 1857 and, after her death, Josephine Burthe née Pages.

Distinctions:
Fellow of the Royal Society 1877; President of the American Chemical Society 1882; Vice-President of the Chemical Society 1888–90; Honorary LLDs from William and Mary College, University of Mississippi, Princeton, John Hopkins, and the University of Pennsylvania.

J.W. Mallet had an early advantage by being able to use his father's extensive library. Private lessons in chemistry from James Apjohn further stimulated his interest, so that by the time he was an undergraduate in Trinity College he had already published a paper. He spent two years studying chemistry under the German chemist Friedrich Wöhler at Göttingen, where he was awarded his PhD. He became an excellent analytical chemist.

In 1853 he went to America: he was appointed Professor of Analytical Chemistry at Amherst, Massachusetts in 1854. In the following year he became chemist to the State Geological Survey in Alabama, and then Professor of Chemistry at the University of Alabama. After the start of the Civil War, he enlisted as a trooper in the Confederate Cavalry, soon becoming aide-de-camp to General Rhodes. During most of the war, he was superintendent of the ordnance laboratories with responsibility for gun-powder manufacture. After the Civil War, he was Professor of Chemistry in the University of Louisiana, undertaking a geological search for petroleum in that State. In 1868 he transferred to the University of Virginia, where he designed the first course on industrial chemistry offered in the USA. Because of his son's failing health, he moved to Austen to oversee the building of a chemistry department in the newly-founded University of Texas in 1882 (in 1931 the University named its science library after him). After the death of his son, he moved to the Chair of Chemistry at the Jefferson Medical School in Philadelphia. In 1884, he returned to the University of Virginia, retiring in 1908.

Further Reading:
F.E. Vandiver: *John William Mallet and the University of Texas*, Southwestern Historical Quarterly, **53**, 422–42, 1950.
Desmond Reilly: John William Mallet, 1832–1912, *Endeavour*, **12**, 48–51, 1953.

William J. Davis, University Chemical Laboratory, Trinity College Dublin.

Born: Ross Carbery, Cork, 4 November 1837.
Died: Margate, Kent, 24 October 1916.

Family:
Hicks was the second of three children of George Hicks, a flax worker, and Gillian Coakley.
Married: Emma Sarah Robertson, a milliner, on 19 June 1862 (she died 19 February 1899); three children.

Addresses:
Ross Carbery, Cork
London
Margate, Kent

Distinctions:
Knight of the Order of St Gregory the Great 1898.

J.J. Hicks from The Optician, *6 October 1898 (courtesy of Islington Local History Library)*

When young, James Joseph Hicks was sent to London, to stay with his mother's relatives. He attended school until the age of 14. In 1852 he was apprenticed to Louis Casella, a well-known maker of thermometers and glass instruments, who was of Italian descent and was based in Hatton Garden. By 1860, Hicks was Casella's foreman, and already designing and improving instruments. He appears to have started in business on his own account at about the time of his marriage. As he prospered and the family grew, they moved into various properties in the north London suburbs.

Hicks specialised in the trade to which he had been apprenticed: philosophical glassware and meteorological apparatus. An astute businessman, who made full use of advertising, marketing and overseas contacts, his company eventually employed over 300 men in his London factories. He had a tenacious grasp of patent and design rights, and he doggedly prosecuted those who infringed them. In 1882 he fought and won the largest patent law suit ever heard in the United States Circuit Court over the shape of thermometer bore. His success was in part due to his exploitation of growing overseas markets, and by keeping an immense throughput of innovations, patents, and designs, while maintaining his near-monopoly of a short-lived but essential instrument, the clinical thermometer. By 1914, Hicks had produced over thirteen million of them.

The business made a wide variety of instruments, principally meteorological in nature and for the brewing trade. A staunch Catholic all his life, Hicks visited Pope Leo XIII in 1898, and was made a Knight of the Order of St Gregory the Great. By 1910, he prepared to retire by successfully amalgamating the business with that of W.F. Stanley of Holborn. He died in 1916, and was buried at Ramsgate. His personal estate was valued at over £36,000.

Further Reading:
Anita McConnell: *King of the Clinicals: the Life and Times of J.J. Hicks 1837–1916*, William Sessions Ltd, York, 1998.

A.D. Morrison-Low, National Museums of Scotland, Edinburgh.

82 JOHN DUNLOP Veterinary Surgeon and Inventor

Born: Ayrshire, Scotland, 5 February 1840.
Died: Dublin, 24 October 1921.

Family:
Married: 1876. Children: One son and one daughter.

In 1845 a young Scotsman named Robert William Thompson took out a patent for an air filled tyre, under the title of 'Improvements in Carriage Wheels', but appears to have done little either to improve or extensively apply the idea, which was basically to intercept vibration before it reached the rim of the wheel. Dunlop was therefore not the first to think of the pneumatic tyre. He was, however, responsible for inventing the first practical form of it.

John Boyd Dunlop 1840-1921

Born in the village of Dreghorn, Ayrshire, John Boyd Dunlop attended Dreghorn parish school, where he proved himself to be a keen and diligent student, capable of assisting in the school as a pupil teacher. Although always suffering in his childhood from rather delicate health, and very prone to chills throughout his life, he was encouraged by his parents to pursue his studies at Irvine Academy under a Dr White, and qualified as a veterinary surgeon at the early age of nineteen.

Moving to Ireland, he set up a most extensive and successful veterinary practice in Gloucester Street, Belfast, with a dispensary and a staff of twelve horseshoers. Always interested in the solution of practical problems, he became fascinated in particular with the improvement of the vehicle wheel.

Bicycles and tricycles of more efficient design were becoming increasingly popular during the last quarter of the nineteenth century, but the rough road surfaces of the day limited their use. With the solid rubber tyres which vehicles of all kinds had, travel could be pretty uncomfortable. Dunlop's son, John, was a keen cyclist, so Dunlop considered the possibility of replacing the solid tyre with an air-filled tube. Towards the end of 1887 he made a tube out of sheet rubber, with a small air supply tube, and fixed it to a wooden disc by means of a linen strip. The tube was then inflated with his son's football pump. In his yard in Gloucester Street he compared the speed and length of run of this wheel with the solid-tyred front wheel temporarily removed from his son's tricycle. The air-filled tyre proved to be faster than the solid one and more efficient. Dunlop then made two wooden rims with air inflated tyres and fitted them to his son's tricycle on the evening of 28 February 1888, with most successful results.

Northern Bank £10 banknote commemorating John Dunlop

Following further experiments with the inflated tyre on tricycles made by Edlin and Sinclair, Belfast, Dunlop applied for a provisional patent, specification No. 10,607, on 23 July 1888 (accepted on 7 December 1888) describing:

> 'An improvement in Tyres of Wheels for Bicycles, Tricycles, or other road cars . . .'
> 'A hollow tyre or tube made of India-rubber and cloth . . . said tube or tyre to contain air under pressure or otherwise and to be attached to the wheel . . .'.

This was followed in October 1888 by an amendment, which embodied the same basic principle of a tube filled with compressed air.

The first experimental bicycle pneumatic tyre ever made by Dunlop was presented by him to the Royal Scottish Museum, Edinburgh and is still on exhibition.

At the Belfast Queen's College Sports, held on 18 May 1889, W. Hume won the cycle race on an Edlin bicycle fitted with Dunlop's pneumatic tyres, an event which was to revolutionise road transport. Many improvements followed but the principle always remained the same.

The new tyres were manufactured first by Edlin and Sinclair of Belfast, later by the Pneumatic Tyre and Booth's Cycle Agency in Dublin, of which Dunlop became a director. He subsequently went to live in Dublin and died there in his 82nd year following a slight cold.

Further reading:

J.B. Dunlop: *The History of the Pneumatic Tyre*, Dublin, no date.
Dictionary of National Biography.

Alfred Montgomery, Department of Local History, (Industrial Archaeology), Ulster Museum, Belfast (retired).

Born: Dublin, 1 July 1840.
Died: Cambridge, 25 November 1913.

Family:
Married: Francis Elizabeth Steele, 1868.
Children: Four sons and two daughters, including the biographer W. Valentine Ball. Nephew of Mary Ball.

Addresses:

1840–1854	3 Granby Row Upper, Rutland Square, Dublin
1865–1867	Birr Castle, Co. Offaly
1868	43 Wellington Place
1874–1892	Dunsink Observatory

Distinctions:
Member of Council of the Royal Irish Academy 1870 (Secretary 1877–1880; Vice President 1885–1892); Fellow of the Royal Society 1873 (Council Member 1897–1898); President of the Royal Astronomical Society 1897–1899; President of the Mathematical Association 1899–1900; President of the Royal Zoological Society, Ireland 1890–1892; Cunningham Medal of the Royal Irish Academy (for mathematical research) 1879; Knighted 1886.

Son of a distinguished Cork naturalist, Robert Stawell Ball received his early education at a school kept by Dr J. Lardner Burke in North Great George's Street, Dublin, and at Abbotsgrange, Tarvin, near Chester. In 1857 he entered Trinity College, Dublin where, in a distinguished career, he gained a scholarship, the Lloyd Exhibition, a University Studentship, two gold medals, and prizes in three successive examinations, 1863–1865.

In the latter year, he accepted a post as tutor to the younger sons of William Parsons, the third Earl of Rosse, on the understanding that he would be allowed to observe with the great six-foot reflector at Birr Castle, then the largest telescope in the world. Between 1866 and 1867 he made observations with this instrument of the positions of many faint nebulae, correcting his measurements for instrumental errors to an accuracy not previously achieved by other users of this telescope. In 1867, on the recommendation of the Earl of Rosse, he was appointed Professor of Applied Mathematics at the then new Royal College of Science in Dublin. A compendium of his excellent lectures to College students on Experimental Mechanics was published in 1871.

In 1874 he was appointed Royal Astronomer of Ireland and Andrews Professor of Astronomy in the University of Dublin at Dunsink Observatory. In this dual capacity, he sought to develop the Dunsink tradition of measuring stellar distances, using a large sample of stars rather than special objects. Although the method he adopted was later found to be inappropriate for the task, his findings served to identify special problems in making extensive sky surveys and anticipated the later development of more accurate investigative methods. In 1892 he was appointed Lowndean Professor of Astronomy and Geometry at Cambridge but the sad circumstance of the deterioration of his eyesight from 1883, culminating in the loss of his right eye in 1897, gradually brought a halt in this period to his activity as a visual observer.

The work for which he is chiefly remembered, his classic researches on screw motions, was developed over more than thirty years in a series of important communications, contributed in great part to the Royal Irish Academy from 1871. He developed a powerful geometrical method to treat the problem of small movements in rigid dynamics, investigating in particular the behaviour of rigid bodies having different degrees of freedom. In the case where there are two degrees of freedom, he demonstrated that the cylindroid shown in the figure represents the cubic surface focus of the screw axis for all possible twists. Thereafter, he took a special interest in exploring the detailed properties of this kind of surface. In the course of his investigations he made independent discovery of certain theorems concerned with the theory of linear complexes in line geometry, a topic which, in his day, was only in its infancy, and he is now ranked among the leaders of nineteenth century mathematics for his contributions to the geometry of motion and force.

Of genial temperament, he was an outstanding public lecturer and his popular works on astronomy (thirteen volumes published between 1877 and 1908), including *The Story of the Heavens,* and a university textbook *A Treatise on Spherical Astronomy,* enjoyed a considerable vogue.

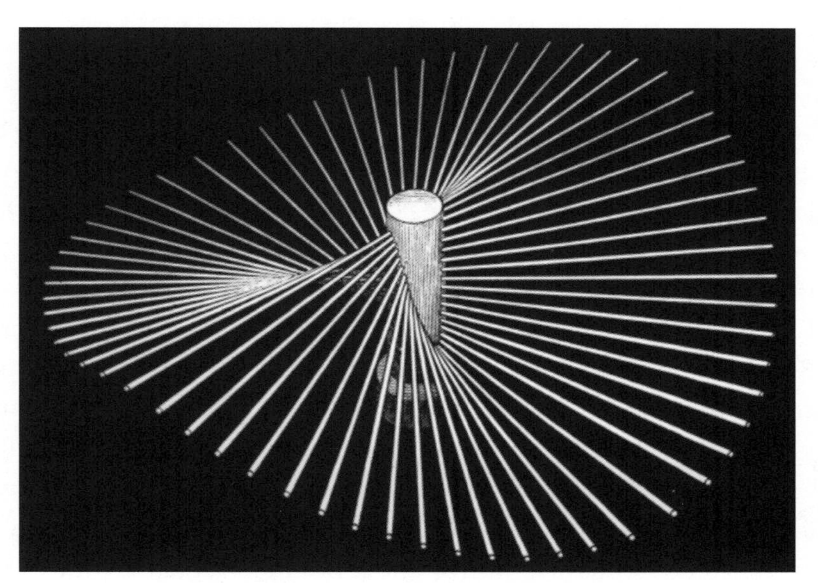

Ball's cylindroid: a model for screw motions

Further reading:

W. Valentine Ball (ed.): *Reminiscences and Letters of Sir Robert Ball,* Cassell and Co. Ltd, London, 1915.

Sir Robert S. Ball: *A Treatise on the Theory of Screws,* Cambridge University Press, 1900.

O. Henrici: The Theory of Screws, *Nature* No. 1075, **42**, 127–132, 1890. (Review of *Theoretische Mechanik Starrer System* by H. Gravelius, published Berlin, Reimer, 1889, an important German treatise based mainly on Ball's work.)

Susan McKenna-Lawlor, National University of Ireland, Maynooth, Co. Kildare.

84 LAURENCE PARSONS Fourth Earl of Rosse, Astronomer and Engineer 1840–1908

Born: Birr Castle, Co. Offaly, 17 November 1840.
Died: Birr, 29 August 1908.

Family:
Eldest son of William Parsons, third Earl of Rosse and Mary née Field. Other surviving brothers: Randal, Richard Clere and C.A. Parsons.
Married: Hon. Frances Cassandra Hawke, only child of Lord Hawke, 1870.
Children: William Edward, fifth Earl, died of war wounds 1918; Geoffry L.; Richard; and Muriel.

Address:
Birr Castle, Co. Offaly

Distinctions:

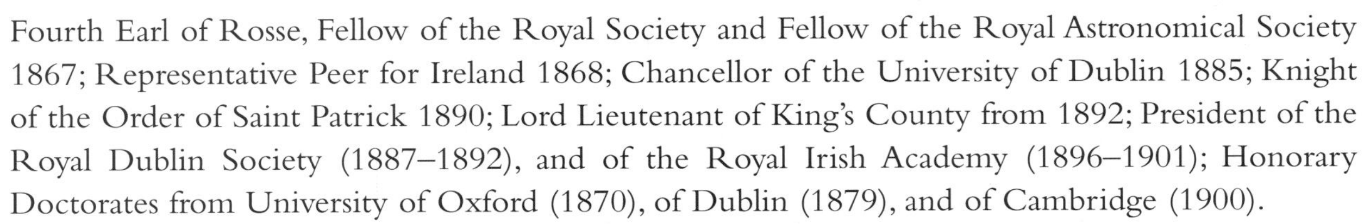

Fourth Earl of Rosse, Fellow of the Royal Society and Fellow of the Royal Astronomical Society 1867; Representative Peer for Ireland 1868; Chancellor of the University of Dublin 1885; Knight of the Order of Saint Patrick 1890; Lord Lieutenant of King's County from 1892; President of the Royal Dublin Society (1887–1892), and of the Royal Irish Academy (1896–1901); Honorary Doctorates from University of Oxford (1870), of Dublin (1879), and of Cambridge (1900).

Laurence Parsons could just recall the completion of the tube for the giant six foot telescope built by his father in the workshops at Birr in 1844. As Lord Oxmantown, he grew up at Birr Castle and was educated at home, where he was encouraged to use his father's workshops. After he graduated from Trinity College Dublin in 1864 he became committed to the work of the Observatory, paying for astronomical assistants from his own pocket. In 1867 he described a hydraulic device for driving an equatorially mounted telescope in the *Monthly Notices of the Royal Astronomical Society.* In 1868 his account of observations made at Birr, published by the Royal Society, contained an excellent engraving of the Great Nebula in Orion, M42. In 1868 using the three foot diameter telescope, he began the research for which he is best known, into the heat radiated by the moon in the infra-red. His original estimate for the temperature of the moon's surface, 500°F, was adjusted downwards in 1877 to 197°F, which is not far from modern observations. During his lifetime his results were not accepted, which disappointed him. In 1875 he had an equatorial mount built for the three foot telescope but, unfortunately, the massive mirror could not be orientated with sufficient accuracy to allow satisfactory photographic time exposures. He was shy by nature but determined in pursuit of an objective. He remained close to his brother Charles, the inventor of the steam turbine. He participated in the trials of Charles' experimental vessel, *Turbinia,* and served as Chairman of the Marine Steam Turbine Company, formed in 1894, and of its successor, the Parsons Marine Steam Turbine Company, until his death.

Fnrther reading:

P. Moore: *The Astronomy of Birr Castle,* Birr, 1971.
W.G. Scaife: *From Galaxies to Turbines,* Institute of Physics, Bristol, 1999.

W. Garrett Scaife, Fellow Emeritus, Trinity College Dublin.

85 JOHN PHILIP HOLLAND Submarine Pioneer, Inventor

Born: Liscannor, Co. Clare, c24 February 1841.
Died: Newark, New Jersey, 12 August 1914.

Family:
The second of four boys, his father, John Holland, was a Coast Guard officer and his mother was Mary Scanlon of Killaloe, Co. Clare. Married: Margaret Foley of Paterson, New Jersey, the daughter of Irish immigrants, on 17 January 1887. They had five children, John Jr, Robert Charles, Joseph Francis, Julia and Margaret.

Addresses:

c1841–1853	Castle Street, Liscannor, Co. Clare
1853–1860	Limerick City
1860–1873	North Monastery, Cork and other Christian Brothers' schools in Portlaoise, Enniscorthy, Drogheda and Dundalk
1873–1874	Boston, Massachusetts
1874–1883	Paterson, New Jersey
1883–1904	185 Court Street, Newark, New Jersey
1904–1914	38 Newton Street, Newark, New Jersey and summers at Fleet's Point, Cutchogue, Long Island

Distinctions:
Honorary MSc from Manhattan College, Riverdale, New York, 1905; Order of Merit, Rising Sun Ribbon, from Emperor Hirohito, 1908.

As the son of a Coast Guard riding officer whose job it was to check the local coastline for shipwrecks and smugglers, Holland had an early introduction to the sea. He was educated at Ennistymon Christian Brothers' School and Synge Street CBS in Limerick, following his father's transfer in 1853. Prevented from pursuing a career at sea because of short-sightedness, Holland joined the Christian Brothers Order in 1858 and moved to North Monastery CBS in Cork, the first of a number of teaching posts in various Christian Brothers' schools. A gifted and innovative teacher of mathematics, physics and music, he was the first person in Ireland to introduce the tonic sol-fa (the system of syllables used in singing). Initially interested in the mechanics of flight, Holland turned his attention to submarine design after reading about Confederate attempts to sink Union Navy warships using submersibles during the American Civil War.

In 1873 Holland left the Christian Brothers (with ongoing ill-health he had not taken his final vows) and emigrated to America. Settling in Paterson, New Jersey, he devoted more and more of his time to submarine design. His specialist knowledge became known to the leadership of Clann na Gael and the Fenian Brotherhood, who were quick to see the potential of an underwater vessel as an offensive weapon. Throughout his life, Holland steadfastly believed that the development of a successful submarine would act as a deterrent to war, and could have peaceful uses.

With financial assistance from the Fenian Brotherhood, *Holland No. 1*, a 14ft, 2.5 ton petrol-engined craft, was successfully launched and tested in 1878. A second submarine, dubbed *The Fenian Ram*, a 31ft, 19 ton petrol-engined craft was launched in 1881. A third submarine was planned and a 16ft,

Advertisement for Holland submarines, 1905

1 ton prototype constructed. Known as *Boat No.3*, this craft and *The Fenian Ram* were stolen from Holland's boatyard and towed away by the leadership of Clann na Gael and the Fenian Brotherhood in an internal feud. *Boat No.3* sank while under tow and the *The Fenian Ram* was abandoned in a shed.

Overcoming these earlier disappointments and others, such as the failure of his fourth submarine, named *The Zalinski Boat* after her backer, in 1885, and political intrigue, when his designs were accepted then rejected by successive US Navy Department administrations, Holland formed the John P. Holland Torpedo Boat Co. in 1893. This company was awarded a US Navy Department submarine contract, and work on Holland's fifth design began. Launched in 1895 and known as the *Plunger*, this 85ft, 168 ton craft possessed steam and electric engines for surface and underwater running.

Holland's next craft, known as *Holland No.6* was launched in 1897 and represented the pinnacle of his submarine achievements. It was this design, and variants of it, which was purchased by the US Navy and, in the years that followed, by the navies of the Japanese, Russian and British Empires. With a length of over 53ft, the 63 ton electric-powered *Holland No.6* was armed with two pneumatic guns and one torpedo tube.

By the beginning of the twentieth century, Holland's name had become synonymous with submarine excellence, but this was of no benefit to Holland himself, as his company had been taken over by the Electric Boat Company. Holland was forced to resign from this company in 1904 amidst legal wrangling over ownership of his various patents. Subsequent fighting through the courts, although successful, left him penniless and, discouraged, he retired from submarine construction in 1907.

In declining health, Holland returned to his earlier interest in powered flight, even drawing up a design for a steam-powered aircraft with moving wings. He died from pneumonia at his family home on 14 August 1914, and was buried at Totowa Cemetery, New Jersey.

Some forty days after Holland died, a submarine, *U-9*, of the Imperial German Navy torpedoed and sank three Royal Navy cruisers, the *Aboukir, Cressy,* and *Hogue* off the Heligoland coast. At no cost to herself, a submarine of just 450 tons, manned by 26 men, had sunk 36,000 tons of enemy shipping with a loss of over 1,400 lives. As a weapon of war, the submarine had arrived.

Further reading:

Frank T. Cable: *The Birth and Development of the American Submarine*, Harper, New York, 1924.
John de Courcy Ireland: *Ireland and the Irish in Maritime History*, Glendale Press, Dun Laoghaire, 1986.
Richard Knowles Morris: *John P. Holland 1841–1914: Inventor of the Modern Submarine*, New Edition, University of South Carolina Press, Columbia, 1998.
Pat Sweeney: John Holland 'Father of the submarine', *Maritime Journal of Ireland*, **43**, 9–12, Summer 1998.

Timothy Collins, Centre for Landscape Studies, National University of Ireland, Galway.

Born: Bandon, Co. Cork, 1841.
Died: Burton-on-Trent, England, 1907.

Addresses:
Bandon, Co. Cork
Burton-on-Trent, England

Distinctions:
First brewing chemist in the world; awarded Longstaff Medal of the Chemical Society 1884; Fellow of the Royal Society 1885.

Cornelius O'Sullivan's early education was at Denny Holland's private school. Later he attended the Cavendish School, endowed by the Duke of Devonshire who held extensive land-holdings in west Cork, where Thomas Lordan taught him. His interest in science may have been sparked off through contact with a local tannery owned by Richard Clear, JP. He attended evening classes in 1861–1862 in Bandon by professors from Queen's College, Cork. These were under the Science and Art Department of the Committee of the Council of Education, then intent on bringing science to many rural areas in Ireland. O'Sullivan distinguished himself by being placed in the first class in inorganic and organic chemistry at the 1862 examination. On the strength of this, he was awarded a Scholarship to the Royal School of Mines in London, where he came under the influence of A.W. Hofmann, perhaps the greatest teacher of chemistry in the nineteenth century. When Hofmann left London to become Professor at Berlin University in 1865, Cornelius went with him as *Privatdocent* to teach and to continue his studies in organic chemistry.

The first firm of brewers to employ a chemist was Bass & Co., at Burton-on-Trent, and on Hofmann's recommendation O'Sullivan was appointed assistant brewer and chemist there in 1866. Here he stayed for the rest of his life, developing exact methods of carbohydrate analysis. He was the first person to put brewing, which until then had been a craft industry, on a scientific basis. He published widely on the chemistry of brewing, on enzymic action, and on the complicated chemistry of gums. When awarding him the Longstaff Medal in 1884, W.H. Perkin, President of the Chemical Society, said of O'Sullivan: 'The methods you have used for the purpose of getting an insight into the complicated structures of these compounds [carbohydrates] by gradually breaking down the molecule and examining the resulting products, have thrown much light on their constitution'.

Cornelius went on to become Head Brewer at Bass in 1894. His last paper on 'Gum Tragacanth' was published in 1901, six years before his death.

Further Reading:
H.D. O'Sullivan: *The Life and Work of Cornelius O'Sullivan, FRS*, Guernsey, 1934.

William J. Davis, University Chemical Laboratory, Trinity College, Dublin.

Born: Cork, February 1842.
Died: London, January 1907.

Courtesy of the British Astronomical Association

Family:
She came from a professional family; her father John William Clerke was a classics graduate of TCD; her mother Catherine née Deasy, a gifted musician, was sister of Lord Justice Deasy. Her brother Aubrey St John became a barrister, and her sister Ellen was both a poet and a writer on literary and scientific subjects. Agnes did not marry.

Addresses:

1840–1861	Skibbereen, Co. Cork
1861–1863	Dublin City
1863–1877	Cobh, Co. Cork
1877–1907	68 Redcliffe Square, London, England

Distinctions:
Actonian Prize of the Royal Institution for her written contributions to Astronomy 1862; Honorary Membership of the Royal Astronomical Society 1903.

Agnes Clerke studied at home and was first drawn to astronomy through the influence of her father, who taught the Clerke children how to make observations with a four-inch telescope. By the time she was fifteen she had formed the intention to write a *History of Astronomy* and began already to draft several of its chapters. Due to delicate health, Agnes spent much of each year from 1867 to 1876 in Italy where she became an accomplished linguist, while also studying literature and music (once playing on the piano for the Hungarian composer Franz Liszt).

When the family settled in London in 1877 she became a contributor on various topics to the prestigious *Edinburgh Review* and an article on the *Chemistry of the Stars*, which appeared in 1880, contains the germ of her important first book *A Popular History of Astronomy during the Nineteenth Century*. In the light of the success of this volume, which ran to four editions between 1885 and 1902, she was invited to visit the Royal Observatory at the Cape of Good Hope to learn the art of night observing. Thereafter, she went on to write many erudite books as well as multitudinous articles which filled a special niche through not only communicating intelligibly what was new to those outside the field of astrophysics, but also through providing professional astronomers with high quality reviews of important, ongoing, international developments in that field – based on her in-depth astronomical knowledge and extraordinary fluency in many languages. In this regard she was described in 1984 by the historian B. Osterbrock as 'the chief astronomical writer of the English speaking world'.

Further reading:
S. McKenna-Lawlor: *Whatever Shines Should be Observed*, Samton Limited, Dublin, 1998, and references therein.

Susan McKenna-Lawlor, National University of Ireland, Maynooth, Co. Kildare.

Born: Belfast, 23 August 1842.
Died: Watchet, Somerset, 21 February 1913.

Osborne Reynolds – portrait by the Hon. John Collier, 1904 (courtesy of the University of Manchester)

Family:
His father, the Rev. Osborne Reynolds, was Principal of Belfast Collegiate School and later Headmaster of Dedham Grammar School in Essex, where his son was educated. Married: Daughter of Dr Chadwick of Leeds in June 1868 (died 1869). Second marriage to a daughter of Rev. H. Wilkinson, rector of Otley, Suffolk, in 1881, two years after the death of his son by his first wife. By his second marriage, he had three sons and a daughter.

Distinctions:
Fellow of the Royal Society 1877, Royal Medal 1888; Honorary Fellow of Queen's College, Cambridge 1882; Telford Premium, Institution of Civil Engineers 1885; LLD University of Glasgow 1884.

Reynolds graduated as a mathematician from the University of Glasgow in 1867 and, a year later, was appointed Professor of Engineering at Owens College, Manchester, later to become the Victoria University of Manchester. His original researches led, during the next thirty years, to the publication of many papers of fundamental importance.

His considerable mathematical ability was supplemented by an almost uncanny insight into underlying physical principles and by an ability to design simple experiments with which to test his ideas. With these gifts and considerable inventive skill and engineering knowledge, he was able to make important contributions in many branches of physics and engineering. Reynolds was born into a generation in which the field of engineering science was only beginning to be explored. He felt that developments and improvements in engineering could only be undertaken with confidence if experience and intuition were supplemented by fundamental knowledge and controlled experiment. He laid the foundations for much of the subsequent work on the theories of turbulence, heat transfer, lubrication, scaling laws, hydrodynamic similarity and many other matters. His experiments on the origins of turbulence, the scaling of estuary models and the determination of the mechanical equivalent of heat were classics of their kind.

He is remembered widely for the *Reynolds Number*, a dimensionless quantity indicating the degree of turbulence of flow past an obstacle. The Royal Medal was awarded to Reynolds for 'his investigations in mathematical and experimental physics, and on the application of scientific theory to engineering'.

Further Reading:
D.M. McDowell and J.D. Jackson (eds): *Osborne Reynolds and Engineering Science Today,* Manchester University Press, Manchester, 1970.

Ronald Cox, Centre for Civil Engineering Heritage, Trinity College, Dublin 2.

89 JAMES EMERSON REYNOLDS Chemist

Born: Booterstown, Co. Dublin, 8 January 1844.
Died: Kensington, London, 17 February 1920.

Family:
Reynolds' father, James, was an apothecary who practised medicine in the Blackrock area of County Dublin. He was also a playwright. His sister was a dressmaker.
Married: Janet Elizabeth Finlayson, the daughter of Prebendary Finlayson of Christchurch Cathedral, Dublin, on 2 February 1875.
Children: His son, Alfred John, who graduated in Engineering at Trinity College, became a Captain in the British Army. He had one daughter called Marion Janet Elizabeth.

Addresses:

1844–1868	5 Booterstown Avenue, Blackrock, Co. Dublin
1869–1876	146 Leinster Road, Rathmines
1877–1879	52 Upper Leeson Street
1880–1886	62 Morehampton Road
1887–1900	70 Morehampton Road
1900–1903	Burleigh House, Burlington Road
1904–1920	3 Inverness Gardens, Kensington, London

Distinctions:
Professor of Chemistry, Trinity College Dublin 1875–1903; Fellow of the Royal Society 1880, Vice President 1901–1902; MD (h.c.) 1876; DSc (h.c.) 1891; President of The Society of Chemical Industry 1891; President of The Chemical Society 1901–1903; Manager of the Royal Institution of Great Britain 1904–1920.

Reynolds seems to have acquired his first scientific knowledge from assisting his father and by self teaching. He never attended a systematic course of instruction. He became devoted to chemistry from an early age, fitted up a home laboratory and published his first paper at the age of seventeen. Having obtained a licentiateship of the Edinburgh College of Physicians and Surgeons in 1865 he practised for a short time but soon was devoting himself full time to chemistry, acquiring his first professional laboratory when he was appointed Keeper of Minerals at the National Museum in Dublin. His forte was chemical analysis and in 1868 he was appointed Analyst to the Royal Dublin Society. In 1870 he became Professor of Chemistry at the Royal College of Surgeons. Reynolds had a thriving practice as an analyst before he became Professor of Chemistry at Trinity College in 1875.

One of his outstanding achievements was the synthesis of thiourea in 1869. The compound urea has a special historical significance in chemistry. The German chemist, Wohler, had shown that this organic compound, a typical animal product, can be made from the purely inorganic substance, ammonium cyanate, by heating. Thus a unifying link was established between two branches of chemistry which had been thought of previously as distinct and separate. Urea was much

H_2N / H_2N — C=S

Thiourea

studied in the nineteenth century especially in the large German schools of chemistry. All attempts to synthesise the corresponding compound thiourea, in which the oxygen atom is replaced by suphur, had failed. In 1869, by using a carefuly designed procedure, Reynolds succeeded in making it while working at the laboratory of the Royal Dublin Society. The original specimen is displayed in the Chemistry Department, Trinity College.

The characteristic feature of the nineteenth century elemental atom was enshrined in its atomic weight (in the twentieth century the characterising feature was to become the atomic number). Reynolds made important contributions to the study and measurement of atomic weights. He was quick to see the significance of the Periodic Law of the Elements. He correctly placed beryllium in the periodic table, albeit on slim evidence. Reynolds laid considerable stress on students carrying out practical work and was one of the first to introduce quantitative work into the early training of students of chemistry. He was a strong advocate for the reformation of the mediaeval standards of culture within the university and argued that a man of modern culture 'must have a perception of the great natural laws under which he lives, and of the means by which they can be utilised for the benefit of mankind'.

EXPERIMENTAL CHEMISTRY

FOR

JUNIOR STUDENTS.

BY

J. EMERSON REYNOLDS, M.D., F.R.S.

VICE-PRESIDENT CHEMICAL SOCIETY OF LONDON,
PROFESSOR OF CHEMISTRY, UNIVERSITY OF DUBLIN.

PART I.—INTRODUCTION

SECOND EDITION.

LONDON:
LONGMANS, GREEN, AND CO.
1883.

Title page of Experimental Chemistry

In 1903, at the height of his prestige as an academic chemist, Reynolds left Dublin to work at the Royal Institution in London. Here he carried on the investigations into silicon compounds which he had started in Dublin.

Further reading:

W.J. Davis: In Praise of Irish Chemists, *Proceedings of the Royal Irish Academy,* **77B**, 309–316, 1977.
J.E. Reynolds: *Experimental Chemistry for Junior Students,* Longmans Green, London, 1897.

William J. Davis, University Chemical Laboratory, Trinity College, Dublin.

Born: Ballylough, County Antrim, 1844.
Died: 6 July 1933.

Courtesy of the Ulster Folk and Transport Museum

William Acheson Traill was born near Bushmills in north Antrim in 1844. Following in the path of his older brother, Anthony, he studied at Trinity College, Dublin, where he gained a Master's Degree in Engineering.

After leaving college, he spent twelve years working for the Geological Survey of Ireland. He was responsible for producing many maps of the rocks and soil lying just below the surface of Ireland. Some of his best known maps show the basalt of the Antrim Plateau, the high country that covers most of the area between the Bann Valley and the Sea of Moyle.

At the Giant's Causeway on the north Antrim coast the basalt cooled to make large many-sided pillars. Even in the eighteenth and nineteenth centuries, the Causeway, with its unusual rock formations, was an attraction for tourists.

For some time, Traill and his brother had been interested in providing railways in County Antrim. By 1881 he had returned north from Dublin to be Chief Engineer to the Giant's Causeway, Portrush & Bush Valley Railway and Tramway Company Limited. The main railway line from Belfast had reached Portrush in 1855. But visitors who wanted to see the Giant's Causeway then had to walk for nearly sixteen kilometres, or travel over a rough road in horse-drawn carriages.

Traill's plan was to develop a means of transport to carry passengers along the north Irish coast, and also to bring iron ore from mines inland to the harbour at Portrush. At this time many small iron ore mines had been opened in north east County Antrim. Already a six-kilometre narrow-gauge railway carried iron ore from Glenariff to a pier near Cushendall.

By the 1870s many tramways had been built in Britain and Ireland, but most used horses to pull the trams. Some carried locomotives driven by energy from steam, like George Stephenson's *Rocket*. Traill's exciting new idea was to use electrical energy to make his trams move, and to get the energy he needed from the flowing water of the river Bush. His tramway was one of the first in the world to be driven by electricity, certainly the first to get its electrical energy from moving water.

In a hydroelectric generator the turning movement, which is converted into electrical energy, comes from letting water flow through the blades of a turbine. In the station at Walkmills, near Bushmills, the water from the river was collected and allowed to drop through the turbine, providing about 75 kilowatts of power. Electrical energy from Walkmills was also used to provide heat and light for the Traill family's Causeway Hotel.

The Giant's Causeway, Portrush & Bush Valley Railway and Tramway Company's light, narrow-gauge railway began regular operation on 28 January 1883. It ran for thirteen kilometres, carrying passengers and goods between Portrush and Bushmills. It was extended to the Giant's Causeway in

1887. The rails were almost one metre apart, and in some places the track climbed gradients of 4%, one in twenty-five. In the early days of 1883, the trams were driven by steam engines, but the hydroelectric power station was built in time for the formal opening of the tramway on 28 September 1883 by Lord Spencer, Lord Lieutenant of Ireland.

The tramway operated at about 300 volts. This voltage was provided between the ground rails and another, live, rail carried on posts about thirty centimetres above the ground. At this voltage the live rail was dangerous but, at the formal opening, Lord Kelvin held, behind his back, a wire leading to the generator while shaking hands with his shocked guests!

The dangers of electricity were not taken as seriously then as now, though in Portrush town the tram was driven by steam because, in a busy town, a live rail might have electrocuted residents. However, after a cyclist who fell against the live rail was killed in 1895, the voltage was reduced to a safer level. Later the government insisted that the live rail must be at least 2.5 metres above the ground, so the rail was replaced by an overhead wire, and these were erected all the way from Portrush to the Causeway. This also had the advantage that the system could operate at a higher voltage, and so provide more energy for running the trams.

For over fifty years the trains ran regularly from Portrush to the Giant's Causeway. The tramway had originally been built to carry both goods and passengers. But not enough money could be made from goods transport so, after 1890, it carried passengers only. The tramway closed for the winter on 30 September 1949 – 66 years and two days after its official opening by Lord Spencer – and never reopened. In the years since then its track and buildings have disappeared, but one of its carriages is on display at the Ulster Folk and Transport Museum.

Traill did not live to see his tramway closed. In 1933, preparations were made to celebrate the golden jubilee of the first use of electrical energy for transport in September 1883. But early that year, he became very ill, and the celebrations were cancelled. William Traill died on 6 July 1933, with his tramway still at the height of its success.

Further reading:

Martin Brown: *William Traill*, The Blackstaff Press, Belfast, 1995.

Martin Brown, Bangor, Co. Down.

91 WALTER NOEL HARTLEY Chemist and Spectroscopist

Born: Lichfield, England, 3 February 1846.
Died: Aberdeenshire, Scotland, 11 September 1913.

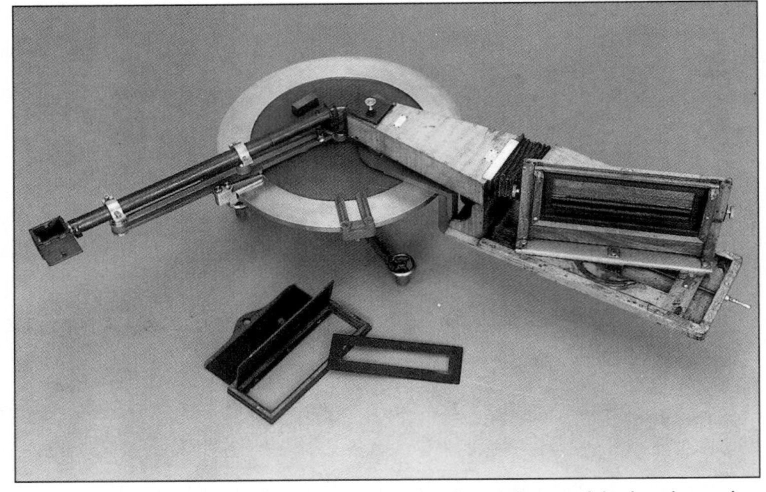

Hartley's ultra-violet spectroscope: the prism table is signed 'Yeates & Son Dublin' (courtesy of the Science Museum, London)

Family: Son of Thomas Hartley, a portrait painter, and Caroline Lockwood.
Married: Mary Laffan of Blackrock, Co. Dublin, a novelist and author 1882.
Children: His son, W.J. Hartley became a lecturer in agricultural bacteriology in University College, Cardiff.

Addresses:

1883–1888	2 Tobernia Terrace, Monkstown, Co. Dublin
1889–1907	36 Waterloo Road, Dublin
1908–1912	10 Elgin Road, Ballsbridge, Dublin

Distinctions:
Professor of Chemistry, Royal College of Science for Ireland 1879–1911; Fellow of the Royal Society 1884; Longstaff Medal of the Chemical Society 1906; Knighted 1911; Honorary DSc Royal University of Ireland 1912.

When Walter Hartley went to Edinburgh University he intended to study medicine, but he abandoned it for chemistry. His chemical education was spread over the years 1864–71 in Germany, Manchester and London. During the later part of this period, he worked in the Royal Institution, where he performed the first research for which he was to become well-known. A French chemist, Bastian, said that he had created a living organism by heating for several hours a solution of inorganic chemicals (sodium phosphate and ammonium tartrate); in other words, he claimed to have achieved the spontaneous generation of life from lifeless matter. If true, this was an astonishing result. With brilliant experimental technique, taking particular care to prevent air from coming into contact with the solution, Hartley demonstrated that it was extremely unlikely that spontaneous generation had taken place, and far more probable that living organisms had entered the solution from the atmosphere.

In 1871 Hartley became Senior Demonstrator at King's College, London, and there he began his career as a spectroscopist. After moving to Dublin in 1879 to the Royal College of Science for Ireland, he took on heavy teaching duties, and most of his research was carried out during the college's holidays. The college premises were then situated in St Stephen's Green.

The science of spectroscopy had been created by the great German scientists Bunsen and Kirchhoff in 1859–60. It immediately became an important technique, but there was little understanding of the relationship between the lines in the spectra of different elements. Hartley was the first to establish that relationships do indeed exist between the wavelengths of spectral lines of the elements related to their positions in the classification called the Periodic Table. He published his results in 1883, though the discovery is usually associated with the names of Balmer (1884) and Rydberg (1885), who produced more mathematical formulations. This was perhaps Hartley's most important discovery: the detailed understanding of spectra was one of the paths which led to Quantum Theory.

Over the three decades in which Hartley worked in Dublin, the main topic of his investigation was the relationship between the structure and the spectra of a wide variety of organic compounds. The importance of this subject had first been noticed by Sir George Gabriel Stokes, Professor of Mathematics at Cambridge and son of the Rector of Skreen, Co. Sligo. Like so many of the scientists who worked in Ireland in the nineteenth century, Hartley was concerned with the practical application of scientific research. Typical of his work in organic spectroscopy was a study of dyes of particular relevance to the Irish textile industry. Derivatives of coal tar, such as benzene and naphthalene, could be converted into dyes with brilliant colours, even though benzene and naphthalene were themselves colourless. Hartley explained that the colours of the dyes were caused by the vibrations within the dye molecules: a molecule that could vibrate with a particular frequency could absorb light of that frequency (our present-day interpretation of these 'vibrations' is the movement of electrons between atomic energy levels. The electron was unknown when Hartley began his spectroscopic work.) He also pointed out that the apparently colourless benzene and naphthalene were in fact absorbing radiation with a higher frequency (or shorter wavelength) than dyes – in other words, they were absorbing the ultra violet.

Hartley's other industrial research covered subjects such as brewing, distilling, the thermochemistry of the Bessemer process of steelmaking, and chemicals that might cure potato blight.

Although spectroscopy was the single method which Hartley deployed in almost all of his researches, he was interested in applymg it to the whole natural world. Like that great Ulsterman, William Thomson, Lord Kelvin, he called himself a naturalist. The Scottish chemist and oceanographer J.Y. Buchanan sent Hartley fragments of shell from the ocean bed. Preserved in spirit, green colouring matter in the shells transferred itself to the spirit. Hardey demonstrated that the colouring was chlorophyll. He subsequendy showed that a deep-sea animal, the sea cucumber, contains a variety of chlorophyll. These observations explained the intensely green colour of the very cold water which is found on certain ocean shores in the tropics: the cold water was identified as having come from the depths of the ocean. It had flowed to the coast to balance the warm surface water being blown offshore by trade winds.

Some of the spectroscopic apparatus used by Hartley is preserved at St Patrick's College, Maynooth. The Science Museum, London, has an ultraviolet spectroscope (see figure), made in part by Yeates of Dublin, which was presented by Hartley's widow in 1913.

Further reading:

J.Y. Buchanan: Obituary of W.N. Hartley, *Journal of the Chemical Society,* **105**, 1207–16, 1913.
W.N. Hartley: On the Colouring Matters Employed in the Illuminations of the Book of Kells, *Scientific Proceedings of the Royal Dublin Society,* **4**, 485–90, 1885.

John Burnett, National Museums of Scotland, Edinburgh.

92 HENRY CHICHESTER HART Botanist, Explorer and Philologist

Courtesy National Library of Ireland

Born: Raheny, Co. Dublin, 29 July 1847.
Died: Carrablagh, Co. Donegal, 7 August 1908.

Family:
Third son of Sir Andrew Searle Hart, mathematician and Vice-Provost of Trinity College Dublin.
Married: 1. Edith Donnelly, at Swords (d. 1901), 1887. 2. Mary Cheshire 1907.
Children: two daughters by his first marriage.
The family's native county was Donegal – 'where my family has been settled since Elizabethan times'.

Addresses:
1847 Glenvar, Raheny, Dublin
Glenalla, Letterkenny, Co. Donegal
Carrablagh, Portsalon, Co. Donegal

Distinctions:
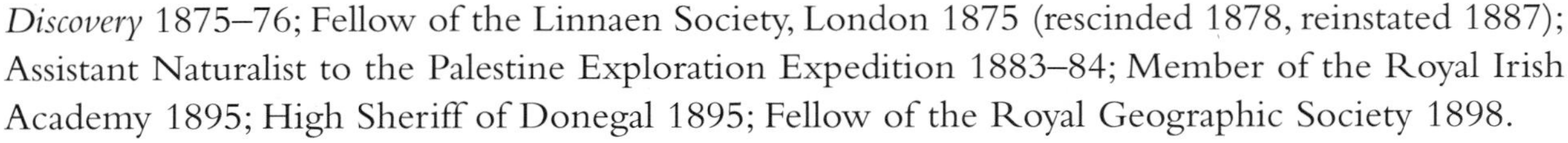
Naturalist to the British Polar Expedition on *HMS Discovery* 1875–76; Fellow of the Linnaen Society, London 1875 (rescinded 1878, reinstated 1887); Assistant Naturalist to the Palestine Exploration Expedition 1883–84; Member of the Royal Irish Academy 1895; High Sheriff of Donegal 1895; Fellow of the Royal Geographic Society 1898.

Henry Chichester Hart was educated at Portora Royal School, Enniskillen, and Trinity College Dublin. He is known primarily as a botanist, but was distinguished in three spheres of activity – physical, scientific and literary. Described as a man of 'magnificent physique', his physical prowess and stamina were acknowledged by contemporaries.

Hart is prominent amongst those who laid the foundations of plant distribution studies in Ireland. Both Hart and R.M. Barrington were 'part of a small band of able botanists whom A.G. More marshalled for the purpose of exploration in connection with the 1898 edition of *Cybele Hibernica*.' From an early age Hart took an interest in flora, influenced probably by his mother, who 'delighted in gardens and wild flowers'. At 'about seventeen years of age', he began the study of the plants of his native county. It was an ongoing project, culminating in the publication in 1898 of the *Flora of the County Donegal*. In the intervening years several papers were published, such as: 'Flora of North-West Donegal', 'Flora of the Croaghgorm Range', 'Report on the Flora of South-West Donegal', 'Botanic Excursions in West Donegal'. The *Flora of the County Donegal,* an early county flora, set a pattern for those that followed. It is unusual, however, in that a sixth of the volume is taken up with extensive climatic data. Hart, who had a deep interest in climate, maintained meteorological equipment at his Donegal residence and was critical of the *Meteorological Atlas of the British Isles* (1883) for the 'very slender knowledge' it provided for Ireland.

Hart's *Flora is* also noteworthy for an appendix setting out plant names and plant-lore accumulated during his travels throughout the county, in the course of which he exercised his ear for dialect. Robert Lloyd Praeger, in his monumental work 'Irish Topographical Botany' (1901), reported that, for information on the north-west of the country, 'I rely almost entirely on the *Flora of Donegal*'.

It was but natural that Hart, with his scientific training and immense vitality, should seek to journey elsewhere in quest of plants. Aided by grants from the Royal Irish Academy, he surveyed islands, river valleys, stretches of coast-line and several mountain ranges. In 1875 he published 'A List of the Plants Found on the Islands of Aran, Galway Bay'. He noted 372 species, of which 176 had not been previously reported. He also provided 'unambiguous data for the distribution of species on the three islands'. In 1880 Hart wrote 'On the Botany of the Galtee Mountains..', in which he paid particular attention to the saxifrages, contrasting them with those specimens he had found on Aranmore, Co. Donegal. In 1882 'On the Botany of the River Suir' was published and in 1883 'Notes on the Flora of the Mayo and Galway Mountains'. In 1890 he summed up his knowledge of upland plants in a paper 'On the Range of Flowering Plants and Ferns of the Mountains of Ireland'. The Report was the result of many years of field work. In it Hart set out the altitudinal ranges of alpine species in western mountains and in the Galtees, the Mournes and the Wicklow range. He reported new locations for alpines and a great deal of altitudinal data for plants of all types.

This stamp, designed by Frances Poskitt, was issued by An Post in 1988 in the Series Endangered Species of Vegetation (courtesy of An Post)
Saxifraga rosacea Moench subsp. hartii (D.A. Webb) D. A. Webb

Hart also botanised outside Ireland – in the Arctic, in Sinai and in Palestine. Polar and Palestine plants collected by him are in the herbaria of Kew, the Natural History Museum, London and the National Herbarium at Glasnevin. His specimens from Donegal and the mountain ranges of Ireland are in the National Herbarium. Some fossil material, and a number of fish and invertebrates are in the National Museum of Ireland. A species first noted in west Donegal by Hart has been named in his honour *Saxifraga hartii* D. A. Webb (see figure). As a botanist Hart had great flair and ability. One notes in his papers the ease with which he assessed plant performance and distribution, climatic and altitudinal data, while keeping in mind the folklore aspect of plants. He also made a notable contribution to zoological studies, publishing papers on mammals, insects, molluscs and birds.

In 1899 Hart read a paper to the Philological Society 'Notes on the Ulster Dialect, chiefly Donegal'. Valuable manuscript notes are in the Royal Irish Academy. He was an authority on the works of Shakespeare and spent his later years editing the Arden series of plays – *Othello, Merry Wives of Windsor, Love's Labours Lost* and others. Hart died at his home Carrablagh on 7 August 1908. His friend R.M. Barrington reported that he was buried at Glenalla in a spot chosen by himself amidst the beautiful scenery and surroundings of his ancestral home.

Further reading:

R.M. Barrington: Henry Chichester Hart, *Irish Naturalist,* **17**, 249–54, 1908.
H.C. Hart: On the Botany of the River Suir, *Scientific Proceedings of the Royal Dublin Society,* **4**, 63–73, 1883.
H.C. Hart: Report on the Flora of the Wexford and Waterford Coasts, *Scientific Proceedings of the Royal Dublin Society,* **4**, 117–46, 1885.
H.C. Hart: *The Flora of Howth,* Dublin, 1887.
M. Traynor: *The English Dialect of Donegal – A Glossary Incorporating the Collections of H.C. Hart,* Dublin, 1953.

In the presentation of this article I received help from G.V. Hart, S.C.; Michael Hewson, National Library of Ireland; Gina Douglas, The Linnaean Society of London; Alan Eager, Royal Dublin Society; Michael Mitchell, University College Galway; Eric Pembrey, Dublin.

Mary J.P. Scannell, former head of Herbarium, National Botanic Gardens, Glasnevin, Dublin.

93 WILLIAM SPOTSWOOD GREEN Marine Scientist

Born: Youghal, 10 September 1847.
Died: West Cove, Co. Kerry, 22 April 1919.

Family:
Only son of Charles Green, JP, of Youghal.
Married: Belinda Beatty, daughter of a fishery owner in Waterville, 22 June 1875.
Children: Five daughters and one son, Charles Green (1876–1938), inspector of fisheries.

Addresses:

1874–1877	Kenmare, Co. Kerry
1877–1889	Carrigaline, Co. Cork
1889–1914	5 Cowper Villas, Dublin
1914–1919	West Cove, Co. Kerry

Distinctions:
Fellow of the Royal Geographical Society 1886; Honorary Member of the Royal Dublin Society 1888; Member of the Royal Irish Academy 1895; Companion of the Bath 1907.

William Spotswood Green was the founder of modern fisheries research in Ireland. Born in Youghal, he developed a love for the sea and fishing as a child. In 1859 he entered Rathmines School in Dublin, where the headmaster, C.W. Benson, was an enthusiastic naturalist. A school leaver in the 1860s, however, had no prospect of formal training or of a career in biology.

While an undergraduate in Trinity College Dublin, he made the first of a number of expeditions to study glaciers in the Alps and in Norway. Ordained in the Church of Ireland ministry, Green's first curacy was in Kenmare, after which he moved to Carrigaline. Both were coastal parishes and he established himself as an expert in marine science. In 1881 Green was advised, on grounds of ill health, to spend the winter away from Ireland. His reaction was characteristic of his phenomenal courage and energy. He set off for New Zealand to climb Mount Cook, an unconquered peak with unexplored glaciers.

Green's formal connection with marine research began in 1885 when the Royal Irish Academy initiated studies on the continental shelf off the south-west coast. His reputation as an expert on sea fishing and boat handling led to an invitation to organise the first expeditions. The steamer *Lord Bandon* was chartered for short periods in the summers of 1886 and 1887 and carried out dredging and trawling down to 2000 metres. The deepest dredging was a failure, but faunal specimens were collected down to 594m. They included a sea anemone new to science, which was named in his honour *Paraphellia greenii.*

The Royal Dublin Society in 1887 commissioned Green to make a report on the state of the fishing industry. This was to be the groundwork for a scheme to develop the fisheries and it led directly to a series of exploratory cruises, including one more funded by the Royal Irish Academy, which took place in May 1888. The *Lord Bandon* was chartered again, but under the new name of *Flying Falcon.* This time successful hauls were made down to 2322m.

Later in the same year Green set off for Canada to study the Selkirk Glaciers in the Rocky Mountains. After the expedition he made a study tour of North American fishing ports on behalf of the Royal Dublin Society. His observations led to important innovations in the Irish fishing industry.

His next expedition was one mounted by the British Museum to collect deep sea fish off the southwest coast in 1889. This was Green's final voyage as a country clergyman. The following year, he was appointed one of the three inspectors of Irish fisheries. Marine exploration continued, this time jointly sponsored by the Royal Dublin Society and the Government. In 1891 Green not only organised the expedition, but skippered the steam yacht *Harlequin* from Southampton to Cobh, up the West coast and round to Dublin. A team of scientists took part in the voyage, and their reports form the basis of scientific information on marine life in Irish coastal waters.

As inspector of fisheries, Green continued to organise research cruises. The most spectacular were two voyages to Rockall in June 1896 that yielded valuable scientific data in spite of foul weather. During these years, Green supervised the work of a number of marine scientists and eventually, in 1900, a permanent government research body was set up in conjunction with the Department of Agriculture and Technical instruction for Ireland. In that year he was appointed Chief Inspector of Fisheries. Under his supervision, the Department purchased its own vessel for research and patrol, the first of two ships named *Helga*. The post of Chief Inspector was partly administrative, partly scientific, and established the necessity for a biological basis in national fisheries management.

Green retired in 1914 to a seaside village in Kerry where he died in 1919. He was buried in Sneem, in sight of the ocean. He published scientific papers on mountain glaciers and marine archaeology besides his fisheries work. R.L. Praeger, who went on several of the cruises, wrote with deep affection of Green as a man whose dedication to work was equalled only by his sense of humour and the desire to help others.

Further reading:

A.E.J. Went: William Spotswood Green, *Scientific Proceedings of the Royal Dublin Society* **B2,** 17–35, 1967.

R. L. Praeger: *The Way That I Went,* Dublin, 1939.

C. Moriarty: The Reverend W.S. Green: A Pioneer in Fisheries Research and Development, in A. McIntyre (ed.), British Marine Science and Meteorology: the History of their Development and Application to Marine Fishing Problems, *Buckland Occasional Papers,* ***2***, 155–168, 1996.

Christopher Moriarty, Marine Institute Laboratories, Abbotstown, Dublin 15.

Born: Dublin, 1848.
Died: London, 1915.

Courtesy of the Royal Astronomical Society

Family:
Her father, John Murray, was a solicitor at King's Inns in Dublin. Her brother Robert became a senior legal figure in the independent Irish State formed in 1921 and was known as the 'Father of the Free State Bar'.
Married: (Sir) William Huggins, an internationally acclaimed astronomical spectroscopist. Children: None.

Addresses:

1848–1875 23 Longford Terrace, Monkstown, Dublin
1875–1913 Tulse Hill Observatory, London
1913–1915 Moore's Garden, Cheyne Walk, Chelsea, London

Distinctions:
Honorary membership of the Royal Astronomical Society and presented with a portrait of William Huggins by the Royal Society 1903; recipient, jointly with her husband, of the Actonian Prize of the Royal Institution for Scientific Writing; styled Lady Huggins when a knighthood was bestowed on William Huggins, with special personal mention in the Honours list of that year 1897; awarded an honorary pension by the Royal Society in recognition of her services to Science.

Educated at home and at a private school in Brighton, Margaret Lindsay was much interested from an early age in science, especially in astronomy. She carried out simple experiments in physics and chemistry in her youth and, from the time of her marriage (in 1875) to the distinguished spectroscopist William Huggins, who owned a small private observatory at Tulse Hill near London, she became her husband's sole assistant. From assistant she graduated to scientific co-author and, among fourteen scientific papers published with William Huggins based on observations of the spectra of a variety of astronomical and laboratory sources, twelve appeared in the prestigious *Proceedings of the Royal Society* and two in the *Astrophysical Journal*. Further, in 1899 William and Margaret jointly produced their important *Photographic Atlas of Representative Stellar Sources*. Later they jointly edited a special edition of all the papers published at their observatory. This latter volume now constitutes a contemporary record of the foundation, through the work carried out at Tulse Hill, of that branch of science known today as Astrophysics.

Margaret's interests were multifaceted and, in addition to publications in her own name and with her husband concerning astrophysics, she also published on the astrolabe, on violins and violin makers and on archaeo-astronomy. Her recreations included landscape painting, wood carving, botany and gardening. She was particularly interested in furthering the education of women and, in this spirit, bequeathed many of the Tulse Hill archives, as well as important personal effects, to Wellesley Women's College in the United States.

Further reading:

S. McKenna-Lawlor: *Whatever Shines Should be Observed*, Samton Limited, Dublin 1998, and references therein.

Susan McKenna-Lawlor, National University of Ireland, Maynooth, Co. Kildare

95 HENRY NEWELL MARTIN Physiologist 1848–1893

Born: Newry, Co. Down, 1 July 1848.
Died: Burley-in-Wharfedale, Yorkshire, 27 October 1893.

Family:
He was the eldest of a family of twelve, his father being at that time a Congregational minister, but afterwards becoming a schoolmaster. His father came from the south of Ireland, his mother from the north.
Married: Mrs Pegram, widow of a Confederate Officer, in 1879; Mrs Martin died in 1892; no children.

Addresses:

1864–1870	London
1870–1876	Christ's College, Cambridge
1876–1893	Baltimore, Maryland
1893	Cambridge and Yorkshire

Distinctions:
DSc Cambridge – the first to take that degree in physiology; MB London; Fellow of the Royal Society 1885; founder member of Physiological Society 1876, and American Physiological Society 1887, of which he was the first Secretary-Treasurer.

Henry Martin received his early education chiefly at home and matriculated at the University of London before he was sixteen. In 1864 he was apprenticed to Dr McDonagh in the Hampstead Road on the understanding that he could attend classes in the nearby Medical School of University College London. As a part-time student there he was 'discovered' and befriended by Michael Foster, lecturer in physiology. In 1870 he moved with Foster when the latter was appointed to the first Professorship of Physiology at Cambridge, where Martin's energy and talents helped popularise the growing School of Natural Science in the University. Although his future in Cambridge was assured, on the recommendation of T.H. Huxley he moved to the newly-founded Johns Hopkins University in Baltimore as first Professor of Biology, and of Physiology on the establishment of its Medical School.

Spontaneous breathing is unconsciously produced by automatic discharge of nerve cells in the brain stem. In Baltimore, Martin showed that there were 'two distinct [respiratory] centres, one for inspiration and one for expiration, each with its own set of muscles'. With Martin's work on the isolated heart-lung preparation (summarised in the Croonian Lecture), 'the study of the mammalian heart [was] made possible to an extent never before attainable'. In 1893 he resigned and returned to Cambridge in an unsuccessful attempt to restore his health.

Further reading:
Dictionary of Scientific Biography.
W.B. Fye: *Journal of the History of Medicine,* **40**, 139, 1985.

Caoimhghín Breathnach, Department of Physiology, University College, Dublin 2.

96 WILLIAM EDWARD WILSON Astronomer and Physicist 1851–1908

Born: Green Island, near Belfast, 19 July 1851.
Died: Streete, Co. Westmeath, 6 March 1908.

Photo of William Wilson by Lafayette, about 1900

Family:
Only son of John Wilson and Frances (née Nangle).
Married: Caroline Granville of Biarritz in 1886.
Children: One son and two daughters.
A nephew, Kenneth E. Edgeworth, predicted in 1943 a source of comets beyond Neptune which is known as the Edgeworth-Kuiper Belt.

Address:
Daramona, Streete, Co. Westmeath

Distinctions:
Member of the Royal Irish Academy 1888; Fellow of the Royal Society 1896; Honorary DSc University of Dublin 1901; High Sheriff for Co. Westmeath 1894.

Educated at home due to delicate health, Wilson at nineteen joined the 1870 solar eclipse expedition to Oran in Algeria. The following year he had a 12-inch Grubb reflector installed at the family estate of Daramona. Ten years later this telescope was replaced by a 24-inch Grubb reflector with an adjacent laboratory, darkroom and workshop. An improved mounting and clock drive installed in 1892 allowed Wilson to take a series of superb celestial photographs during the following decade.

In 1895–6 the 24-inch telescope was used for pioneering measurements of stellar brightness. G.M. Minchin with the help of Wilson and G.F. Fitzgerald detected ten stars photoelectrically, the brightest being Betelgeuse. Minchin's selenium cells were first tested on planets by W.H.S. Monck and S.M. Dixon at 16 Earlsfort Terrace, Dublin, in 1892.

Wilson's crowning achievement was his estimation of the surface temperature of the Sun. The observations were made in 1894 with the assistance of P.L. Gray using equipment loaned by the Royal Society. The final value of 6590°C compares favourably with modern estimates. Sunspots were photographed regularly with a Grubb 4-inch refractor. In August 1898 Wilson took a series of 400 photographs of a sunspot in almost four hours: this was probably the first application of cinematography to astronomy. He photographed the solar corona at the successful joint RIA-RDS solar eclipse expedition to Plasencia, Spain in 1900. Other projects included a search for a trans-Neptunian planet, uses of X-rays, the invention of a sunshine recorder, and an investigation of the effects of high pressures on an electric arc and on radioactive substances.

Further reading:
Dictionary of National Biography and obituary notice in *Proceedings of the Royal Society*, **83–A**, iii–vii, 1910.

Ian Elliott, Dunsink Observatory, Dublin 15.

97 GEORGE FRANCIS FITZGERALD Physicist

Courtesy Trinity College, Dublin

Born: Monkstown, Co. Dublin, 3 August 1851.
Died: Dublin, 22 February 1901.

Family:
His father was Bishop of Cork and later of Killaloe. His uncle was the famous Irish mathematician and physicist George Johnstone Stoney, 1826–1911. His father-in-law, John Hewett Jellett, held the Chair of Physics at Trinity College Dublin.
Married: Harriet Mary Jellett 1885.
Children: Five daughters and three sons.

Addresses:
The Rectory, Kill-o'-the-Grange, Monkstown, Co. Dublin
19 Lower Mount Street, Dublin
7 Ely Place, Dublin

Distinctions:
Fellow of Trinity College Dublin 1877; Member of the Royal Irish Academy 1878; Erasmus Smith Professor of Natural and Experimental Philosophy, Trinity College 1881–1901; Secretary of the Royal Dublin Society 1881–1889; Fellow of the Royal Society 1883; Royal Society Medal 1899.

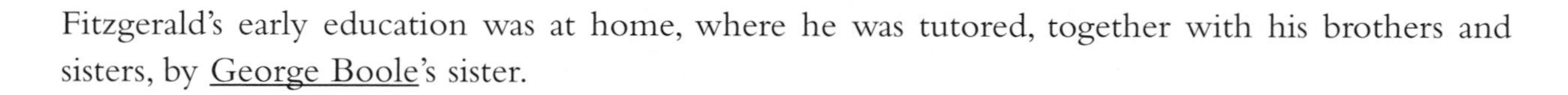

Fitzgerald's early education was at home, where he was tutored, together with his brothers and sisters, by George Boole's sister.

J. Clerk Maxwell published his celebrated *Treatise on Electricity and Magnetism* in 1873. One of the scientists who took Maxwell's electromagnetic theory seriously was George Francis Fitzgerald. He did much to spread and advance knowledge of Maxwell's work. When Hertz succeeded in producing electromagnetic waves it was Fitzgerald who called attention to the experiment at the 1888 meeting of the British Association at Bath, thereby ensuring that its importance was realised. Today he is remembered mainly for the theory which bears his name: the Fitzgerald-Lorentz Contraction.

The late nineteenth-century physicists assumed the existence of an all-pervading ether as the medium for the transmission of light. The famous Michelson-Morley experiment (1881) was an unsuccessful attempt to prove conclusively the existence of the universal ether framework by means of a sensitive optical device. Fitzgerald proposed (1892) that the failure of the experiment arose from the fact that a moving body contracts in the direction of its motion. He suggested that, just as a ship pushing through a calm sea undergoes a small but definite contraction because of the pressure on its bow, so a body moving through the ether undergoes a similar contraction. Consequently, he argued, the contraction cannot be measured because measuring rods shrink in the same proportion.

The Fitzgerald-Lorentz Contraction (Lorentz, a Dutch physicist, gave mathematical support to

Fitzgerald took a practical interest in flight. In 1895 he tested his glider in Trinity College Park. It flew well without him but carried him only a few yards.

Off the ground – just about!

Fitzgerald's theory) was a significant step towards Einstein's theory, which visualised an identical contraction, although the concept of an ether was abandoned.

Fitzgerald was also the first to suggest a method of producing radio waves, and so helped to lay the basis of wireless telegraphy.

Until his appointment to the Professorship of Physics, there had been in Dublin no teaching in practical physics. Obtaining possession of a disused chemical laboratory, Fitzgerald began teaching classes in experimental physics. He took his responsibilities as a teacher very seriously and it was largely owing to his efforts that technical education became established in Ireland. This led to the founding, in 1887, of the Kevin Street Technical School (now part of the Dublin Institute of Technology).

Fitzgerald's versatile mind and wide-ranging interests – he even tried to produce a flying-machine – combined theoretical and practical knowledge. He clearly saw science as a co-operative effort, where one person's ideas would stimulate and add to another's. His remarkable enthusiasm, generosity of spirit and unstinting helpfulness brought him all manner of onerous tasks, resulting in continuous overwork which undoubtedly contributed to his untimely death at the age of 50.

Further reading:

J. Larmor (ed.): *The Scientific Writings of the Late G.F. Fitzgerald,* Dublin, 1902.
Proceedings of the Royal Society, **75**, 152–160, 1905 (Obituary notice).
B.J. Hunt: Practice vs Theory: The British Electrical Debates, 1888–1891, *ISIS,* **74**, 341–355, 1983.
E. Whitaker: G.F. Fitzgerald, *Scientific American*, **185** (5), 93–98, 1953.

Sheila Landy, Science Library, The Queen's University of Belfast.

Born: Copenhagen, 13 February 1852.
Died: Oxford, 14 September 1926.

Family:
Father: John Christopher.
Married: Katherine Tuthill of Kilmore, Co Limerick, 1875.
Children: Three sons and one daughter.

Addresses:
1882–1916 Armagh Observatory

Distinctions:
Gold Medal, Copenhagen (1874); Gold Medal – Royal Astronomical Society (1916), President RAS (1923–1925); DSc Belfast, Hon. MA Oxford.

From his schooldays in Copenhagen, J.L.E. Dreyer showed unusual ability in history, mathematics and physics – the subjects which were to form the background to his later work.

In 1874 he accepted the position of Assistant to Lord Rosse at Birr, where the giant six-foot Leviathan, at that time the largest telescope in the world, was at his disposal. Here he initiated a comprehensive survey of star clusters, nebulae and galaxies. From 1878–1882 he became assistant at Dunsink Observatory before moving to Armagh, where he became Director in 1882.

Financially, Armagh Observatory was destitute, with no prospect of replacing its ageing instruments. Though Dreyer obtained a new 10-inch refractor by Grubb, the lack of funding for an assistant precluded him from a continuation of traditional positional astronomy. Instead he concentrated on the compilation of observations made earlier, namely *The Second Armagh Catalogue of Stars*, and what became his most important contribution to astronomy, *The New General Catalogue of Nebulae and Clusters of Stars* (NGC). In this catalogue, which to this day remains the standard reference used by astronomers the world over, he listed 7840 objects. This he followed with two supplementary *Index Catalogues* (1895, 1908) which contained a further 5386 objects. It is the order in which they appear in these catalogues that defines the name of many prominent galaxies, nebulae and star clusters.

Throughout his life, Dreyer had a fascination for the early development of his subject and, in particular, the work of his fellow countryman, Tycho Brahe, on whose work Kepler based his theories of planetary motion. Dreyer's account of the life and work of Tycho, published in 1890, is the standard biography in English. Subsequently, he began his *magnum opus*, a comprehensive version in Latin of the complete works of Tycho, eventually to fill 15 volumes. In 1906 he published *The History of the Planetary System from Thales to Kepler*, another classic of historical astronomy.

Further reading:
J.A. Bennett: *Church, State and Astronomy in Ireland*, Armagh Observatory, 1990.

John Butler, Armagh Observatory, N. Ireland

Born: 13 Connaught Place, London, 13 June 1854.
Died: At sea, off Kingston, Jamaica, 11 February 1931.

Family:
Youngest son of William Parsons, third Earl of Rosse, and Mary née Field. Other surviving brothers, Laurence, fourth Earl of Rosse, Randal and Richard Clere.
Married: Katherine Bethell of Rise in Yorkshire, 10 January 1883.
Children: Rachel Mary, born 25 January 1885, died 1 July 1956; Algernon George, born 19 October 1886, killed in action, 26 April 1918.

Addresses:
Birr Castle, Co. Offaly
Holeyn Hall, Wylam upon Tyne and Ray, Kirkwhelpington, Northumberland
1 Upper Brook Street, London

Courtesy of The Science Museum, London

Distinctions:
Companion of the Bath 1904; Knight Commander of the Order of the Bath 1911; Order of Merit 1927; Fellow of the Royal Society 1898 – medals – Rumford 1902, Watt 1904, Franklin 1920, Kelvin 1926, Copley 1928, Bessemer 1929; Honorary Doctorates from eleven universities.

When Charles Parsons was born in 1854, his father, who had built the giant six foot telescope at Birr and had revealed the spiral nature of some of the nebulae, was President of the Royal Society. Charles grew up at Birr Castle in Parsonstown (Birr, Co. Offaly) and was educated at home by young tutors who were also employed as assistants in his father's observatory. He was encouraged to use his father's workshops. With the family, he sailed in the yacht *Titania* as far afield as Spain.

In 1871, like his brothers, he entered Trinity College Dublin, but transferred to the University of Cambridge, graduating in 1877. He became a premium apprentice in the Elswick Works of Sir William Armstrong at Newcastle upon Tyne. In 1881 he joined his brother at Kitson and Co. in Leeds, where he perfected a quite novel high speed rotary steam engine, designed to power an electric dynamo for lighting. He also experimented with a novel form of torpedo which was powered by a gas turbine.

Having married in 1883, he invested £20,000 in Clarke Chapman as a junior partner in 1884. In that year, and with great rapidity, he solved the problems associated with the creation of the world's first practical steam turbine driven generator. This ran at what is still an extraordinarily fast speed of 18,000 rpm, and generated 6 kW. Dissatisfied with slow progress, he left his partners in 1889 to establish his own firm at Heaton in Newcastle, and for a time lost the right to use his 1884 patents. Nevertheless by 1891 his alternative design, fitted with a steam condenser, was proved to be more economical in steam consumption than conventional engines. Before his death his firm had built a turbo-alternator rated at 50,000 kW.

In 1893 he established a syndicate to fund the building of a high speed vessel to demonstrate the potential of the turbine for marine use. After many set-backs and reverses, his 40 ton 100 ft long *Turbinia* reached a world record speed of 32.6 knots (37.5 mph). It was demonstrated to great effect at the Spithead Naval review for Queen Victoria's diamond jubilee in 1897. Progress was now very rapid. The giant turbines of the liner *Mauretania*, which helped her to reach 26 knots on her trials in 1907, developed 76,000 horse power (57 MW) to drive four propellers. Turbines completely ousted steam engines in warships and high speed liners. It was an extraordinary achievement to harness a turbine which works best at high speeds of rotation, to screw propellers which require much lower speeds. In 1910 he fitted gearing to the *Vespasian*, a merchant vessel of 4,350 tons. This allowed a high speed turbine to drive a propeller at 72 rpm, and convincingly demonstrated the advantages of the combination. In the course of this work he also invented the 'creep' method of gear cutting.

While at Clarke Chapman, Parsons introduced the idea of moulding plate glass sheets to a parabolic shape suitable for use in searchlight mirrors. During the first world war his firm became one of the largest manufacturers of mirrors in any country. After the war he took over the firm of Sir Howard Grubb. A new factory was built in Newcastle for Sir Howard Grubb, Parsons and Co. Many large telescopes were built there for observatories around the world.

In the early years Parsons built electric arc lamps, and sought to improve the quality of the carbon rods used in them by generating high pressures and temperatures in a special hydraulic press. Later this led him to attempt to transform carbon into diamond. Although he was able to sustain pressures of 15,000 atmospheres and temperatures well in excess of 2,000°C in his press, he had ultimately to admit defeat. He spent £20,000 on the venture.

In 1901 Parsons began work on a valve for an acoustic amplifier, carrying out the delicate handiwork involved as a form of relaxation at a time of intense business activity. The Auxetophone, as it was known, was patented and successfully employed with gramophones and stringed instruments.

Parsons licensed manufacturers in Europe, America and Japan to build turbines to his designs which were continually being refined at Heaton. Although he was faced with many competitors, he remained pre-eminent in the engineering development of steam turbines during his lifetime. He could be irascible and difficult, but he did have a reputation for fair dealing in business and for concern for his employees. Despite receiving so many honours he always retained his natural modesty.

Further reading:

A. Richardson: *The Evolution of the Parsons Steam Turbine*, London, 1911.
Rollo Appleyard: *Charles Parsons – His Life and Work*, London, 1933.
R.H. Parsons: *The Development of the Parsons Steam Turbine*, London, 1936.
W.G. Scaife: The Parsons Steam Turbine, *Scientific American*, **252**, 132–139, 1985.
W.G. Scaife: *From Galaxies to Turbines*, Institute of Physics, Bristol, 1999.

W. Garrett Scaife, Fellow Emeritus, Trinity College Dublin.

100 AUGUSTINE HENRY Plant Collector and Dendrologist

Born: Dundee, Scotland, 2 July 1857.
Died: Dublin, 23 March 1930.

Family:
Son of Bernard and Mary (née McNamee) Henry of Tyanee, Portglenone, County Derry.
Married: 1, Caroline Orridge, 20 June 1891: she died in Denver, Colorado, in September 1894;
2, Alice Brunton, 17 March 1908: she died in Dublin in 1956. There were no children.

Address:
47 Sandford Road, Ranelagh, Dublin

Distinctions:
Fellow of the Linnean Society 1888; Veitch Memorial Medal 1902; Victoria Medal of Honour 1906; Silver Medal, Société Nationale d'Acclimatation de France 1923.

Dr Henry's portrait in the former Forestry Herbarium at the National Botanic Gardens, Glasnevin, Dublin; on the table is a copy in original wrappers of Elwes' and Henry's seven-volume Trees of Great Britain and Ireland

Augustine Henry was born in Scotland, but his parents soon returned to the family home, near Portglenone, on the west side of the River Bann in County Derry. There Augustine spent his childhood. He was a bright lad, leaving school (Cookstown Academy) to enter Queen's College (now University College) Galway, where he studied natural history and medicine, gaining his BA in 1877. Shortly after completing his master's degree at Queen's College Belfast, Henry was persuaded by Sir Robert Hart, the Ulster-born diplomat and civil servant, to join the Chinese Imperial Maritime Customs Service as a doctor. In the summer of 1881, aged 24, Augustine Henry set sail for China: he was to live in remote parts of that vast country for most of the succeeding twenty years.

At first he was stationed at Yichang, a port situated 1000 miles inland from Shanghai on the Chang Jiang (River Yangtze), where he treated the sick and also undertook duties as a customs officer. Henry soon became fascinated by the native medicines and it was this interest that led him to undertake the collection of plant specimens for study and identification at the Royal Botanic Gardens, Kew.

Because few Europeans had collected the plants of central China, Augustine Henry's specimens proved to be new and exciting – during his years in Yichang he discovered more than five hundred new species, about twenty-five new plant genera, and one entirely new plant family. Many of his discoveries proved to be attractive, hardy plants suitable for European gardens, and Henry sent seeds and bulbs to Kew for cultivation there. Among his plant introductions was the orange-blossomed lily, *Lilium henryi,* which inhabits the limestone gorges near Yichang. Later he collected in southern

China too, but his discoveries there are not suitable for cultivation even in the mildest parts of Ireland, and they are less familiar to us. Henry did collect in the field but he also employed native Chinese collectors to gather, dry and press specimens. The principal set of his Chinese plants is now in Kew.

He tried to encourage others to collect plants, especially those of economic and medicinal value, and published a small handbook to assist collectors. In it he predicted that the Chinese gooseberries (species of *Actinidia*) had considerable potential as an edible fruit – the misnamed 'kiwi fruit' is one of these.

After completing his time in China, Dr Henry took up the study of trees (dendrology), beginning by entering the famous French School of Forestry at Nancy. In 1903, Henry Elwes, a wealthy English landowner and gardener, invited Henry to collaborate in writing a book on the native and exotic trees cultivated in British and Irish gardens. The magnificent seven-volume book, illustrated with photographs, was completed in 1913, the same year that Henry was appointed Professor of Forestry at the Royal College of Science (later University College) Dublin. He was to remain there until he retired in 1927.

Augustine Henry promoted the idea of planting marginal land with fast growing coniferous trees, principally those from western North America. He is thus the 'father of Irish forestry'. He also did pioneering work on the production of vigorous artificial hybrid trees, especially poplars. His forestry herbarium, containing specimens cultivated in Ireland and Britain, is now at the National Botanic Gardens, Glasnevin.

Many botanical collectors are commemorated by having their names incorporated into the scientific names of plants that they found. Augustine Henry is no exception. There are too many to list but these plants mark his contribution to the discovery of the Chinese flora: *Rhododendron augustinii* (a blue flowered rhododendron), *Parthenocissus henryana* (a red-leaved vine), *Acer henryi* (a maple), and *Hypericum augustinii* (a St John's wort). His first wife Caroline is also commemorated in the lovely primrose, *Carolinella henryi* (now *Primula henryi*).

Dr Henry was also an unwitting agent of a famous botanical hoax – a made-up plant concocted by one of his Chinese collectors fooled botanists at Kew and was named *Actinotinus sinensis.*

Principal publications:

Notes on Economic Botany of China, Shanghai, 1893; facsimile reprint, Kilkenny, 1986.
With Henry J. Elwes: *The Trees of Great Britain and Ireland,* Edinburgh, 1907–13.
Forests, Woods and Trees in Relation to Hygiene, London, 1919.

Sources and further reading:

S. Pim: *The Wood and the Trees, A Biography of Augustine Henry,* Second revised edition, Kilkenny, 1985.
E.C. Nelson: Some Botanical Hoaxes and Chinese puzzles, *Kew Magazine,* **3**, 178–85, 1987.
E.C. Nelson: The Joys and Riches O' Kathay: Augustine Henry and the Trees of China, *Irish Forestry*, **52**, 75–87, 1995.

E. Charles Nelson, Outwell, Wisbech, England.

101 JOSEPH LARMOR Mathematician and Physicist

Born: Magheragall, Co. Antrim, 11 July 1857.
Died: Holywood, Co. Down, 19 May 1942.

Courtesy of the Royal Society

Family:
Son of Hugh Larmor and Anna Wright. Larmor himself never married.

Distinctions:
Senior Wrangler, Cambridge University 1880; Secretary of the Royal Society 1901–12; Knighted 1909; Member of Parliament for the University of Cambridge 1911–22; President of the London Mathematical Society 1914; De Morgan Medal of the London Mathematical Society 1914; Royal Medal of the Royal Society 1915; Copley Medal of the Royal Society 1921; Honorary Freeman of the City of Belfast; Honorary Member of the Royal Irish Academy, the American Academy of Arts and Sciences, and the Accademia dei Lincei; Honorary Degrees from the Universities of Dublin, Oxford, Belfast, Glasgow, Aberdeen, Birmingham, St Andrews, Durham and Cambridge.

Like many talented Irish men of science, Joseph Larmor achieved his greatest successes outside Ireland. He was educated at the Royal Belfast Academical Institution and Queen's College Belfast but, after taking his degree at Queen's, he went on to take another degree at St John's College Cambridge, as was common for promising students from provincial universities. At the final mathematical examination in Cambridge in 1880, he won the top prize, an event celebrated in Belfast as it was the second year in a row that a student from Belfast had been crowned 'senior wrangler'. (Andrew Allen [1856–1923] had obtained this distinction in 1879.) Larmor then returned to Ireland as Professor of Natural Philosophy at Queen's College Galway, a position he held for only five years. He soon returned to Cambridge to take up a new mathematics position and, on the death of Sir George Gabriel Stokes in 1903, he was appointed to the prestigious Lucasian Chair of Mathematics.

Larmor is best known for his contributions to the theory of electromagnetism, in particular the electron theory of matter. From the beginning of the nineteenth century, scientists had attempted to find a theoretical model of the aether, the strange medium in which light waves were believed to travel. When James Clerk Maxwell showed in the 1870s that light is an electromagnetic wave, the search was on for a theory of the aether that would explain not only the phenomena of light, but also those of electricity and magnetism. Irish scientists showed a particular interest in aether theories. In the 1880s, George Francis Fitzgerald, a professor at Trinity College Dublin, adapted a model of the aether created in the 1830s by James MacCullagh, another Dublin mathematician. Fitzgerald showed that the model could mathematically account for much of optics and electromagnetism, though its properties were difficult to visualise mechanically. Larmor took up this model with a view towards understanding the relationship between electromagnetic fields and charged bodies. He soon found that small point-like strains or twists in the aether would act like charged particles. He called these strains 'electrons', a term coined by the Irish physicist George Johnstone Stoney. These were not solid particles, rather they were like knots in the aether, able to move about according to the laws of electromagnetism, but having no independent existence. Soon after Larmor's discovery, J.J. Thomson found experimental evidence that negatively charged subatomic particles exist, and Larmor identified these as electrons. Larmor published his collected papers on electromagnetism in 1900 in a famous book entitled *Aether and Matter*, though one physicist thought it might have more appropriately been titled 'Aether and *no* Matter', because of the way it made matter simply a property of the aether.

Larmor's electron theory of matter also predicted that, under certain conditions, matter might contract slightly due to its interaction with the aether. This idea, based on a suggestion from Fitzgerald, was to become a key concept in the special theory of relativity. Larmor's work therefore, while rooted in the classical physics in which he had been trained, eventually led to the breakdown of classical physics and the rise of relativity theory and quantum mechanics. Sir Arthur Eddington described him as 'one who re-kindled the dying embers of the old physics to prepare the advent of the new'. Larmor, however, remained a conservative in physics, and never reconciled himself to the new theories, complaining that all true scientific progress ceased around 1900. His conservatism extended to other realms as well. He once opposed the installation of baths in college saying, 'We have done without them for four hundred years, why begin now?'

Larmor saw himself as part of an Irish scientific tradition. He adored the principle of least action as it had been developed by Sir William Rowan Hamilton, and utilised the work of MacCullagh, Kelvin, Stokes and Fitzgerald. In fact, Larmor was involved in editing the collected works of a number of Irish scientists, including Sir George Gabriel Stokes, James Thomson, Lord Kelvin, and George Francis Fitzgerald. In many ways Larmor represented the culmination of the great flowering of Irish mathematical physics in the nineteenth century. Though he spent most of his career in England, he returned to Ireland most summers and moved there permanently after his retirement from the Lucasian chair. His commitment to the Union of Ireland with Great Britain led him to serve in Parliament as a member for Cambridge University from 1911 to 1922.

Further Reading:

Jed Z. Buchwald: *From Maxwell to Microphysics*, University of Chicago Press, Chicago, 1985.
Arthur S. Eddington: Joseph Larmor, *Obituary Notices of the Fellows of the Royal Society*, **4**, 197–207, 1944.
Joseph Larmor: *Aether and Matter*, Cambridge, 1900.
Andrew Warwick: Frequency, Theorem and Formula: Remembering Joseph Larmor in Electromagnetic Theory, *Notes and Records of the Royal Society of London*, **47**, 49–60, 1993.

David Attis, Program in the History of Science, Department of History, Princeton University, Princeton, New Jersey USA.

102 JOHN JOLY Geologist, Physicist and Mineralogist

Born: Hollywood, Co. Offaly, 1 November 1857 (uncertain).
Died: Dublin, 8 December 1933.

Courtesy of the Royal Dublin Society

Addresses:
Apart from his early childhood in Hollywood, John Joly, who never married, lived mostly in Trinity College; he also had a house, 'Somerset', in Temple Road, Dartry, near Trinity Hall, which he bequeathed to his colleague Professor H.H. Dixon.

Distinctions:
Fellow of the Royal Society 1892; Professor of Geology in Trinity College Dublin (a post he held until his death) 1897; Member of the Royal Irish Academy 1897; Commissioner of Irish Lights 1901; President of Section C (Geology) of the British Association for the Advancement of Science 1908; Royal Medal 1910, Boyle Medal 1911, Murchison Medal 1923; President of the Royal Dublin Society 1929 (Council 1889, 1897, Hon. Secretary 1897–1909; Vice-President 1912); Hon. Member of Russian Academy of Sciences 1930; Also LLD (Michigan), ScD (Cambridge), ScD (National University of Ireland).

John Joly, son of a County Kildare Church of Ireland rector and of Belgian and Italian descent, entered Trinity College Dublin in 1876. After a wide range of studies, including classics and modern literature, he graduated in Engineering in 1882, and was then appointed Assistant to the Professor. In 1897 he was appointed Professor of Geology, a post he held until his death.

His interests were many. A keen walker, traveller, yachtsman and alpinist, he found scientific fascination all around him. His major early scientific work was in mineralogy, for which he devised apparatus ('meldometer') for observing melting and sublimation behaviour of minerals. He also measured their specific heat capacities. His famous steam calorimeter, which later played a part in the kinetic theory of gases, was originally devised for mineralogical use.

He attempted to calculate the age of the Earth (a matter of keen dispute at the turn of the century) from the quantity of salt in the oceans but, after the discovery of radioactivity, he realised that much better estimates might be obtained for the ages of rocks from the radioactivity they contained. When slices of certain rocks are examined under a microscope, tiny circles called pleiochroic haloes can be seen; Joly realised that these were due to the damage caused in the rock structure by the products of radioactive decay, mainly alpha particles, from minute inclusions of elements such as thorium incorporated in the solidifying rock. Careful measurements, and a knowledge of half-lives, enable ages of rocks to be found which, when combined with stratigraphy, can lead to lower limits, at least, of the Earth's age.

Joly also showed that the presence of radioactive elements in the Earth could account for much of the heat produced in the Earth: calculations based on the cooling of an originally molten planet by Kelvin and others had suggested that the Earth could not be nearly so old as we now know it to be. Much of the extra heat, and indeed much of the heat and energy revealed in geological processes, Joly showed could come from radioactive decay.

With Dr Walter Stevenson, of Dr Steevens' Hospital, Dublin, Joly pioneered the use of radioactivity in the treatment of cancer. In 1914 he persuaded the Royal Dublin Society to purchase a supply of radium salts, and to set up a Radium Institute. Here the radon gas produced in the radioactive decay of radium was collected in fine hollow tubes which could be surgically implanted in tumours, destroying them.

In 1894 Joly invented a system of colour photography based on taking and viewing photographs on plates ruled with many thin lines in three colours. The method produced excellent transparencies. His interest in this field continued with a quantum theory of colour vision in the 1920s. Also, on the biological side, he devised, with H.H. Dixon FRS, Professor of Botany in Trinity College Dublin, an adhesion-cohesion theory for the ascent of sap in trees.

As a Commissioner of Irish Lights, he made contributions to signalling and safety at sea.

As with any other scientist, some of his theories have been superseded or replaced. Nevertheless, Joly must be considered one of Ireland's leading scientists and innovators, as well as one who has, through his work in Trinity College Dublin and the Royal Dublin Society on the organisation of science and its teaching, had a considerable influence on later generations.

Further reading:

K.C. Bailey: *A History of Trinity College Dublin 1892–1943,* Dublin University Press, 1947.
J. Joly: *The Surface History of the Earth,* Oxford University Press, 1930.
J. Joly: History of the Irish Radium Institute, *Royal Dublin Society Bicentenary Souvenir, 1731–1931,* Dublin, 1931.

Adrian Somerfield, formerly of St Columba's College, Dublin.

103 SYDNEY YOUNG Physical Chemist

Born: Farnworth, Lancashire, 29 December 1857.
Died: Clifton, Lancashire, 7 April 1937.

Family:
Third son of a Liverpool merchant. His brother Alfred was a mathematician and was elected a Fellow of the Royal Society in 1934.
Married: Grace Martha Kimmins (1896), sister of educationalist Charles Kimmins.
Children: One of his twin sons was killed at Ypres in 1915; the other became headmaster of Lincoln School and later of Rossall.

Addresses:

1906–1916	12 Raglan Road, Ballsbridge, Dublin.
1917–1930	13 Clyde Road, Ballsbridge, Dublin.

Distinctions:
Professor of Chemistry, University College Bristol 1887–1903; Fellow of the Institute of Chemistry 1888; Fellow of the Royal Society 1893; Founder Fellow of the Institute of Physics 1893; Professor of Chemistry, Trinity College, Dublin 1903–28; President of the Chemical Section, British Association 1904; President of the Royal Irish Academy 1921–26.

Distillation as a method of concentration or separation of liquids has a very ancient history. Alchemical writings show that it has engaged chemists' attention from earliest times and its often central role in chemical processes continues in the present day. We owe much of our present knowledge to Sydney Young, a chemist who spent over half his professional life in Dublin.

Sydney Young received his chemical training at Owens College, Manchester, under the guidance of Sir Henry Roscoe. Later he spent some time working with the German organic chemist Rudolf Fittig in Strasbourg where, as well as learning organic chemistry, he acquired the glass-blowing skills which later he found so useful in his work on distillation. His first work as a student at Manchester was to investigate the discovery that ice cannot be liquefied by great heat so long as it is kept at low pressure under vacuum. (This finds modern utility in the process we now call 'freeze drying'.) His Dublin colleague John Joly was to show that the inverse was true, that is, that ice melts at a lower temperature when under pressure.) So started an interest in the relationships between the states of matter, solids, liquids and gases which was to remain a continuing theme of his scientific researches.

Young was a pioneer in the separation and specification of pure organic chemicals. Changes of state or phase or of composition are mirrored by the heat changes that accompany them, the study of which is usually called thermochemistry. The rules of this behaviour are enshrined in the laws of

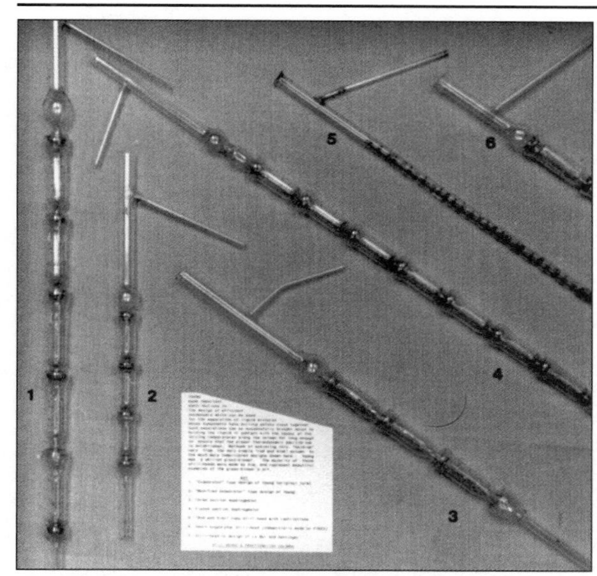

Glass fractionating columns made by Sydney Young (from a display in the Chemistry Department, Trinity College Dublin)

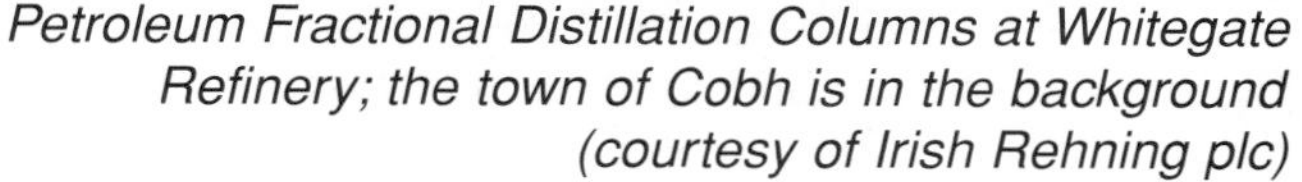

Petroleum Fractional Distillation Columns at Whitegate Refinery; the town of Cobh is in the background (courtesy of Irish Rehning plc)

thermodynamics, a subject that was being developed when Young was carrying out his early work. Thermodynamics makes evident the conditions under which separation of substances of similar boiling point can be carried out in a process called fractional distillation. By careful determinations of critical and other physical constants, and the relationship between temperature and vapour pressure, Young clarified crucial thermodynamical relationships for solids and liquids. From 1885, the potential for extracting paraffin hydrocarbons from petroleum led Young to devise improved techniques of fractional distillation. He developed a new method of dehydrating ethyl alcohol using benzene in an 'azeotropic mixture'; in 1902 he published this method in a paper entitled 'The Preparation of Absolute Alcohol from Strong Spirit'. This was immediately before he came to Dublin in 1903.

At the beginning of the twentieth century Young was regarded as the foremost authority in the world on distillation. He continued his research in Dublin but, because of his increased teaching and administrative duties, he carried out scientific work at a lesser pace than before. He held a number of positions on government and scientific committees. He was a founder member of the Irish Section of the Royal Institute of Chemistry. His text-book *Stoichiometry* and a new and enlarged edition of his book on distillation were published while he was professor in Trinity College. He also contributed articles on distillation, thermochemistry and thermometers to *Thorpe's Dictionary of Applied Chemistry*.

Further reading:

S. Young: *Practical Distillation*, London, 1903.
S. Young: *Stoichiometry,* London, 1908; 2nd edition 1918.
S. Young: *Distillation Principles and Processes*, London, 1922.

William J. Davis, University Chemical Laboratory, Trinity College, Dublin.

104 ALEXANDER ANDERSON Physicist and University Administrator

Born: Coleraine, 12 May 1858.
Died: Dublin, 5 September 1936.

Family:
Son of Daniel Anderson, Camus, Coleraine.
Married: Emily, daughter of William Binns, National Bank, Galway.
Family: Three daughters, Elsie, Helen, Emily, and son Alexander.

Addresses:

1891–1895	Tariffa Lodge, Maunsels Road, Galway
1895–1899	Sea Road, Galway
1899–1934	President's Residence, University College Galway
1934–1936	7 Pembroke Park, Dublin 4

Portrait of Anderson in presidential robes by Charles Lamb RHA

Distinctions:
Queen's College Galway – BA 1880 (with gold medal), MA 1881, Professor of Natural Philosophy 1885–1934, President 1899–1934; Sixth Wrangler, University of Cambridge 1884; Fellow of the Royal University of Ireland 1886; Fellow of Sidney Sussex College, Cambridge 1891; Honorary Doctorates – University of Glasgow 1901, National University of Ireland 1909; Member of the Royal Irish Academy 1909.

Like many a northerner of his time Alexander Anderson sought his university education in Galway. In 1877 he began a distinguished career and an association of more than fifty years with that College. He took first place in an open scholarship to Sidney Sussex College in the University of Cambridge where he studied physics and mathematics under J.J. Thomson. He was the Goldsmith's Exhibitioner in 1882 and was sixth Wrangler in the Tripos examination in 1884. He returned to Galway in 1885.

Shortly after his appointment as Professor, succeeding Joseph Larmor, he overcame obstacles and secured apparatus in order to introduce practical physics into the curricula of science, medical, and engineering students. Anderson's interest in the practical applications of physics is illustrated by the fact that his department was providing a medical radiography service in Galway from 1898. He was also involved in industrially sponsored research as the Eastman-Kodak Company provided a fellowship for the study of X-ray photography around 1899. This activity attracted world-wide attention in 1904 when the College authorities were taken to court in what was probably the first instance of litigation in these islands on the injurious effects of ionizing radiation. A child had been brought for X-ray of a knee in which a piece of a needle was thought to be embedded. No part was found but the tissue was damaged and scarred by the X-ray exposure. Many well known physicists of the time, like Lord Kelvin and Sir Oliver Lodge, came to Dublin as expert witnesses at the trial. The verdict was in favour of the College and the child recovered without permanent disability.

Physics Laboratory in Queen's College Galway about 1902

Anderson excelled in both the theoretical and practical aspects of physics. His research covered a wide range of topics from the focometry of lenses to theoretical studies in electrostatics and magnetism. It was published in over 30 papers, mainly in the *Philosophical Magazine*. While his Anderson's bridge method of measuring the self inductance of a circuit is probably his best known contribution to practical physics, his constant volume method of measuring the viscosity of a gas and the pendant drop method of determining the surface tension of a liquid were also included in textbooks of practical physics.

He also participated in contemporary problems in science and took a special interest in Einstein's general theory of relativity. He submitted a paper 'On the Advancement of the Perhelion of a Planet (Mercury) and the Path of a Ray of Light in the Gravitational Field of the Sun' to the *Philosophical Magazine* (**39**, 626–628, 1920). In it, after deriving the necessary formulae, he wrote as follows '...if the mass of the sun were concentrated to a sphere of radius 1.47 kilometres, the index of refraction near it would become infinitely great, and we should have a very powerful condensing lens, too powerful indeed, for the light emitted by the sun itself would have no velocity at the surface. Thus.....it will be shrouded in darkness, not because it has no light to emit, but because its gravitation field will become impermeable to light'. This seems to be the first published suggestion of the possibility of 'black holes' which is correct in the relativistic sense. Anderson's bold and prescient description of Einsteinian black holes deserves wider recognition and may become his main claim to fame.

Besides his work of teaching and research in experimental and mathematical physics, Professor Anderson proved himself as an able university administrator. During his term as President there were many new developments both in Ireland and in the position of the Queen's College – its incorporation in the new National University of Ireland, the establishment of the Irish Free State and the beginning of the Gaelicisation of the College being some of the more revolutionary changes. Throughout all these changes his tact and broadminded attitude avoided many a difficulty and solved many a problem. When, about 1925, the very existence of the College was at stake, his great reputation and skilful direction were, to a large extent, responsible for averting the danger.

On his retirement he lived in Dublin, where he died on 5 September 1936. He was buried in Dean's Grange Cemetery.

Further reading:

Tom O'Connor: Natural Philosophy/Physics, Chapter 4 in *From Queen's College to National University*, Tadgh Foley (ed), Four Courts Press, Dublin, 1999.

Thomas C. O'Connor, Department of Physics, National University of Ireland, Galway.

ROBERT FRANCIS SCHARFF Naturalist 1858–1934

Born: Leeds, England, 9 July 1858.
Died: Worthing, Sussex, 13 September 1934.

Courtesy of The Irish Naturalists' Journal

Family:
Father: Edward Scharff. Married Alice (née Hutton), 1889, two sons; Jane (née Stephens), 1920, one daughter. His second wife was Assistant Naturalist at the National Museum and an expert on sponges.

Addresses:
Knockranny, Bray, Co. Wicklow
24 Broadwater Road, Worthing, Sussex

Distinctions:
Vice-President of the Royal Irish Academy, 1903–06, 1909–11, 1919–21; Prize of HM Emperor Nicholas II of Russia, International Congress of Zoology, 1895.

Scharff was educated at University College, London, and at the universities of Heidelberg, Edinburgh, Bordeaux and St Andrews. He was appointed Assistant Keeper of the Natural History Collections at the National Museum, Dublin, in 1887 and served as Keeper from 1890 to 1921. He was acting director of the Museum for part of that time, and a member of the Advisory Committee on Fisheries Research. He was chair of the committee which organised the Clare Island Survey of 1911–1915, perhaps the most intensive natural history survey of part of Ireland. In 1905 he unsuccessfully tried to retain the type specimens of new species discovered in the seas off Ireland for the National Museum, but was overruled by the Natural History Museum in London (the type specimen is the basis for the first scientific description of a species).

Scharff wrote descriptions of many groups of animals, but his most important work was on the geographical distribution of species. There are many cases where the naturalist has difficulty explaining why species live where they are actually found. Thus some species of animals found in Ireland are also found in Spain – but nowhere in between. Scharff believed that in all such cases there had once been a 'land bridge' connecting the two areas, across which the animals had migrated, the bridge subsequently having been sunk beneath the ocean by geological forces. He postulated such bridges across most of the world's oceans, in defiance of geological theories, and was called a 'reckless bridge builder' by one American critic. Some of the connections are now explained by continental drift.

Further reading:
R.F. Scharff: *The History of the European Fauna,* London, 1899.
R.F. Scharff: *European Animals,* London, 1907.
P.J. Bowler: *Life's Splendid Drama,* Chicago, 1996.
R.L. Praeger: *Some Irish Naturalists,* Dundalk, 1949.

Peter J. Bowler, School of Anthropological Studies, The Queen's University of Belfast.

RICHARD DIXON OLDHAM Geologist and Seismologist 1858–1936

Born: Rugby, England, 31 July 1858.
Died: Gwalia Hotel, Llandrindod Wells, 15 July 1936.

Family:
Son of Thomas Oldham and his wife, L.M. Dixon of Liverpool. He was unmarried.

Addresses:
1879–1903 India
1903–1936 Isle of Wight, Kew, Hyères in France, and Llandrindod Wells

Distinctions:
Lyell Medal 1908, President 1920–1922, Geological Society of London; Fellow of the Royal Society 1911; Honorary Fellow Imperial College of Science 1931.

Although born in England, Oldham was paternally of Irish descent. Educated in Rugby and the Royal School of Mines in London, he joined the Geological Survey of India in 1879, and followed in the footsteps of his father (its first Superintendent [later termed Director] from 1850–1876), and his uncle Charles Oldham (1831–1869). Richard Oldham served as Director from 1896–1897. He retired in 1903 suffering from the tropical disease, sprue.

In India he carried out extensive field work, in the Himalaya, the desert of Rajputana, where he prospected for coal, in the Andaman Islands and in Burma. He authored over 70 papers on Indian geology, including the second edition of the *Manual of the Geology of India* (1893).

On 12 July 1897 a catastrophic earthquake shook Assam and many lives were lost. Oldham took charge of an investigation that led him to important seismological conclusions, which remain his greatest contribution to geology. While the existence of seismic waves was already known, Oldham, in 1900, distinguished three types produced by earthquakes: now known as P (compressional), S (shear), and L (Love)-waves. The former two pass through the Earth while the latter travels around it. In 1906 Oldham showed from arrival patterns of waves from various earthquakes that the Earth had a core (its existence had been suspected), which he thought was liquid. It was later shown to be composed of iron and nickel, having a solid centre surrounded by a liquid portion.

On his retirement he lived on the Isle of Wight and subsequently at Kew. Later he spent the winter in the south of France, where he studied the changes which affected the Rhone delta since Roman times, and the summer in Wales.

Further reading:
Dictionary of National Biography: missing persons.
C. Davison: Richard Dixon Oldham, *Obituary Notices of Fellows of the Royal Society,* **2**, 111–113, 1936.
S.G. Brush: *Nebulous Earth*, Cambridge University Press, 1996.

Patrick N. Wyse Jackson, Department of Geology, Trinity College, Dublin 2.

107 THOMAS PRESTON Physicist

Born: Ballyhagan, Kilmore, Co. Armagh, 23 July 1860.
Died: Dublin, 7 March 1900.

Family:
Preston's wife Katherine, who survived him by more than half a century, became principal of Alexandra College. His son George was the co-discoverer of Guinier-Preston zones in metallurgy. He had two other children, one of whom died in infancy.

Address:
Bardowie, Orwell Park, Rathgar, Dublin

Distinctions:
Member of the Royal Irish Academy 1897; Fellow of the Royal Society 1898; Honorary DSc 1898; Boyle Medal of the Royal Dublin Society 1899.

Thomas Preston enrolled in Trinity College Dublin in 1881, the year in which G.F. Fitzgerald became Professor, and he was one of the many who benefited from the great man's leadership in the period leading up to the end of the century, shortly after which both men suffered untimely deaths. This was the more so in Preston's case, for he was at the very height of his powers. But one need not speculate on mere potential: Preston recorded many outstanding achievements in the last decade of his life. These include the discovery of the Anomalous Zeeman Effect and the writing of two authoritative textbooks. *The Theory of Light* (1890) and *The Theory of Heat* (1894) are exceptionally thorough and professional treatments of their subjects. A wealth of informative detail is carried along by a magnificent prose style and beautiful illustrations. Preston produced an expanded second edition of *Light* in 1895 and the world-wide popularity of both books was such that they survived as recommended texts throughout half of the present century, with further editions revised by other hands.

This work was begun while Preston was still associated with Trinity, and continued after he became Professor of Natural Philosophy at University College Dublin in 1891. The post carried with it fellowship of the Royal University and, in due course, access to the new spectroscopic facilities which were installed by W.E. Adeney. Using these, together with a magnet borrowed from W.F. Barrett (Royal College of Science), Preston embarked on an investigation of the Zeeman Effect in 1897. In this he was undoubtedly encouraged by Fitzgerald's theoretical interest, but his investigations were as single-handed as they were single-minded. Despite competing demands on his time (he was, among other things, inspector of Science and Arts), he succeeded first in surpassing Zeeman's original observations and then in discovering that the true effect was more complicated: hence the term Anomalous Zeeman Effect. Preston's publication established priority over many other distinguished physicists who had been drawn to this fashionable topic. He went on to establish empirical rules for spectral lines, which are still associated with his name, and was still working in that area when he died of a perforated ulcer, presumably brought on by overwork.

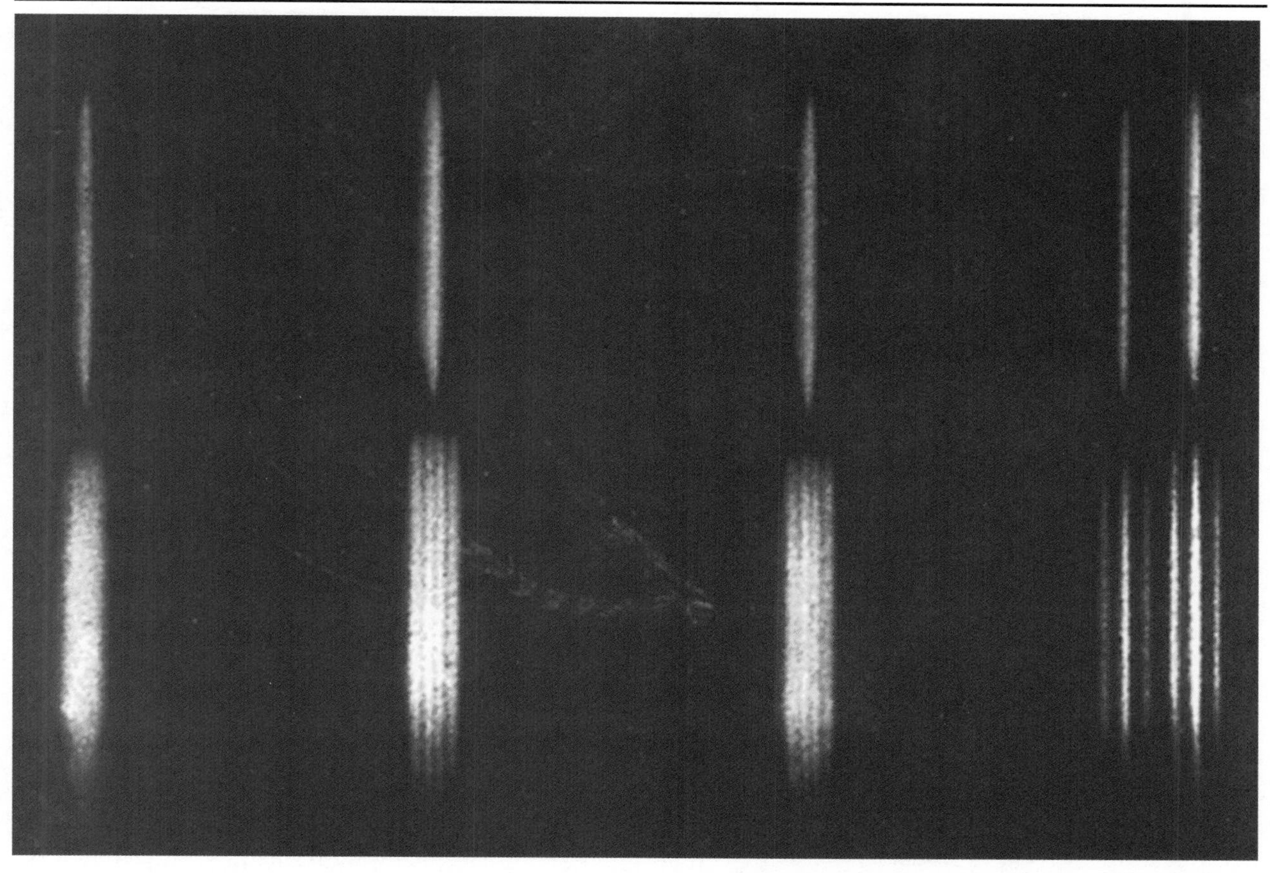

This photograph is taken from Preston's paper to the Royal Dublin Society in December 1897. It shows spectrum lines of cadmium and zinc photographed in zero magnetic field (top) and a strong magnetic field (bottomJ. The two lines on the right hand side show a triplet splitting of the type reported by Zeeman and anticipated by Lorentz's electron theory. The next two lines show a distinct quartet structure. This was the first reported example of the Anomalous Zeeman Effect. Preston's work demonstrated the inaccuracy of classical theory, and it was not until the 1920s that a proper explanation of Zeeman splitting was produced with the introduction of the concept of electron spin and the development of quantum mechanics.

Although much praised and honoured at the time, Preston's achievements have been insuffciently recognised since then. In particular it is noteworthy that Zeeman and Lorentz received the Nobel Prize for their pioneering work in the area in which Preston seized the lead so dramatically in 1897.

Further reading:

D. Weaire and S. O'Connor: Unfulfilled Renown: Thomas Preston (1860–1900), and the Anomalous Zeeman Effect, *Annals of Science,* **44**, 617–44, 1987.

Denis Weaire, Department of Physics, Trinity College Dublin.
Sé O'Connor, Department of Physics, University College Dublin.

THOMAS RANKEN LYLE Physicist 1860–1944

Born: Coleraine, 26 August 1860.
Died: South Yarra, Australia, 31 March 1944.

Family:
Married: Jane Ranken. Children: one son, three daughters.

Address:
Walsh Street, South Yarra, Melbourne, Australia

Distinctions:
ScD (Dublin) 1905; Fellow of the Royal Society 1912; Knighted 1922.

Rugby International

Thomas Lyle (courtesy of the Royal Society)

Thomas Lyle arrived in Trinity in 1879 from Coleraine Academical Institute. To his outstanding academic performance he quickly added athletic distinction, progressing from the College Third XV to the international rugby team in a single season and becoming champion sprinter of Ireland. He graduated in 1883, sweeping the board with first places in mathematics and experimental science and two gold medals. While waiting for a Fellowship he taught at the Catholic University and made a study of alternative lighting methods for the Commissioners of Irish Lights.

In 1889 Lyle was appointed to the Chair of Natural Philosophy at Melbourne University, with which Trinity was closely associated at that time. The devoted service which made him a father-figure of physics in Australia is well remembered there, and memorialised by a medal awarded for distinguished Australian research in mathematics and physics. In difficult circumstances he undertook experimental research in electrical technology. He went on to become an important government advisor on electrical power, and other matters of scientific, technical and educational policy.

In his research he introduced the method of analysis of circuits by complex functions. Although priority for this approach must be given to others unknown to him, he used it to great effect in a series of original investigations. A great all-rounder, who loved to work with his hands as much as his head, he devoted the first part of his retirement to making high-quality diffraction gratings.

Further reading:
Australian Dictionary of National Biography, Melbourne University Press.

Denis Weaire, Department of Physics, Trinity College, Dublin 2.

109 THOMAS JOHNSON Botanist 1863–1954

Born: Barton-upon-Humber, Lincolnshire, 27 February 1863.
Died: Dublin, 9 September 1954.

Family:
Son of George Johnson and Mary Young.
Married: Bessie Stratton Rowe.
Children: two sons and a daughter, all medical practitioners.

Addresses (in Dublin):

1892–1895	5 Pulteney Terrace (later 75 St Lawrence Road)
1896–1898	Gilford Road (initially at 12, later at 2)
1909–1918	13 Palmerston Park
1919–1930	'Glenmore', 63 Orwell Park
1940–1954	64 Terenure Road East

Distinctions:
DSc University of London; Member (1893), Vice-President (1904–05) of the Royal Irish Academy.

Thomas Johnson was educated at the Normal School of Science, South Kensington (now part of Imperial College), the premier British institution for innovative laboratory-based education in the sciences, pioneered by its then Dean, T.H. Huxley. As Demonstrator in Botany in the School from 1885 until his appointment as Professor of Botany in the Royal College of Science for Ireland in 1890, Johnson was in the vanguard of those, inspired by Huxley's reforms, who spread modern teaching of biology throughout Britain and Ireland. For some thirty years, he was an innovator in teaching, research and administration of plant sciences in Ireland. In 1891 he was requested to organise the herbarium of the new Science and Art Museum, and served as Keeper of the Botanical Collections until 1923 – the founding director of the National Herbarium. (His erstwhile assistant Matilda Knowles succeeded him as a curator.) After the transfer (1900) of both College and Museum to the Department of Agriculture and Technical Instruction for Ireland, Johnson established (1901) Seed Testing and Plant Diseases Laboratories, the first official provision for seed testing in Britain and Ireland. (His former assistant George H. Pethybridge [winner of the Boyle Medal in 1921] succeeded him as their director in 1908.) Johnson greatly stimulated the scientific study of plant pathology in Ireland.

Over a notably long period (1887–1951), Johnson published many scientific papers, chiefly on plant pathology, fossil plants, and algae. These disparate themes diminished the long-term scientific impact he might otherwise have achieved by more focused research. And as official Government botanist in Ireland, he published numerous pamphlets on diverse aspects of economic botany and botanical education. His significant role in the development of botany in Ireland has been little appreciated since Irish independence. When the Royal College of Science was transferred to University College Dublin in 1926, Johnson remained as Professor of Botany, conjointly with Joseph Doyle, until his retirement in 1928.

James White, Department of Botany, University College Dublin.

110 FREDERICK THOMAS TROUTON Physicist 1863–1922

Born: Dublin, 24 November 1863.
Died: Downe, Kent, 21 September 1922.

Courtesy of the Royal Society

Family:
Married: Annie Fowler, 1887.
Children: Four sons and three daughters.

Addresses:
Until 1902 Caerleon, Ballybrack, Co Dublin
1914–1922 Tilford, Surrey: The Rookery, Downe, Kent

Distinctions:
ScD University of Dublin 1892; Member of the Royal Irish Academy 1894; Fellow of the Royal Society 1897.

From a wealthy Dublin family, Trouton was educated at Dungannon Royal School and Trinity College, Dublin. He performed brilliantly in his undergraduate work, simultaneously taking degrees in engineering and science. Even before he graduated he had, on the one hand, taken a leading part in surveying for a railway and, on the other, enunciated the relationship between latent heat and molecular weight known as 'Trouton's rule'.

On graduation in 1884 he immediately became an assistant to Trinity's Erasmus Smith's Professor of Natural and Experimental Philosophy (Physics), G.F. FitzGerald, and the two enjoyed a close collaboration and friendship. Trouton's wide experimental interests there included the nature of electromagnetic waves, the possibility of detecting the earth's motion through the ether, the thermodynamics of steam evaporation from salt solutions, and the conduction of electricity in liquids. In lighter vein, he and FitzGerald composed in an atmosphere of 'hilarious levity' a letter published in *Nature* proposing to coin new words analogous to 'resistivity', like 'diffusivity', 'emissivity', 'expansivity', 'frictivity', and so on. After FitzGerald's untimely death in 1901, one of his other assistants was elected to the professorship, whereas on the same day Trouton was appointed to a fixed-term lectureship. Trouton's response was to become Quain Professor of Physics at University College London next year. There he resumed his studies of the ether wind with H.R. Noble and, later on, A.O. Rankine, all, of course, with null results. Other interests included measurements on high-viscosity materials, and the adsorption of water vapour by materials for hygrometers for measuring atmospheric humidity.

He was forced to retire in 1914 through an illness which eventually led to permanent paralysis in his legs. He was remembered for his integrity, friendliness, lively imagination touched often with 'Irish whimsicality', and love of athletics and gardening.

Further reading:
Obituaries in *Nature*, **110**, pp. 490–491, 1922, and *Proceedings of the Royal Society, A*, **110**, pp. iv–ix, 1926.

Eric C. Finch, Department of Physics, Trinity College, Dublin 2.

111 MATILDA KNOWLES Botanist

Born: Cullybackey, Co. Antrim, 31 January 1864.
Died: Dublin, 27 April 1933.

Family:
Daughter of William James Knowles and Margaret Spotswood Cullen; one of six children – two (William and Mary) died in childhood, Elizabeth died in 1894 (aged 33), and two (Catherine Casement and Margaret) survived her. Her father (1832–1927) was the land agent of the large Casement estates in Co. Antrim and secretary (1878–1920) of the Antrim County Land, Building and Investment Company. Author of some 60 papers (nine in the *Proceedings of the Royal Irish Academy*) chiefly on the archaeology of Ulster, he was elected (1883) a member of the Royal Irish Academy for his notable collections and writings. He discovered the celebrated Neolithic axe factory at Tievebulliagh (near Cushendall) and excavated numerous prehistoric coastal settlements. Through his extensive social and business contacts, he purchased and amassed a huge private collection of antiquities of 'astonishing extent and richness', subsequently sold in a four-day sale at Sotheby's in 1924. He has been dubbed 'the father of Ulster antiquities'. Her sister Margaret, trained as an artist, illustrated many of her father's papers and those of other archaeologists closely linked to her father in the Academy.

Addresses:

c.1885–1902	16 Flixton Place (now 4 Cullybackey Road), Ballymena
1895–1896	7 Ranelagh Road, Dublin
1910–1933	College Park Chambers, 11 Nassau Street, Dublin

'One of the finest pieces of work ever carried out in any section of the Irish flora' was R.L. Praeger's description of *The Lichens of Ireland* by Matilda Cullen Knowles *(Proceedings of the Royal Irish Academy*, **39B**, 173–434) on its publication in 1929. Although little is known of her early life, Matilda Knowles and her sisters were encouraged to study natural history by their father, attending meetings of both the Belfast and Ballymena Naturalists' Field Clubs, and accompanying him on his many collecting trips to coastal sites in the north and west of Ireland. Her earliest publications included reports on kitchen middens in County Clare where she deputised at a field meeting for her father, who was secretary of the Sandhills Committee of the Royal Irish Academy. Through him she met and befriended the leading archaeologists and biologists of the period, including George Coffey, A.C. Haddon and R.L. Praeger.

With her sister Catherine Casement, she came to Dublin in October 1895 and both registered as occasional students in the Royal College of Science for Ireland, taking a variety of courses in natural sciences (one in botany) during the academic year. Unlike her sister, however, Matilda did not re-attend for a second year, and appears to have returned to Ballymena. In June 1902, on the recommendation of Professor Thomas Johnson, she was appointed as a temporary assistant for six months in the Botanical Section of the Science and Art (later National) Museum; this remained her official position for at least a decade, her contract being renewed half-yearly. She worked closely with Johnson in organising the recently established (1891) Herbarium, of which he was the Keeper, and in 1910 they published *Hand-list of Irish Flowering Plants and Ferns*. During these years, there was an active programme of foreign acquisitions for the Herbarium and their incorporation kept the assistants busy.

In June 1909, Knowles became involved in the Clare Island Survey (sponsored by several learned institutions in Britain and Ireland) as an assistant to the foremost British lichen expert Annie Lorrain Smith. Initially she collected the larger foliose forms and sent them to Smith. In 1910, she accompanied Smith, Praeger and others to Mayo, collecting lichens, many of them new to Ireland. Thereafter she committed herself almost completely to the study of Irish lichens. Professor Johnson opposed this commitment at the expense (in his opinion) of the daily routines of the Herbarium; happily, however, for Irish lichenology, he was overruled by the Director of the Museum who had a high opinion of her. Knowles spent time at the Natural History Museum (London) studying under Smith, and became 'thoroughly interested in the lichens', as she afterwards recollected. For her part, Smith praised Knowles for her 'exceptional ability as an observer and collector' – skills she had doubtless long-since acquired from her indefatigable father. Devoting all her spare time during 1910–1913 to field work and identification, her first substantial paper on lichens (*The Maritime and Marine Lichens of Howth* was published in 1913 (*Scientific Proceedings of the Royal Dublin Society*, **14**, 79–143). This was instantly, and remains, a classic in Irish botanical literature. It was the first detailed ecological study of coastal-cliff lichens in Britain or Ireland, and received international recognition when its results were reprinted in detail in A.L. Smith's (1921) *Lichens*, the outstandingly authoritative monograph of its time on the subject. Knowles acknowledged her indebtedness to Smith by naming one of the new lichen species she discovered at Howth *Verrucaria lorrain-smithii*. Following directly on publication of this paper, which established her reputation as a lichenologist, she was invited by R.L. Praeger, on behalf of the Fauna and Flora Committee of the Royal Irish Academy, to prepare a systematic monograph on the lichens of Ireland, in the style of his own *Irish Topographical Botany* (1901). Subsequently she devoted much of her time to the collection and study of lichens throughout Ireland, and with the assistance of many amateur and professional naturalists assembled a huge collection for the National Herbarium. Her voluminous data, compiled under Praeger's guidance, resulted in *The Lichens of Ireland* (1929), her most notable intellectual legacy. This records the distribution of some 800 species in Ireland, several of which she had herself added to the list.

When Thomas Johnson retired as Keeper of the Herbarium in 1923, Knowles was officially appointed assistant Keeper of the Botanical Section of the National Museum. Her meticulous and dedicated curatorship is still reflected in the excellent condition of the accessions made during her period of management of the National Herbarium, an enduring part of her legacy to Irish science.

She died in 1933, and within two days her sister Margaret (who appears to have come from her home in Ballycastle to nurse her) also died. They are buried together at Dean's Grange Cemetery.

Praeger eulogised Matilda Knowles: *A woman of very fine character: courageous, humorous, sympathetic, a staunch companion, a fearless critic, a loving friend.... She was a wise counsellor of the young and in matters scientific her breadth of view and tolerance of divergent opinion rendered more valuable her wide knowledge relating to Systemic Botany.*

Further reading:

A bibliography of publications of Matilda Knowles may be found in M. Scannell: Inspired by Lichens, in *Stars, Shells and Bluebells*, Women in Technology and Science (WITS), Dublin, 1997.

James White, Department of Botany, University College Dublin.
Helena Chesney, 35 Colenso Parade, Belfast BT9 5AN.

112 ANDREW CLAUDE DE LA CHEROIS CROMMELIN Astronomer 1865–1939

Born: Cushendun, Co. Antrim, 6 February 1865.
Died: London, 20 September 1939.

Courtesy of the Royal Astronomical Society

Family:
Third son of Nicholas de la Cherois Crommelin.
Married: Letitia Noble in 1897.
Children: Two sons and two daughters.
A sister was married to John Masefield, Poet Laureate.

Address:
1891–1939 Blackheath, London SE.

Distinctions:
Lindemann Prize of the Astronomische Gesellschaft; Honorary DSc Oxford; Walter Goodacre Medal and Gift of the British Astronomical Association 1937.

Andrew Crommelin was descended from Louis Crommelin, the Huguenot founder of the Ulster linen industry. He was educated at Marlborough College and Trinity College Cambridge where he graduated in 1886. Even as a child he was a keen observer of the stars and he joined the Royal Astronomical Society before leaving university. After teaching for a time at Lancing College, he was appointed an Assistant at the Royal Observatory, Greenwich, in 1891. His duties at Greenwich involved routine observations of the Moon, of occultations of stars and of comets and the reduction of the observations. His chief interest was the calculation of the orbits of comets and asteroids. As Director of the Comet Section of the British Astronomical Association, he issued regular notes about comets for more than forty years and thereby gained a world-wide reputation. In 1906 Crommelin pointed out that the existing predictions of the coming date of the perihelion passage of Halley's Comet differed by nearly three years. In collaboration with P.H. Powell, he simplified the method of calculation and predicted the perihelion date in April 1910 to within three days of the actual time. The recorded appearances of Comet Halley were also traced back to 240 BC. For this work the authors were awarded the Lindemann Prize and honorary DSc degrees.

Crommelin was keenly interested in predicting eclipses of the Sun and the Moon. He took part in six solar eclipse expeditions, the most important being to Brazil in 1919 as part of Eddington's plan to measure the deflection of starlight by the Sun's gravitational field. The results confirmed the General Theory of Relativity and helped make Albert Einstein a household name.

Crommelin played an active part in the affairs of both the British Astronomical Association and the Royal Astronomical Society. He was President of the BAA from 1904 to 1906 and President of the RAS from 1929 to 1931.

Further reading:

Obituary in *Monthly Notices of the Royal Astronomical Society*, **100**, 234–236, 1940.

Ian Elliott, Dunsink Observatory, Dublin 15.

113 ROBERT LLOYD PRAEGER Natural Historian

Born: 'Woodburn', Holywood, Co. Down, 25 August 1865.

Died: Rock Cottage, Craigavad, Co. Down, 5 May 1953.

Robert Lloyd Praeger (left) with Francis Joseph Bigger (who succeeded Praeger as Secretary of the Belfast Naturalists' Field Club when he moved to Dublin) on a Field Club excursion in 1893

Family:

The second of four brothers and one sister, his father Willem Emil Praeger was a Dutch-born linen exporter, and his mother, Belfast-born Maria Patterson, was the daughter of Robert Patterson, the well-known mid-nineteenth century naturalist.

His older brother Willem became the first Professor of Ecology at the University of Kalamazoo, Michigan, and his sister Sophia Rosamond was a noted sculptress and writer of children's books.

Married: Hedwig Magnusson, daughter of the artist Christian Carl Magnusson of Schleswig-Holstein, 1901. No children.

Addresses:

1866–1893	'The Croft', Holywood, Co. Down
1894–1895	31 Upper Baggot Street Dublin
1895–1900	'Delbrook', Dundrum, Co. Dublin
1901–1921	'Lisnamae', Zion Road, Rathgar, Dublin
1922–1952	19 Fitzwilliam Square, Dublin
1952–1953	Rock Cottage, Craigavad, Co. Down

Distinctions:

Praeger was twice elected President of the Belfast Naturalists' Field Club; he received their Gold Medal in 1927; he is one of the few people to have twice received the Gold Medal of the Royal Horticultural Society; at various times he was elected President of the Royal Irish Academy, the British Ecological Society, the Royal Zoological Society of Ireland, The Geographical Society of Ireland, the Bibliographical Society of Ireland and First President of the Library Association of Ireland; honorary doctorates were conferred on him by the three Irish universities (Queen's University Belfast, Trinity College Dublin and the National University of Ireland); he was elected an Associate of the Linnean Society of London, Honorary Life Member of the Botanical Society of the British Isles and First President of An Taisce, the National Trust for Ireland.

From the 1880s to the 1950s Robert Lloyd Praeger dominated the subject of Irish natural history. As a child he developed an early interest through his membership of the Belfast Naturalists' Field Club. Graduating from Queen's College Belfast in 1886 as an engineer, he began to study the fossil shells then being unearthed in the construction of the Alexandra Dock in Belfast. Praeger's interest in post-glacial geology resulted in the publication of papers on the raised beaches of the north east of Ireland, which gave information on the climate in Neolithic times. Changing careers, he joined the staff of the National Library of Ireland in Dublin as an Assistant Librarian in 1893.

Active in fieldwork all his life, he literally surveyed the whole of Ireland in his spare time in the 1890s, publishing the results in *Irish Topographical Botany* in 1901. Other major work on the Irish flora included *A Tourist's Flora of the West of Ireland* in 1909 and *The Botanist in Ireland* in 1934. All of these floras detail what he had seen during his field trips and are still useful sources of reference for botanists. In other areas of endeavour, Praeger studied the major cave systems of Ireland with R.F. Scharff in the 1900s and later surveyed many archaeological sites for the Royal Irish Academy with R.A.S. Macalister in the 1920s.

A gifted organiser, Praeger spent much of his time planning the fieldtrips of the Belfast Naturalists' Field Club and later the Dublin Naturalists' Field Club. In 1905 he organised the Lambay Survey, a biological survey of the island, the success of which led to the famous Clare Island Survey of 1909–1911. To date, this is the largest multi-disciplinary survey of a specific area ever undertaken in these islands, with almost two hundred people involved over a three year period. The results, which were published in a special three-part volume of the *Proceedings of the Royal Irish Academy* between 1911 and 1915, have served as base-line studies for researchers ever since. Praeger retired from the National Library as Librarian in 1923, but continued his botanical fieldwork, which now included work on the floras of other European countries. Throughout the 1930s and 1940s, he continued writing up the results of his travels, assisting others with similar work and becoming the Grand Old Man of Irish natural history. Following the death of Hedwig in 1952, Praeger moved to the home of his sister Sophia Rosamond in Craigavad, Co. Down. He died there on 5 May 1953 and was buried beside his wife in Dean's Grange Cemetery Dublin.

Praeger was a prolific writer, publishing many hundreds of papers in learned journals, articles popularising the study of Ireland's flora and fauna and a number of highly readable books. His best known, partly autobiographical, book *The Way that I Went* was published in 1937. *Some Irish Naturalists*, published in 1949, is a biographical reference book. *The Natural History of Ireland* published in 1950 and *The Irish Landscape* in 1953, were the result of his long acquaintance with the Irish countryside. An authority on Irish botany, Praeger was also recognised for his contributions to Irish archaeology, quaternary geology, phytogeography, ecology, history, librarianship, travel and zoology.

Further reading:

Timothy Collins: *Floreat Hibernia: a Bio-Bibliography of Robert Lloyd Praeger 1865–1953,* Royal Dublin Society, Dublin, 1985.

Robert Lloyd Praeger: *A Populous Solitude,* Hodges, Figgis, Dublin, 1941.

Robert Lloyd Praeger: *The Way That I Went: an Irishman in Ireland,* Hodges, Figgis, Dublin; Methuen, London, 1937 (second edition 1939, third edition 1947). Reprinted many times owing to its continuing popularity, the most recent being a facsimile reprint of the first edition by the Collins Press, Cork, 1997.

Criostóir MacCarthaigh and Kevin Whelan (eds): *New Survey of Clare Island, Volume I: History and Culture Landscape*, Royal Irish Academy, Dublin 1999. This new survey follows up Praeger's original work, with forthcoming volumes dealing with the geology, archaeology, flora and fauna of the island.

Timothy Collins, Centre for Landscape Studies, National University of Ireland, Galway.

Courtesy of the Royal Society

Born: Ballystockart, Comber, Co. Down, 2 May 1866.
Died: Douglas, Isle of Man, 14 August 1934.

Family:
He came from farming stock; his father was a farmer and flax mill owner.
Married: Elizabeth Campbell, 1892.
Children: Three daughters.

Address:
Inisfail, 6 Bushy Park Road, Terenure

Distinctions:
Member of the Royal Irish Academy 1904; Fellow of the Royal Society 1909; Hon. DSc Queen's University Belfast 1919.

Orr received his early training in Mathematics at the local national school, and later at Methodist College, Belfast. Following a scholarship in mathematics to the Royal University of Ireland in 1883, he graduated at that University in 1885, from Queen's College, Belfast, a constituent college of the Royal University. He proceeded to St John's College, Cambridge, where he obtained first place in Part II of the mathematical tripos. In 1891 he was elected into a fellowship at St John's College. That same year, at the age of 25, he was appointed Professor of Applied Mathematics at the Royal College of Science for Ireland in Dublin. When this College, essentially a technical institution, was absorbed by University College Dublin in 1926, he was offered and accepted an equivalent position there as Professor of Pure and Applied Mathematics, from which he retired in 1933, one year before his death.

His first paper dealt with Bessel functions (introduced by the German astronomer to determine motions of bodies under mutual gravitation), a subject to which he returned, twenty years later. His privately printed book *Notes on Thermodynamics for the Use of Students* (1909) is a model of its kind, well written, clear, well argued. He is best remembered for his great work on the stability of the parallel shear flows of a liquid. In this connection the equation he derived in 1907 – the Orr-Sommerfeld equation – is still the subject of research.

Prof. Orr was a first class cyclist. So it is hardly surprising that he made a study of the dynamics of the bicycle.

Michael Hayes, Department of Mathematical Physics, University College Dublin.

115 RALPH ALLEN SAMPSON Astronomer 1866–1939

Born: Skull, Co. Cork, 25 June 1866.
Died: Bath, England, 7 November 1939.

Courtesy of the Royal Astronomical Society

Family:
Fourth child of James Sampson, a metallurgical chemist.
Children: A son and four daughters.
A daughter, Professor Peggy Sampson, was a celebrated cellist.

Addresses:
1895–1910 Observatory House, Durham
1910–1937 Royal Observatory, Edinburgh

Distinctions:
Fellow of the Royal Society 1903; Gold Medal of the Royal Astronomical Society 1928; Honorary ScD Durham; Honorary LLD Glasgow.

When Ralph Sampson was five, the Sampson family moved from Skull to Liverpool, where they endured hardship. Ralph eventually entered St John's College, Cambridge, and graduated as third wrangler in the Mathematical Tripos of 1888. After lecturing in mathematics in King's College London, he returned to Cambridge in 1891 and developed the theory of radiative equilibrium in a star's interior. In 1893 Sampson went to Newcastle-on-Tyne as professor of mathematics in Durham College of Science and, two years later, he was appointed to the chair of mathematics in Durham University. There he undertook his greatest work, the dynamical theory of the four large satellites of Jupiter. This entailed critical study of thousands of observations of eclipses of the satellites, and required high mathematical ability. His *Tables of the Four Great Satellites of Jupiter* were published in 1910 and the general theory appeared in 1921. For this research he was awarded the Gold Medal of the Royal Astronomical Society.

In 1910 Sampson was appointed Astronomer Royal for Scotland and Professor of Astronomy in the University of Edinburgh. He took a keen interest in matters connected with the determination of time and, in particular, he encouraged the development of the Shortt free pendulum clock which became standard equipment in many observatories. He collaborated with E.A. Baker in constructing a photo-electric photometer which was used for measuring photographs of stellar spectra and for deriving colour temperatures of stars. He also studied ways of improving the optical performance of reflecting telescopes.

Sampson served as President of the Royal Astronomical Society (1915–1917) and, in 1921, was elected a corresponding member of the Bureau des Longitudes in Paris. He was a true pioneer who explored several important areas of astronomy and astrophysics with distinction.

Further reading:
H.A. Brück: *The Story of Astronomy in Edinburgh from its Beginnings until 1975*, Edinburgh University Press, 1983.

Ian Elliott, Dunsink Observatory, Dublin 15.

116 ERNEST WILLIAM MacBRIDE Zoologist 1866–1940

Born: Belfast, 12 December 1866.
Died: Alton, Hampshire, 17 November 1940.

Courtesy of the Royal Society

Family:
Father: Samuel MacBride; Mother: Mary Jane (née Browne).
Married: Constance Harvey, 1902, two sons.

Addresses:
Cambridge; Montreal; London; Alton, Hampshire.

Distinctions:
Fellow of the Royal Society 1905; Fellow of the Zoological Society 1909.

MacBride was educated at Queen's College, Belfast and the Universities of London and Cambridge. In 1897 he was appointed Professor of Zoology at McGill University in Montreal. In 1909 he was appointed Lecturer and in 1914 Professor of Zoology at Imperial College, London, where he remained until his retirement in 1934. He was on the Council of the Linnean Society and the Zoological Society of London, and served on several government commissions investigating fisheries.

MacBride's main area of research was the embryology of invertebrates, especially the echinoderms (starfish, sea urchins etc.). He showed that the larvae of some echinoderms suggest that the origin of this group is closely linked to that of the vertebrates (animals with backbones). He remained loyal to several theories that were popular in the late nineteenth century, but which were discredited by the rise of modern genetics. He believed that the development of the embryo accurately recapitulated the evolutionary history of the species. He also supported the evolutionary theory known as Lamarckism or the 'inheritance of acquired characters' (this supposes that a character acquired during adult life, such as a weightlifter's large muscles, can be transmitted to the offspring). MacBride defended the Austrian Lamarckian Paul Kammerer in the 1920s, and is described as the 'Irishman with a heart of gold' in Arthur Koestler's account of this episode. Koestler's comment is particularly inappropriate because MacBride came from Ulster Protestant stock and considered his racial origins to be quite separate from the native Irish. He was a member of the Eugenics Society, which called for selective breeding of the human race, and he argued that the Irish component of the British population should be sterilised. He believed that the races had evolved distinct characters because they originated in different locations – the Irish population had migrated from a Mediterranean homeland and was less hardy than the Anglo Saxons.

Further reading:

P.J. Bowler: E.W. MacBride's Lamarckian Eugenics, *Annals of Science,* **41**, 245–260, 1984.
A. Koestler: *The Case of the Midwife Toad,* London, 1971.
New Dictionary of National Biography.

Peter J. Bowler, School of Anthropological Studies, The Queen's University of Belfast.

117 ANNE L. MASSY Marine Scientist and Pioneer Conservationist 1867–1931

Born: Wicklow 1867.
Died: Dublin, 16 April 1931.

Family:
Although she lived on into the twentieth century in Dublin and was a distinguished scientist, Anne L. Massy's family background is almost unknown. She was living in Enniskerry in 1885 and this was probably the place of her birth. She moved to Malahide and then to Howth.

Addresses:

1885	Coolakeigh, Enniskerry, Co. Wicklow
1905–1910	9 St James's Terrace, Malahide, Co. Dublin
1926–1931	Galteemore, Baily, Co. Dublin

Anne L. Massy had a world-wide reputation as an expert on the classification of molluscs, particularly the Cephalopda, the group containing octopus and squid. However, she first came to notice at the age of twenty as an ornithologist and the first person to discover the redstart nesting in Ireland. It remains an extreme rarity.

In 1901 she was employed by the Fisheries Branch of the Department of Agriculture and Technical Instruction, one year after the establishment of its scientific service. By that time systematic marine exploration had been in progress for more than twenty years on the Irish coast and large collections of zoological specimens awaited identification. Her first published results appeared in 1907 and, from then until 1930, she published papers nearly every year in the leading international journals, including the Department's own Scientific Investigations series. The reputation she established led to her being invited to work on material from all over the world. This included major collections held by museums in India and South Africa, and marine animals from the Terra Nova and Discovery expeditions to the Antarctic. She was still working on the Discovery collections at the time of her death.

Her ornithology continued over the years and, in 1904, she was one of the founders of the Irish Society for the Protection of Birds. She remained a leading member and, when support for the Society had become problematical in 1926, she volunteered to serve as Secretary. Her success in reversing its fortunes led to its continued existence until the 1960s when it merged with other bird conservation organisations. This was a vital period in the development of nature conservation in Ireland because it led to the passing of the Wild Birds Protection Act in 1930. This was the first legislation in the new state aimed purely at conservation, rather than for the maintenance of stocks for hunting.

Further reading:
G.P. Farran, C.B. Moffat: Obituaries, *Irish Naturalists' Journal*, **3**, 215–216, 1931.

Christopher Moriarty, Marine Institute Laboratories, Abbotstown, Dublin 15.

118 JOHN SEALY EDWARD TOWNSEND Physicist

Born: Galway, 7 June 1868.
Died: Oxford, 16 February 1957.

Courtesy of the Royal Society

Family:
Father: Edward Townsend, Professor of Civil Engineering at Queen's College, Galway.
Married: Mary Georgiana Lambert in 1911; she later became an alderman, honorary freeman and twice mayor of Oxford.

Addresses:
Galway; Dublin; Cambridge; Banbury Road, Oxford

Distinctions:
Foundation Science Scholarship in Mathematics at Trinity College Dublin 1888; Double Senior Moderatorship at TCD 1890; Clerk Maxwell Scholar at Cambridge 1898; Wykeham Professor of Experimental Physics at Oxford 1900; Fellow of the Royal Society 1903; Member of the Franklin Institute in Philadelphia 1924; Officer of the Legion of Honour; Corresponding Member of the Institut de France; Knighted 1941.

Townsend was born and educated in Galway, where his father was Professor of Civil Engineering. He entered Trinity College Dublin in June of 1885 and read mathematics, mathematical physics and experimental science, graduating first in mathematics in 1890. He was one of the many bright students of George Francis FitzGerald, who introduced him to Maxwell's theory of electromagnetism and trained him in the latest experimental techniques.

After obtaining his degree, Townsend spent four more years at Trinity. Then in 1895, he became one of the first outside research students at Cambridge University. The Cavendish Laboratory at Cambridge was then one of the most exciting places in the world to study physics. J.J. Thomson, who soon after measured the charge and mass of the electron (and is thus usually credited with its 'discovery'), directed the laboratory, while Townsend's fellow students included such great physicists as Ernest Rutherford, Joseph Larmor, C.T.R. Wilson and Paul Langevin. Their research centred on the recently discovered cathode rays, x-rays and other forms of radiation that seemed to offer insight into the structure of the atom. Most of these rays, discovered in the 1880s and 1890s, were created by the discharge of electrodes in gas filled tubes. Cathode rays, for example, were produced in tubes similar to those that later formed the basis of the visual displays in television sets.

For this reason, Townsend's research at Cambridge focused on the electrical properties of gases, a subject that would occupy him for the rest of his life. He discovered early on that the gases produced when an electric current is passed through a liquid carry electric charges and form clouds when they meet moisture. This not only demonstrated that clouds in the atmosphere are due to the electrification of gases, it also helped inspire his fellow student C.T.R.Wilson to build an early form of particle detector known as a cloud chamber. Townsend himself used the principle to make one of the earliest measurements of the charge of the electron (though he did not yet know that it was the electron whose charge he was measuring). He also developed a collision theory of ionization that became the basis for the particle detector built by Rutherford and Geiger (a precursor of the now common Geiger Counter for measuring radioactivity).

In 1900 Townsend was appointed to the newly created Wykeham Chair of Experimental Physics at Oxford, and he immediately introduced the teaching of electricity and magnetism to Oxford. At first, he had to struggle to find laboratory space but, in 1910, Oxford finally built a real electrical laboratory where Townsend would work until his retirement in 1941. In the new laboratory he built a flourishing research school with students from around the world, some of whom went on to become famous scientists in their own right, such as Henry Moseley and Robert van de Graaff. Townsend seems to have been somewhat isolated scientifically, however. He rarely read scientific publications or attended scientific meetings.

At Oxford, he continued his research on the electrification of gases. He worked, for example, on the problem of the electric breakdown of gases and the diffusion of electrons within a gas. He also discovered (simultaneously with German researchers) that slow electrons could pass through argon gas without interacting with it. This phenomenon, which demonstrated the wave-like properties of electrons, came to be known as the Ramsauer-Townsend Effect, though Townsend had little interest in its quantum mechanical implications.

During the First World War, he carried out research on radio communication for the Royal Naval Air Service with the rank of Major. World War I was the first time that vacuum tubes were used for radios, and Townsend applied his scientific knowledge to improve their functioning. He even volunteered for a dangerous mission to Russia in 1914. It was so dangerous that a colleague at Oxford dropped him a note before he was due to leave reminding him to recommend the colleague as his successor if he were not to return from the mission. Luckily for science, the mission was cancelled, and Townsend was able to continue his research and teaching. He maintained his interest in radio after the war and later worked on the cavity magnetron, a high-frequency transmitter that was essential to the development of radar.

Though he never achieved the same level of fame as some of his fellow students at the Cavendish Laboratory, Townsend's painstaking experimental work and his dedication to teaching made important contributions to the development of a number of areas of physics in the first half of the twentieth century.

Further reading:

Dictionary of Scientific Biography.
Dictionary of National Biography.
A. von Engel: John Sealy Edward Townsend, *Biographical Memoirs of Fellows of the Royal Society*, **2**, 257–72, 1957.

David Attis, Program in the History of Science, Department of History, Princeton University, Princeton, New Jersey USA.

119 WILLIAM O'LEARY Priest Scientist 1869–1939

Born: Dublin, 19 March 1869.
Died: Riverview College, New South Wales, Australia, 16 April 1939.

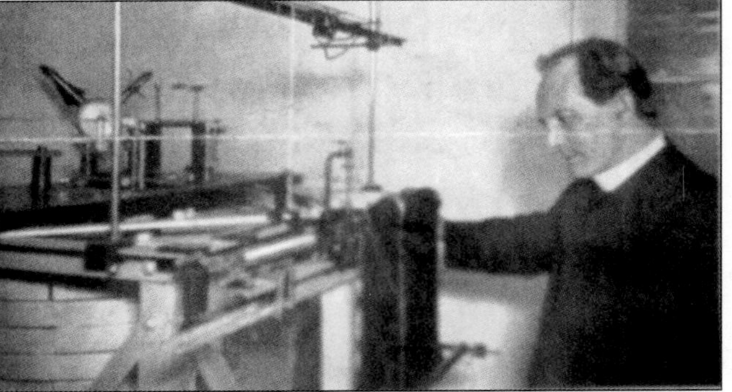
Father O'Leary with a seismograph in Australia

Family:
Son of Dr William O'Leary, Professor of Anatomy at the Royal College of Surgeons, Dublin, and Home Rule MP in the party of Isaac Butt.

Addresses:

1886	Entered Jesuit Order, Dromore
1905–1915	Mungret College, Limerick
1915–1929	Rathfarnham Castle, Dublin
1929–1939	Riverview College, Australia

Distinctions:
Fellow of the Royal Astronomical Society;
President of the British Astronomical Association.

In 1909, three Jesuit Colleges, Stoneyhurst in England, Riverview in Australia, and Mungret near Limerick, set up seismological observatories. The Mungret observatory was in the charge of Father O'Leary. He had constructed a seismograph which was exhibited in London at the Coronation Exhibition in 1911. He also made observations of the upper atmosphere, using sounding balloons, on behalf of the Royal Meteorological Society. When transferred to Rathfarnham Castle in 1915, he designed and worked on a much larger seismograph of inverted pendulum form, probably with the assistance of the Grubb instrument making firm. The 'bob' of about 1800 kg of octagonal iron plates sat on top of a vertical rod about two metres long, at the bottom of which was a circular steel plate to which were attached three steel rods making a small angle to the vertical and fixed to the ground just below the bob. This pendulum was almost unstable; any small movement of the ground relative to the inertia of the bob would cause it to rock with its period of about 16 seconds. Electromagnetic dampers quickly brought it to rest for the next tremor, while levers magnified the movement in two dimensions and recorded them through glass styli on smoked paper on a drum driven by clockwork.

O'Leary also designed a three-dimensional seismograph in 1913 but this was never completed. In 1929 he went to Riverview College. Here he continued seismography, but also turned his attention to the photographic study of variable stars using a blink comparator of his own design.

Further reading:

R.E. Ingram and J.R. Timoney: Theory of an Inverted Pendulum with Trifilar Suspension, *Geophysical Bulletin*, **9**, DIAS, 1954.
R.E. Ingram and P.M. Troddyn: Earthquake Recording at Rathfarnham Castle, *Irish Jesuit Directory and Year Book*, 139, 1938.
Rev. T. MacMahon, Rathfarnham Castle, *Dublin Historical Record*, **41**, 21–23, 1987.

I thank Professor Thomas Murphy, former Director of the Dublin Institute for Advanced Studies (DIAS) for assistance.

Adrian Somerfield, formerly of St Columba's College, Whitechurch, Dublin 16.

120 HENRY HORATIO DIXON Botanist

Born: Dublin, 19 May 1869.
Died: Dublin, 20 December 1953.

Portrait by Phoebe Donovan, 1949 (courtesy of Trinity College Dublin)

Family:
Son of George Dixon, soap-manufacturer, and Rebecca Yeates, daughter of Dublin's leading instrument-maker.
Married: Dorothea Mary Franks, daughter of Sir John Franks, CB 1907.
Children: three sons, two of them scientific fellows of King's College, Cambridge.

Addresses:

1869–1880	30 Holles Street, Dublin
1881–1888	4 Earlsfort Terrace, Dublin
1889–1894	17 Earlsfort Terrace, Dublin
1894–1905	23 Northbrook Road, Dublin
1906–1907	Milverton, Temple Road, Dublin
1908–1934	Clevedon, Temple Road, Dublin
1934–1953	Somerset, Temple Road, Dublin

Distinctions:
Professor of Botany, Trinity College Dublin 1904–49; Fellow of the Royal Society 1908; Boyle Medal of the Royal Dublin Society 1917; Croonian lecturer, Royal Society 1937; President, Royal Dublin Society 1944–47; Member of the Royal Irish Academy 1947; Honorary Fellow, Trinity College Dublin 1950; Honorary President, International Botanical Congress, Stockholm 1950.

Dixon was the youngest son of a large and talented family; two of his brothers also became professors. He entered Trinity College in 1887 and won a foundation scholarship in classics in 1890, but then changed to natural science, in which he graduated with a gold medal in 1891. The change was made under the influence of John Joly, with whom he went on a walking-tour in Switzerland in 1888: this opened his eyes to the beauty and wonder of the natural world and laid the foundations of a life-long friendship.

From 1892 to 1904 Dixon was assistant to E.P Wright, whom he succeeded in the chair of botany in 1904, but he first spent a year at Bonn, where Strasburger was working out the details of the recently discovered process of meiosis. Dixon's studies here were mainly cytological (cytology is the study of the structure and function of cells), but his imagination was also stimulated by Strasburger's demonstration that the sap in trees could rise even when the cells in the stem had been killed. Back in Dublin he considered the implications of this. If the living cells in the wood did not pump up the sap (as had generally been supposed), then it must rise under purely physical forces, root pressure, capillarity and atmospheric pressure all being plainly inadequate. With Joly's help he elaborated a theory that the motive force was primarily the evaporation from the leaves: this produces a tension in the water filling the vessels, and thereby, thanks to the cohesive strength of a bubble-free column of water (far greater than had been hitherto imagined), draws up the water from the roots. Many experiments had to be devised to refute sceptics: they were done with simple but ingenious apparatus and explained with patience and clarity. By about 1914 most of his critics had been silenced, and a few years later the cohesion theory found its way into the textbooks as undisputed orthodoxy.

The School of Botany, Trinity College Dublin

Dixon had a very alert and imaginative mind, and he made many pioneering observations and suggestions which, because he did not follow them up, are often credited to others. These include the explanation of bivalent chromosomes at meiosis, the use of thermocouples for determining osmotic pressure in small volumes of fluid, and the mutagenic effect of cosmic rays. His theory of the transport of organic substances in plants turned out to be mistaken, but it stimulated research in a field in which much still remains to be explained.

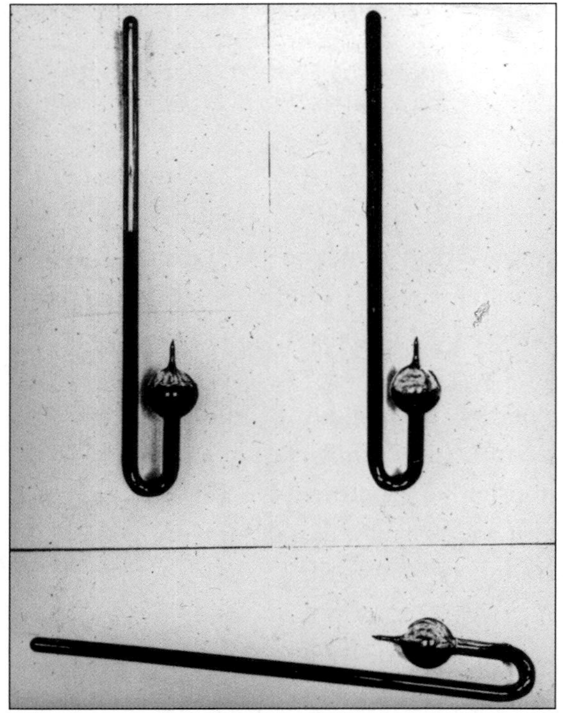

Apparatus devised by Dixon to demonstrate the cohesive strength of water. The J-tube contains only coloured water and water vapour, and if inclined so as to fill the long arm (bottom) and then raised carefully the long arm (right) remains full.

In 1904 botany was still taught in cramped and unsatisfactory rooms originally designed for residence. Thanks to the persistent advocacy of Joly and the generosity of Lord Iveagh, money became available in 1906 for the erection of a proper laboratory. In the design of this building Dixon almost played as great a part as the architect, and with great success. It gave scope for the development of the subject by twentieth-century standards, and even today its only serious defect is that it is too small for the greatly increased numbers of staff and students. In 1910 an annex was added to house the herbarium adequately for the first time in its long history, and it is to Dixon's credit that, although he had no real interest in taxonomy, he devoted the best part of two years to filing and indexing the specimens.

Further reading:

H.H. Dixon: *Transpiration and the Ascent of Sap in Plants*, London, 1914.

W.R.G. Atkins: Henry Horatio Dixon, *Obituary Notices of Fellows of the Royal Society*, **9**, 79–97, London, 1954.

David Webb 1912–1994.

Born: Newry, 26 July 1872.
Died: Cambridge, 21 March 1947.

Portrait of Joseph Barcroft
by R.G. Eves, RA

Family:
He was the second of five children of Henry Barcroft – an inventive industrial manager – and his wife, Anna Richardson.
Married: (1903) Mary Agnetta (Minnie), daughter of Sir Robert Ball, formerly Royal Astronomer of Ireland, Professor of Astronomy and Geometry at Cambridge.
Children: Two sons, Henry FRS (1904–1998) and Robert.

Addresses
1903–1947 Grange Road, Cambridge

Distinctions:
Fellow of the Royal Society 1910, Royal Medal 1922, Copley Medal 1943, Croonian Lecture 1935; Knighthood 1935; Professor 1925–1937; in 1952 Mount Barcroft (3,908 m) in White Mountains, California, was named in his honour when the University of California established a high-altitude laboratory there.

Joseph Barcroft was educated at home, at the Friends' School, Bootham, York, and at Leys School and King's College Cambridge. After taking the Natural Sciences Tripos in 1896, he 'was clearly accepted at once by his seniors as a new recruit of the first rank to experimental physiology'. In 1908 he invented the differential blood gas manometer for his own researches and it rapidly passed into world-wide use. He led three mountain expeditions to study the physiology of life at high altitudes and the results of two of them were incorporated in *The Respiratory Function of the Blood* (1914). This monograph firmly established his reputation as a respiratory physiologist, and after chlorine was used at Ypres on 22 April 1915 Barcroft, in spite of his Quaker beliefs, eventually found 'his bowler among the brass hats' when he became chief physiologist at Porton Down.

Using contrast cine-radiography he studied the dramatic changes in the circulation at birth. *Researches on Pre-Natal Life* (1946) recounted the story of these and other adventures that initiated the modern study of fetal physiology. In *Features in the Architecture of Physiological Function* (1934) he argued that the stability of the environment surrounding the cells of the body is the essential condition for the 'higher' nervous function, that is, for mental activity.

Although Barcroft did all his work outside Ireland, he was always looked upon, and looked upon himself, as an Irishman.

Further reading:
K.J. Franklin: *Joseph Barcroft 1872–1947,* Oxford, Blackwell, 1953.
Dictionary of National Biography; *Dictionary of Scientific Biography.*

Caoimhghín Breathnach, Department of Physiology, University College, Dublin 2.

EDMUND TAYLOR WHITTAKER Astronomer and Mathematician 1873–1956

Sir Edmund Whittaker, F.R.S.

Born: Birkdale, Lancashire, 24 October 1873.
Died: Edinburgh, 24 March 1956.

Family:
Son of a railway engineer and contractor.
Married: Mary Boyd, daughter of the Rev. Thomas Boyd of Cambridge in 1901.
Children: Three sons and two daughters.

Addresses:
1906–1912 Dunsink Observatory, Dublin
1954 48 George Square, Edinburgh

Distinctions:
Second wrangler, mathematical tripos, Cambridge 1895; Fellow of Trinity College, Cambridge 1896; Fellow of the Royal Society 1905, Sylvester Medal 1931, Copley Medal 1954; Andrews' Professor of Astronomy, University of Dublin, and Royal Astronomer of Ireland 1906–12; Professor of Mathematics, University of Edinburgh 1912–46; President, Royal Society of Edinburgh 1939–44; The Cross pro Ecclesia et Pontifice 1935; Member, Pontifical Academy of Sciences 1936; Knight Bachelor 1945.

Edmund Whittaker did much of his work in Cambridge, before coming to Dublin in 1906, and in Edinburgh, after leaving Dublin in 1912. Nevertheless, his years at Dunsink Observatory, which he remembered vividly for all his days, were influential in focusing his attention on the problems of incorporating sound mathematics into his astronomy, both as a tool for interpreting observations and as a mode of expression of physical theories. Although he was himself a mathematician, he saw clearly that progress in astronomy required an approach based on theoretical physics and, during his time as Royal Astronomer of Ireland, he wrote his influential *History of the Theories of the Aether and Electricity* which, in its 1951 second edition, is a standard work on the subject.

At Dunsink, Whittaker undertook responsibility for the practical work of the Observatory, mainly concerned with maintenance of accurate time; and an opportunity was taken to develop satisfactory photometry of stars using the photographic capability of the 15-inch Isaac Roberts reflector. Accurate light curves for bright variable stars were derived, and this work was continued later by Plummer and Martin. It became influential in assisting Eddington, in Cambridge, to develop theories of stellar structure.

Whittaker's period as Royal Astronomer at Dunsink and his influence on de Valera then and later were major factors in determining the course of scientific work in the Dublin Institute for Advanced Studies up to the present day.

Further reading:
D. Martin and others: Whittaker Memorial Number, *Proceedings of the Edinburgh Mathematical Society,* **11**, June, 1958.
W. H. McCrea: Edmund Taylor Whittaker, *Journal of the London Mathematical Society,* **32,** 234–56, 1957.

Patrick A. Wayman 1927–1998.

Born: Kilkee, Co. Clare, 13 August 1874.
Died: Esher, Surrey, 4 April 1943.

Courtesy of the National Library of Ireland

Family:
Father: Thomas, Resident Magistrate in Kilkee, Cahirsiveen. Mother: Anne (née Barry).
Married: Nina Le Mesurier of an old Guernsey family.
Children: One son, two daughters.

Addresses:
2 Belgrave Place, Cork
Also: Kilkee; Cahirsiveen; Calcutta; Esher, Surrey

Distinctions:
Exhibition 1851 Scholarship 1898; DSc, National University of Ireland 1920; Commander of the Indian Empire 1921; Fellow of the Royal Society 1926; President British Mycological Society 1927; President Association of Economic Biologists 1929; Commander of the Order of St Michael & St George 1932; LLD Aberdeen University 1938; Knighted 1939.

Edwin John Butler graduated in Medicine at Queen's College Cork in 1898, but never practised. With Marcus Manuel Hartog (1851–1924), Professor of Natural History, Assistant Director Peradeniya (Ceylon) (1874–1877), a gifted cytologist (cytology is the study of cells), he explored microflora in the College ponds. His interest in *Pythium*, a parasitic fungus had begun. He worked with foremost mycologists in Kew, Paris, Freiburg and the Antibes. In 1901 he was appointed Cryptogamic botanist to the Government of India. He trained mycologists and technicians, set up a herbarium and a culture collection, inspired workers, and advanced knowledge and research. His first posting was to Calcutta, with the Botanical Survey of India – the centre of Systematic Botany. In 1902 he made field collections at Dehra Dun. In 1905 he was appointed Imperial Mycologist at the Agricultural Research Station, Pusa. He studied life in the soil and fungus diseases in potato, wheat, rice, sugar, coconut, rubber and many other tropic crops. His first paper, on potato diseases of India, appeared in 1903, and a classic account of *Pythium* in 1907. Some 40 research papers followed, leading to *Fungi and Diseases in Plants* (1918). In 1931, with G.R. Bisby, he issued *The Fungi of India.* In 1920 Butler was Agricultural Adviser to the Government of India. In July 1920 he founded the Imperial Mycological Bureau in London, later the International Mycological Institute, and was its Director for 15 years. The abstracting journal *Review of Applied Mycology* was initiated with Butler as editor. To-day he is known as 'The Father of Indian Plant Pathology and Mycology'. The Butler Medal was initiated by the Society of Irish Plant Pathologists in 1977.

Further reading:
E.J. Butler: The development of economic mycology in the empire overseas, *Transactions of the British Mycological Society*, **14**, 1–18, 1918.
E.J. Butler: *Fungi and Disease in Plants*, Calcutta, 1918 (reissued 1955).
Dictionary of National Biography, Missing Persons, 1993.

Mary J.P. Scannell, former Head of Herbarium, National Botanic Gardens, Glasnevin.

ARTHUR WILLIAM CONWAY Mathematician, Physicist and University Administrator

Born: Wexford, 2 October 1875.
Died: Dublin, 11 July 1950.

Addresses:

1903–1912 100 Leinster Road, Rathmines, Dublin
1912–1920 Elsinore, Coliemore Road, Dalkey, Co. Dublin
1920–1929 Abbeyview, Coliemore Road, Dalkey, Co. Dublin
1929–1950 Colamore Lodge, Coliemore Road, Dalkey, Co. Dublin

Family:

Son of Myles and Teresa (née Harris) Conway.
Married: Agnes Christina Bingham, daughter of William Bingham of Ballymena, Co. Antrim, 1903.
Children: Teresa Mary, Morgan, Verna, and Orlaith.

Distinctions:

Professor of Mathematical Physics, University College Dublin 1901; Honorary DSc Royal University of Ireland 1908; Fellow of the Royal Society 1915; President of the Royal Irish Academy 1937; Honorary ScD University of Dublin 1938; Honorary LLD University of St Andrews 1938; Member of the Pontifical Academy of Sciences 1939; Honorary Fellow of Corpus Christi College Oxford 1940; President of University College Dublin 1940; President of the Royal Dublin Society 1942.

Arthur William Conway entered University College, 86 St Stephen's Green, Dublin, in 1892 as a resident student and, in 1897, he received the MA degree with first class honours in mathematical science. He then transferred to Corpus Christi College, Oxford, where in 1898 he was awarded the Junior and in 1902 the Senior University Scholarship in mathematics. In 1900 he became a junior fellow of the Royal University of Ireland. Appointed to the chair of mathematical physics at University College in 1901, he was also external lecturer at St Patrick's College, Maynooth, from 1903 to 1910, during the transition period when Maynooth was preparing for association with the National University of Ireland. He was succeeded in this lectureship by his student Eamon de Valera.

On the establishment in 1908 of the National University, with University College as one of its constituent colleges, Conway was appointed Registrar of the College. He occupied this position until he became President of the College 32 years later. In spite of his teaching and administrative duties, Conway pursued scientific research until the end of his life, when he left behind papers that were published posthumously. His research was greatly influenced by the investigations of the nineteenth century Irish mathematician Sir William Rowan Hamilton. This is seen by his interest in quaternions, on which he continued to work from 1900 until his last years, by which time he had come to be acknowledged as the world's greatest authority on the subject. It is also shown by his undertaking and accomplishing the arduous task of joint editor of the first two volumes of Hamilton's mathematical papers, the first on geometrical optics being published in 1931 in collaboration with J.L. Synge, and the second on dynamics in 1941 in collaboration with A.J. McConnell.

In the course of a discussion on elementary particles at a meeting of the Royal Irish Academy about 1945, Conway recalled that during his scientific life the electron had been discovered by J.J. Thomson. This discovery was a crucial point in the history of theoretical physics. Conway was to follow closely developments in atomic models and to apply his mathematical dexterity in elucidating and developing the new theories of atomic structure that were emerging. In the middle of the first decade of the present century, it was generally supposed that the atom is an electrical system having proper vibrations and that the periods of these vibrations give rise to the set of lines in the spectrum of an atom. In 1907 Conway made the bold supposition that each atom produces a single spectral line, so that the production of the spectrum is a collective effect of the atoms in a sample under investigation. This correct theory was put forward six years before the publication of the Bohr theory of the atom.

The first decades of the twentieth century were exciting for both theoretical and experimental physicists. In 1900 Planck proposed that energy is transferred discretely in integral multiples of the unit $h\nu$, where ν is the frequency of radiated energy and h is Planck's constant. Thus began quantum theory. Quantum theory and relativity provided new avenues of research for Conway.

In 1911 he showed that quaternions were especially well suited to express results in special relativity and to lead to new theorems. When later Dirac proposed his relativistic theory of the electron and Eddington his theory of protons and electrons, Conway expressed the four-by-four matrices employed in these theories very simply in quaternion form.

When the Bohr theory of the atom was extended by Sommerfeld and Wilson, Conway applied their quantum theory to show how the Zeeman triplet effect could be explained. He also showed how one could account for the ordinary spectrum of ionised helium and the s-term of orthohelium or parhelium, and could provide the correct value of the ionisation potential.

Immediately after the publication in 1926 by Uhlenbeck and Goudsmit of their paper on electron spin, Conway wrote a paper on the dynamics of a spinning electron. In this he provided a quantum mechanical theory in which the electron is regarded as a uniform rotating sphere that describes an orbit around a fixed nucleus. Then in 1927 he applied the de Broglie-Schrödinger wave mechanics, which he called 'undulating theory', to two-electron orbits and obtained the Rydberg form of series terms.

Conway was a man of genial disposition and with a boyish sense of humour. His university lectures were well ordered and were delivered with great speed. With the help of related textbooks one could build on his lectures to acquire a deep knowledge of the subject matter. When you heard him lecture on Hamiltonian mechanics, or on some other topic to which he had made a personal contribution, he infused enthusiasm. He was admired and respected by his students and in particular by de Valera, who as Taoiseach often sought his advice on academic matters.

Having spent his boyhood near the Wexford coast, Conway had a love of the sea and, during most of his life, he lived on the shore of Dublin Bay. He was a strong swimmer and enjoyed sailing. He was hospitable and, if you called unexpectedly during the evening, you could be sure of a warm welcome.

Further reading:

James McConnell (ed.): *Selected Papers of Arthur William Conway*, Dublin, 1953.

Reverend Monsignor James McConnell, 1912–1999.

125 WILLIAM SEALY GOSSET ('STUDENT') Brewer and Statistician

Student in 1908

Born: Canterbury, 1876.
Died: London, 16 October 1937.

Family:
He was the eldest of four sons and a daughter of Colonel Frederic Gosset, RE, who married Agnes Sealy Vidal in 1875. The Gossets were an old Huguenot family.
Married: Marjory Surtees Phillpot, youngest daughter of the head-master of Bedford School, in 1906. She was captain of the English ladies' hockey team and subsequently played for, and captained, the Ireland team.
Children: One son Harry, who became a physician, and two daughters, Bertha and Ruth.

Addresses:
Woodlands, Monkstown, Co. Dublin
Holly House, Newtownpark Avenue, Blackrock, Co. Dublin

The Dublin brewery which, since 1759, has borne the name Guinness, traditionally brewed porter or stout from malted barley, flavoured with hops and darkened by roasted malt. In the latter part of the nineteenth century its product penetrated markets in Britain and overseas, and the brewery expanded enormously. This demanded a more scientific approach to quality control, and the board set about appointing persons with a scientific background as Brewers, the senior men in charge of the process. In 1899, W.S. Gosset, a scholar of Winchester and of New College, Oxford, and a graduate in mathematics and natural science, was appointed as a Brewer.

Gosset soon became aware of the need to be able to estimate experimental errors, the problem being that often only a small number of samples was available. It took at least a day to obtain the result of changing one variable in the experimental brewery, while in the development of new strains of barley, in which Guinness worked closely with the Department of Agriculture and its cereal station at Ballinacurra, the testing of one 'cross' might take a year. The usual method of repeating each experiment many times was therefore impracticable. In 1904, Gosset wrote a report on 'The Application of the Law of Error to Work of the Brewery', as a result of which he was put in touch with Karl Pearson (1857–1936), the mathematician, who had been building up the large biometric laboratory at University College London and was probably the leading statistician of the day. Pearson had recently introduced the chi-squared goodness-of-fit criterion and founded the journal *Biometrika*.

Pearson's work was on large samples, but in 1906 Gosset spent two terms in London on the application of statistics to small samples, and it was during that period that he laid the foundation for the basis of his most famous paper, 'The Probable Error of a Mean', published in *Biometrika* in 1908. (At that period, Guinness did not allow its staff to publish under their own names, and for this reason Gosset used the pseudonym 'Student' for all of his 22 published papers.)

In this paper, through examining by a mixture of theory and practice how means of small samples were distributed, he produced tables from which could be computed the probability that the population mean would lie within certain numbers of standard deviations of the sample mean for sizes of sample 4 to 10. Ronald Aylmer Fisher (later Sir Ronald, FRS, 1890–1962), a mathematical student at Cambridge, wrote to Gosset giving a more rigorous proof of the distribution and recommending a change from n to n-1 in calculating the variance. When this was done, the statistic became known as Student's t (see box) and the t-test entered the language of statistics. Fisher became statistician to Rothamsted Experimental Station at Harpenden, and it was largely due to his advocacy that Student's t became widely accepted as a test of significance.

A USE OF STUDENT'S t

We wish to construct a 95% confidence interval estimate of the mean yield of barley per square metre μ on a farm. We take a random sample of size n plots each of 1 square metre area, find the yield x in each, and the sample mean $\bar{x}$. Were we to know the population variance σ^2 of the yields per square metre, then based on the normal probability tables $x \pm(1.96\ \sigma/\sqrt{n})$ would with probability 0.95 contain the unknown mean μ. When σ^2 is not known, it is usually estimated by the sample variance

$$s^2 = \frac{\sum(x-\bar{x})^2}{n-1}$$

For large n (above say 30), $\bar{x} \pm(1.96\ s/\sqrt{n})$ constitute an approximate 95% confidence interval for μ. 'Student', however, showed that for small n such interval estimates were much less reliable than 95%, and that in such a situation 1.96 should be replaced by the appropriate t value with n - 1 degrees of freedom. Values of

$$t = \frac{(\bar{x}-\mu)\sqrt{n}}{s}$$

are obtained from the curves and tables for this function which he had calculated.

For example, if n =12, $\bar{x}$ = 325g and s^2 = 125, the appropriate t value for 11 degrees of freedom is 2.201, and the resulting 95% confidence interval for the mean yield per square metre would be

$$\bar{x} \pm 2.201\ s/\sqrt{12} = [318, 332]$$

(I thank Dr Philip Boland, University College Dublin, for assistance.)

Gosset's interests extended to correlation co-efficicents, randomisation of field trials (he had differences here with Fisher, though they remained friends), genetics, and other areas of statistics. However, it would be a mistake to think of him as a 'statistician employed by Guinness', or as a 'brewer with a sideline in statistics'. His brewery work made statistics a necessary tool; where the tool was inadequate he improved it. His was an essentially practical approach; he developed methods that worked, being content to leave mathematics to Fisher. On the other hand, he could test the practical usefulness of the mathematics. In these electronic days, it is easy to forget that the calculations of the vast tables for z and its successor t had all to be done 'by hand' or on what would seem to us primitive mechanical calculators. Much of this was done by Gosset himself at home, though he had some assistance in the Brewery from W.A. Bowie and, from 1922, E. Somerfield.

Gosset was a well-liked person, interested in the outdoor life, golf, gardening, sailing, skiing, walking and fishing; a little eccentric perhaps. He built several boats for fishing, one of which, with a novel design with two rudders, was featured in *The Field* in 1936.

In 1935 he left Dublin to take charge of the new Guinness brewery at Park Royal in London. He was able to supervise the start-up of this enterprise, but he died in 1937 at the early age of 61. His t-test and process of 'Studentisation' are his memorial.

Further reading:

E.S. Pearson and J. Wishart: *'Student's' Collected Papers,* University College, London, 1942.
C. Chatfield: *Statistics for Technology,* 1978.
D. Hoctor: *The Department's Story: A History of the Department of Agriculture,* Dublin, 1971.
J.F. Box: *R.A. Fisher, the Life of a Scientist,* 1978.

Adrian Somerfield, formerly of St Columba's College, Dublin.

ARTHUR WILSON STELFOX Entomologist, Conchologist, Botanist and Naturalist

Courtesy of the trustees of the Ulster Museum

Born: Belfast, 15 December 1883.
Died: County Down, 19 May 1972.

Family:
Son of James Stelfox, MICE, and Jennie MacIlwaine.
Married: Margarita Dawson Mitchell 1914.
Children: Two sons and a daughter. His grandson, Dawson Stelfox, was the first Irishman to climb Mount Everest, in 1993.

Addresses:

1883–1907	Oakleigh, Ravenhill Road, Belfast
1907–1914	Delamere, Chlorine Gardens, Belfast
1914–1920	Ballymagee, Bangor, Co. Down
1920–1956	Clareville Road, Harold's Cross, Dublin
1956–1972	Tullybrannigan Road, Newcastle, Co. Down

Distinctions:
Member of the Royal Irish Academy 1912; Honorary Fellow of the Linnaean Society 1947; He refused two offers of honorary degrees.

Arthur Wilson Stelfox was educated in Campbell College, Belfast, and trained as an architect in London and Belfast. He obtained his Associateship of the Royal Institution of British Architects in 1909, though he practised his profession little. He ran a market garden in County Down for a short period. In 1920 he was appointed to the Natural History Division of the National Museum of Ireland. For several years from 1924 he was Acting Keeper of the Natural History Division. However, he failed to be appointed to the post of Keeper of the Division (on the grounds that he did not have a university degree!) in spite of his international standing in several fields of natural history and in spite of his achievements in the post in an acting capacity. Subsequently he was appointed Deputy Keeper. He retired in 1948 and later went to live in Newcastle, Co. Down.

Stelfox was a keen naturalist from a very early age, and his first publication dates from 1904. He became a leading authority in Britain and Ireland on nonmarine molluscs, then on bees, wasps and ants, then on sawflies; he then began a vast study of the parasitic hymenoptera, such as the ichneumons and braconids. In addition to these main areas, Stelfox was a polymath in natural history: he was an expert in the identification of mammals and birds from bone fragments in cave deposits; he had a wide knowledge of flowering plants, particularly arctic-alpine plants and difficult groups such as sedges and willows; and he carried out genetical studies on several species of snail, involving long-extended breeding studies. He has been described as 'a naturalist of prodigious knowledge, intelligence, versatility and memory'. He used his gifts to stimulate and help many of the young naturalists of his time.

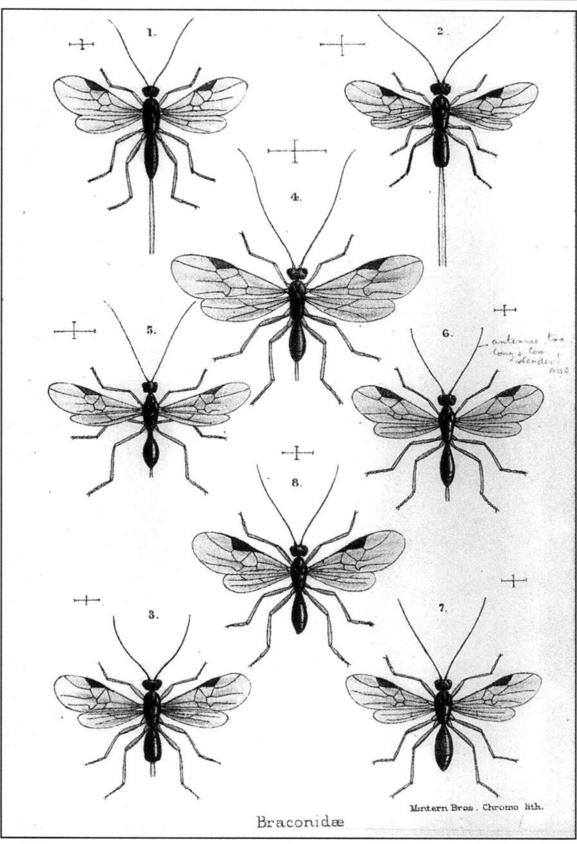

*Plate X of T.A. Marshall's (1889) A Monograph of British Braconidae (*Transactions of the Entomological Society of London*) annotated by Stelfox (courtesy of the National Museum of Ireland)*

His biggest contribution was in the collecting and recording of insects and molluscs throughout Ireland and in many parts of Britain. He appreciated the importance of collecting long series of each species with detailed records on habitat. In the case of the parasitic hymenoptera, his collection contained over 100,000, perhaps 200,000, superbly mounted and catalogued specimens, many new to science: it is one of the greatest collections ever made by one person and has been described as a national treasure. Only a small fraction of the information contained in this collection has been published, and his collections and records will provide material for publication by generations of entomologists. It is a tragedy that the then lack of government interest in the Natural History Division of the Museum, together with unfortunate official decisions affecting himself and some other people whom he respected, led to the conviction that his collections would never get the required degree of study in this country, and he donated all his collections and records outside the state. The parasitic hymenoptera collection went to the Smithsonian Institution in Washington. Happily, duplicate specimens from this collection have been returned recently to the National Museum, and these will enable his work to be carried on by entomologists in Ireland.

Further reading:

Various Authors: Arthur Wilson Stelfox, 1883–1972, *The Irish Naturalists' Journal*, **17**, 286–307, 1973.

N.F. McMillan, A.E. Ellis and L.M.Cook: Arthur Wilson Stelfox, 1883–1972, *Journal of Conchology*, **27**, 520–33, 1972.

B.P. Beirne: Irish Entomology: the First Hundred Years, *The Irish Naturalists' Journal, Special Entomological Supplement,* 1985. This includes a critical review of Stelfox's character and contributions. Many who knew him regard some of it as unfair, and parts of it have been replied to by G.F. Mitchell and N.F. McMillan, *The Irish Naturalists' Journal,* **22**, 122, 1986.

Frank Winder, Department of Biochemistry, Trinity College, Dublin.

127 HARRY FERGUSON Pioneer Aviator and Inventor

Born: Growell, Co. Down, 4 November 1884.
Died: Stow-on-the-Wold, Gloucestershire, 25 October 1960.

Family:
Married: Maureen Watson 1913.
Children: One daughter.

Harry Ferguson, christened Henry George, was the fourth son in a family of eleven children. He left school at the age of fourteen to work on his father's farm at Growell, near Dromore in Co. Down, a job he disliked intensely. His natural mechanical aptitude was soon put to better use when he started to serve an apprenticeship with his elder brother Joe, who had started a car and cycle repair business in Belfast in 1901.

After a brief interest in competitive motor-cycling, in 1909 he became fascinated by the latest invention – aviation. So he built his own aeroplane, and flew the machine successfully later that year. A replica of this early machine may be seen in the Ulster Folk and Transport Museum, Cultra, Co. Down.

From his early years he showed traits of character which were to remain with him all his life. Not only was he very strong willed, but a perfectionist in all that he attempted, with an obsession for punctuality, tidiness and cleanliness.

In 1911 he opened his own motor business in Belfast – Harry Ferguson Limited – with an agency for several cars. At this time he even entered the motor racing field for a short period. The business prospered and in 1913 he married a life-long friend, Maureen Watson, though not with the wholehearted approval of the Watson parents, who considered Harry an agnostic, mad young man. Despite this initial opposition, it was a successful union, and Maureen did everything to encourage her husband with all his various projects throughout his life.

In 1917 Ferguson was approached by the Irish Board of Agriculture for advice on how to improve the yield of food production by means of easier and more efficient mechanical ploughing. He approached the problem with his usual zeal and enthusiasm and soon realised that the heavy cumbersome tractors and the complicated ploughing attachments then in use were at fault. So he designed a plough that could be attached to any type of tractor, and successfully demonstrated it in late December of that year, hitched to the recently developed Ford tractor – the Fordson.

Basically the principle of the Ferguson system was hydraulic control of various implements linked to the tractor.

Harry Ferguson and Henry Ford negotiating the famous 'handshake agreement' (courtesy of the Ulster Folk and Transport Museum)

Believing that a vehicle of his own manufacture was the only way to do full justice to his plough and attachments, he designed a tractor which was completed in 1933, and fortunately this prototype is preserved in the Science Museum. Ferguson next came to an agreement with the David Brown Engineering Company to manufacture the Ferguson tractor, to be known as the Brown Ferguson. Finished in grey, with a range of attachments, it was introduced to the public in 1936.

In 1939 he approached Henry Ford of the Ford Motor Company, who wanted to acquire the Ferguson patents, which Ferguson flatly refused to relinquish. So the Ford Ferguson tractor partnership was formed – a gentleman's agreement sealed with a handshake. Unfortunately, it was an agreement which was to end up in a lengthy lawsuit in 1950.

By this time Harry Ferguson had acquired the country house and small estate of Abbotswood, near Stow-on-the-Wold, where he spent the last years of his life. Despite failing health and insomnia, he still remained active and continued his interest in tractors and farming equipment.

On the morning of 25 October 1960 he died from an overdose of drugs. The coroner's jury returned an open verdict.

Further reading:

Colin Fraser: *Harry Ferguson, Inventor & Pioneer,* London, 1972.
John B. Rae: *Harry Ferguson & Henry Ford,* Ulster Historical Foundation, Belfast, 1980.
Stevan D.T. Patterson: *Fifty Years of the Ferguson 20 Tractor,* Omagh College of Further Education – Publication No. 124177 716905.
Bill Martin: *Harry Ferguson,* Ulster Folk & Transport Museum, Cultra, Co. Down.

Alfred Montgomery, Department of Local History (Industrial Archaeology), Ulster Museum, Belfast (retired).

Born: Vienna, 12 August 1887.
Died: Vienna, 4 January 1961.

Erwin Schrödinger, senior professor at the Dublin Institute for Advanced Studies, 1940–1956

Family:
Son of Rudolf and Georgine Emilia Brenda (née Bauer) Schrödinger
Married: Annemarie Bertel 1920

Addresses:

1940–1956	26 Kincora Road, Clontarf, Dublin
1956–1961	Pasteurgasse 4, Vienna

Distinctions:
Membership of the following scientific academies: Vienna 1928, Prussian 1929, Royal Irish 1931, Madrid 1935, Pontifical 1936, USSR 1940, Lima 1944, Lincei 1947, Royal Society 1949; Honorary doctorates of University of Ghent 1939, Dublin University 1940, National University of Ireland 1940.
Medals: Medaglia Matteuci 1929, Nobel prize for physics 1933, Max Planck 1937.

Erwin Schrödinger was a member of a cultured Viennese family. As a child he derived from his father an interest in botany, philosophy and painting and from his mother a proficiency in the English language, his maternal grandmother having been born at Leamington. His early formal education was chiefly in the ancient classics, and this helped him to become well acquainted with Greek philosophy. From 1906 to 1910 he studied at the University of Vienna, where he obtained an excellent training in theoretical and experimental physics from Fritz Hasenöhrl and Franz Exner. Before coming to Dublin, Schrödinger had held University posts at Vienna, Jena, Stuttgart, Breslau, Zürich, Berlin and Graz. During his stay in Zürich (1921–27) he proposed what became known as the 'Schrödinger equation', which provided a means of applying the quantum theory of Max Planck to physics, chemistry and biology. In spite of his immense influence in spreading the knowledge of quantum theory, Schrödinger appears to have remained at heart a classical, that is pre-quantum, physicist. Schrödinger's publications include sixteen books and about one hundred and sixty papers, many of which were translated into foreign languages. The range of his scientific publications embraces quantum theory, statistical mechanics, Brownian motion, dielectric theory, general relativity, optics. He wrote on interdisciplinary topics; in particular he investigated how physics and chemistry might be applied to biological problems. He was also very much concerned with the cultural value of the natural sciences.

Erwin and Annemarie Schrödinger became Irish citizens in 1948. Schrödinger left Ireland in 1956 to take up a personal chair in the University of Vienna.

Further Reading:
Walter Moore: *Schrödinger, Life and Thought*, Cambridge, 1989.

Reverend Monsignor James McConnell, 1912–1999.

129 J. J. and P. J. NOLAN Physicists 1888–1952 and 1894–1984

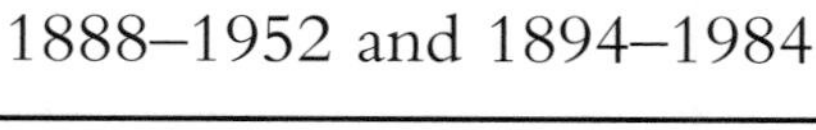

J.J. Nolan

P.J. Nolan

Addresses:

J. J. Nolan

Until 1941	Ros-na-Greine, 14 Avoca Avenue, Blackrock, Co. Dublin
1941–1945	30 Fitzwilliam Place, Dublin
1945	Lugnaquilla, Cowper Road, Rathmines, Dublin

P. J. Nolan

Until 1937	65 St Stephen's Green, Dublin
1937	35 Eglinton Road, Donnybrook, Dublin

In the early years of this century two remarkable brothers entered University College Dublin (UCD) to study for degrees in physics. Both were later to become professors in the College and helped to establish the Atmospheric Physics Research Group which is still thriving today. John J. and Patrick J. Nolan were born in Omagh, Co. Tyrone, where they received their early education and were awarded scholarships to study in UCD.

J. J. Nolan (the elder brother) graduated from UCD in 1909 and joined the staff of the University in the following year. In 1920 he succeeded J. A. McClelland as professor of Experimental Physics. He was appointed Registrar of the College in 1940 and held both posts until his death in 1952.

He published his first paper with McClelland in 1911 on the subject of the electric charge on rain, which signalled the start of a lifelong interest in atmospheric physics in general and atmospheric electricity in particular. His research carried him into many aspects of this branch of physics research and in each case he made a significant contribution to the fund of knowledge on the particular topic. However, it was his work on the equilibrium of ionisation in the atmosphere that many consider to be his most important piece of research, much of which was done in collaboration with his younger brother.

P. J. Nolan entered UCD in 1911 and graduated in 1914. The following year he was awarded the MSc degree and won a travelling studentship. However, because of the war, he was unable to travel abroad and so the first two years of the studentship were spent in UCD. At the end of the war he went to the Cavendish Laboratory in Cambridge and worked under Sir J. J. Thomson and Lord

Rutherford. It is interesting to note that amongst P. J. Nolan's fellow research students in the Cavendish at that time were E.V. Appleton, J. Chadwick, A.H. Compton and G.P. Thomson, all of whom subsequently were awarded the Nobel Prize for Physics.

After his period at the Cavendish, P. J. Nolan returned to UCD and began a long and very fruitful career as a lecturer and later as a professor in the physics department. Probably his greatest contribution in the field of atmospheric physics was his development, with L.W. Pollak, of the photoelectric nucleus counter that bears their names. This instrument, which is used to measure condensation nuclei, is now the standard instrument in use throughout the world. (Condensation nuclei are very small particles – about 10^{-8}m in radius – which are present in the atmosphere and on which water vapour condenses.)

Two incidents from the long career of these two distinguished physicists will show that scientific research can have its lighter moments. The first relates to 1929 when members of the group went to Colwyn Bay in North Wales to investigate the effects of a solar eclipse on atmospheric ions. After all the travel and preparation, the expedition was literally a wash out because of heavy rain. Added to that, local interest caused crowds to gather at the site chosen by the Irish scientists. The ensuing pollution caused by cigarette smoke, car exhausts, etc. made measurements of atmospheric particles valueless.

The second event occurred almost four years later when, early in 1933, Professor J.J. Nolan and his chief technician, Mr Jack Hughes, were making measurements on atmospheric ions and conductivity in a cottage in what was then a rather remote valley at Glencree in the Wicklow mountains. They went up to the cottage well prepared for a stay of some days. However, a heavy snowfall cut off communications. Some people in Dublin, fearful for their safety, convinced Mr Eamon de Valera, the new Head of State, that they were in danger of death from hunger and exposure. He ordered the army out to rescue the professor and his technician. Time passed and having heard nothing, Mr de Valera, accompanied by his sons, Eamon and Vivian, set out to find out what was happening. At Enniskerry they found the army lorry which had returned there after getting stuck on the road up to the cottage. Mr de Valera and his sons pressed on and later reached the cottage. And, far from finding Professor Nolan and Mr Hughes suffering badly from the cold, found them instead in front of a roaring kitchen fire entertaining with a bottle of whiskey the two army officers who had made it up on foot.

Under the guidance of J.J. and P.J. the atmospheric research group gained in reputation and became internationally known as the 'Nolan School'. Together the Nolan brothers published over 80 scientific papers and it is a measure of the quality of their work that papers in this area of research being published today still quote the work of the Nolans, which in some cases had been published over sixty years ago.

Further reading:

Professor J.J. Nolan, *Nature,* **169**, 1036, 1952.
Professor J.J. Nolan, *Studies,* **41**, 317–22, 1952.
P. J. Nolan: The Photoelectric Nucleus Counter, *Scientific Proceedings of the Royal Dublin Society, Series A*, **4** (12), 161–80, 1972.

Tony Scott, Department of Physics, University College, Dublin.

DOROTHY PRICE Physician 1890–1954

Born: Clonskeagh, Co. Dublin, 8 September 1890.
Died: Dublin, 30 January 1954.

Family:
Third child of Jemmeth Stopford, accountant to the Irish Land Commission, and his wife Constance Kennedy, daughter of Dr Evory Kennedy, a Master of the Rotunda Hospital.
Married: Liam Price, DJ, in 1925.

Dorothy Stopford was born at Newstead, Clonskeagh, County Dublin. When she was twelve her father died and the family moved to London where she was educated at St Paul's Girls School. For some years after leaving St Paul's, Dorothy was satisfied to work with the Charity Organisation Society. And when, eventually, she sought admission to Dublin University it was to study medicine rather than history.

At the Meath Hospital she came under the influence of Professor William Boxwell, a grandson of Dr William Stokes. Boxwell interested her in research which gradually eclipsed her political activities – she had supported Ireland's demand for independence. The direction which her enquiries were to follow was determined, after graduation in 1921, by the realisation when working in Kilbrittain, County Cork, that her teachers had taught her very little about tuberculosis in children. Within a few years she had established a medical practice in Dublin and was one of the founders of St Ultan's Hospital for infants.

Almost casually, in the course of practice, Dr Price was introduced to the tuberculin test, a skin test which when positive indicated that a tuberculous infection had occurred. To inform herself more fully about tuberculosis, Dorothy Price learned German in order to read text books in that language, and she visited clinics in Germany, Sweden and Denmark. She was appointed children's specialist to the Royal City of Dublin Hospital, Baggot Street, in 1932 and ten years later published a useful book, *Tuberculosis in Childhood*.

By then it was clear to Dr Price how many Irish children lived in homes where tuberculosis was an inescapable risk. They became infected at an age when resistance was minimal. Hoping to create increased immunity in individuals she had espoused the method of vaccination with an attenuated tubercle bacillus developed in France by A. Calmette and C. Guerin (Bacille Calmette et Guerm, 'BCG'). Her advocacy of this preventive method was not unopposed but her pioneer venture of injecting 35 children with BCG at Ultan's in 1936 led to Dublin Corporation's BCG scheme in 1948. In the following year she became chairperson of a National BCG Committee. The incidence of lethal tuberculosis has since fallen in Ireland due to a number of factors which include the wide application of BCG. Dr Dorothy Price became seriously ill in 1950 and died on 30 January 1954.

Further Reading:

Obituary in *Journal of the Irish Medical Association*, **34**, 84, 1954.
Liam Price: *Dr Dorothy Price*, Dublin, 1957.

J.B. Lyons, Department of the History of Medicine, Royal College of Surgeons in Ireland, Dublin.

131 JOSEPH DOYLE Botanist

Born: Glasgow, 25 March 1891.
Died: Dublin, 11 April 1974.

Joseph Doyle in conversation with Eamon de Valera in the Royal Irish Academy about 1965

Family:
Married: Elizabeth Leonard, a lecturer in the Department of Botany, University College Dublin, 1919.
Children: Three sons and one daughter. His sons all became medical doctors; the youngest, J. Stephen, was Professor of Medicine in the Royal College of Surgeons in Ireland. His daughter, Mary Helen (married M.W. O'Reilly), a lecturer in the Department of Botany, UCD from 1945 to 1962, obtained a PhD on conifer wood structure in 1948. She died in 1987.

Addresses:

1920–1924	20 Victoria Avenue, Donnybrook
1924–1926	40 Marlborough Road, Donnybrook
1926–1933	6 Achill Road, Drumcondra
1933–1956	69 Anglesea Road, Ballsbridge
1956–1966	6 Park Drive, Rathmines
1966–1972	14 St Kevin's Park, Dartry
1972–1974	'Carrigower', Mount Anville Road

Distinctions:
Member (1927), on Council at various periods from 1931 onwards, Secretary (1957–63), President (1964–66), Vice-President (1966–67) of the Royal Irish Academy; Boyle Medal of the Royal Dublin Society 1942; President of the Botany Section, British Association for the Advancement of Science Meeting in Dublin 1957; Vice-President of the Royal Dublin Society 1968–74.

Joseph Doyle graduated in natural science at University College Dublin in 1911 and, after a period of graduate research in Hamburg, joined the staff of the Biology Department in 1913. In 1924 he was appointed Professor of Botany in succession to J. Bayley Butler (who was simultaneously appointed Professor of Zoology in succession to George Sigerson). Thereafter Doyle was responsible for organising an independent Botany Department and directing it until his retirement in 1961, when he was succeeded by Phyllis Clinch.

Over the period 1916–72 he published a long series of research papers on the reproductive biology of conifers, which established his international scientific reputation. Conifers are woody seed plants, widespread in northern and southern hemispheres. Their reproductive biology had, in general outline, been determined in the 19th century, but many details still remained obscure until recent times. In common with all living seed plants, the dominant vegetative plant produces spores, which develop into microscopic structures bearing sex organs: female eggs in ovules and male gametes in pollen grains. Only the pollen grains are released from the spore-bearing plant. They are borne in the air by wind currents to the retained female eggs and, after fertilisation, a complex series of transformations occurs in the young embryo before a new spore-bearing plant is formed and released eventually as a seed. The study of conifer embryogeny engaged Doyle throughout his scientific career. It called for great patience in obtaining the various stages of embryo development: these could only be investigated by the preparation of large numbers of thin sections of material which had been chemically arrested and stained for observation of cellular development.

Doyle and his students in the 1930s devoted attention to pollination mechanisms in conifers: the reception of air-borne pollen by the ovule. In 1945 he summarised his own and pre-existing observations in a review paper 'Developmental Lines in Pollination Mechanisms in the Coniferales' (*Scientific Proceedings of the Royal Dublin Society*, **24**, 43–62), which is still considered to be the definitive statement in modern textbooks. The stages subsequent to pollination particularly engaged Doyle and several collaborators over many years. Making use of the outstanding collections of mature conifers at Glasnevin, Kilmacurragh, Powerscourt, Rostrevor and elsewhere, they described the embryogeny of a series of southern-hemisphere conifers – *Athrotaxis, Callitris, Fitzroya, Podocarpus, Saxegothaea*. This new information supplied an important perspective for existing knowledge on northern-hemisphere genera, especially *Pinus*, which had hitherto provided the staple textbook examples of conifer reproductive biology. Doyle soon came to the conclusion that some of these southern-hemisphere conifers, such as *Podocarpus andinus*, showed a simpler scheme of embryological development than that found in *Pinus*. A masterly summary of the evidence was presented in 'Pro-embryogeny in *Pinus* in Relation to that in Other Conifers – A Survey' (*Proceedings of the Royal Irish Academy*, **62** (13), 181–216, 1963). It is a lasting legacy of his research that the great variability in embryological events found in conifers may now be seen more clearly as modifications of a relatively simple basal plan. Many conifers, during the course of development of a single fertilised egg, show a proliferation of embryos (often dozens); this phenomenon is known as 'cleavage polyembryony'. To this topic, with the experience of over fifty years, Doyle turned in his last two-part paper with his colleague Martin Brennan, SJ: 'Cleavage Polyembryony in Conifers and Taxads – A Survey' (*Scientific Proceedings of the Royal Dublin Society*, **A4** (6, 10), 1971/2). Here they considered the wide array of existing evidence, and concluded that the absence of cleavage of an embryo was the more primitive condition from which cleavage arose independently in various evolutionary lines of conifers. Once again *Pinus*, for long treated as typical of conifers embryologically, was considered to be a specialised and derived type, not a prototype.

Almost all of Doyle's research papers on conifer embryology were published by the Royal Dublin Society, of which he was a devoted member, serving on many sub-committees over a long period. In the first half of this century the Society's *Scientific Proceedings* achieved international eminence botanically for three remarkable series of papers: those of Henry H. Dixon FRS and his collaborators in Trinity College Dublin on plant physiology, of Paul A. Murphy and colleagues (including Phyllis Clinch) in University College Dublin on plant pathology, and of Doyle and his collaborators on conifer reproduction.

Since the 1940s, Doyle had repeatedly urged the Government to establish a National Aboretum. In 1963, supported by the Royal Dublin Society, he published cogent scientific and cultural arguments in its favour. In quite unforeseen circumstances, his vision was subsequently accepted and realised as an appropriate national tribute to the memory of John F. Kennedy.

James White, Department of Botany, University College, Dublin.

JAMES MARTIN Engineer and Inventor 1893–1981

Born: Killinchy Woods, Crossgar, Co. Down, 11 September 1893.

Died: Southlands Manor, Denham, Bucks, 5 January 1981.

Family:
Father: Thomas, farmer, died 1895; Mother: Sarah Coulter.
Married: Muriel Haines, 1942.
Children: Twin sons, two daughters.

Addresses:

1893–1924	Crossgar
1924–1929	London
1929–1981	Denham

Distinctions:
Order of the British Empire 1950; Wakefield gold medal, Royal Aeronautical Society 1951; Commander of the British Empire 1957; Barbour Air Safety Award 1958; Cumberbatch Air Safety Trophy 1959; Royal Aero Club gold medal 1964; Knighted 1965; DSc The Queen's University of Belfast 1970; DSc College of Aeronautics, Cranfield 1975.

James Martin inherited his inventing streak from his father, who used self-designed agricultural implements. Refusing an offer at Queen's University, he went to London in 1924 to establish his own business designing specialised vehicles. In 1929 he formed the Martin Aircraft Works to build aeroplanes, and was joined in 1934 by his friend Captain Val Baker who acted as chief test pilot. Baker was unfortunately killed in 1942 when the MB3, built to replace Spitfires and Hurricanes, crashed. Martin was now determined to find ways of saving pilots' lives. One successful invention used during World War II was the barrage balloon cable cutter which severed the thick steel cable using an explosive device on the aeroplane wing.

Aeroplanes, now faster and often jet-propelled, made escape 'over the side' impossible. In 1944 the Ministry of Aircraft Production sought his help. He conceived the idea in which both pilot and seat were forcibly ejected using a two-stage explosion. The first 'live' shot from a plane was in June 1946. Within a year Martin-Baker ejector seats were fitted in all new British planes. After solving the problem of ground-level ejection in 1955, MB seats were fitted in all US carrier and NATO planes. By February 1983 the 5,000th life was saved with this seat and its use is now universal. Martin continued to develop ejector seats for higher speeds, greater altitudes, vertical take-off, multiple crew escape and even underwater ejection. MB produce escape systems involving canopy jettison, automatic life-rafts and life vest inflation units for more than 400 types of aircraft. His two sons run the present very successful company.

Further reading:

J. Jewell: *Engineering for Life: The Story of Martin-Baker*, Martin-Baker Aircraft Company, Denham, 1979.
W. Garvin and D. O'Rawe: *Northern Ireland Scientists and Inventors*, Blackstaff Press, Belfast, 1993.

Wilbert Garvin and Des O'Rawe, The Queen's University of Belfast.

ERNST JULIUS ÖPIK Astrophysicist 1893–1985

Born: Port Kunda, Estonia, 23 October 1893.
Died: Bangor, Co. Down, 10 September 1985.

Family:
Children: Uno, Helgi, Tiiu, Elina, Inna, Maija.

Addresses:
30 College Hill, Armagh
99 Clifton Road, Bangor, Co. Down

Distinctions:
Member of the Royal Irish Academy 1954; Medals: National Academy of Sciences 1960, Meteoritical Society 1968, American Association for the Advancement of Science 1972, Royal Astronomical Society 1975, Astronomical Society of the Pacific 1976; Honorary degrees: Belfast 1968, Sheffield 1977.

E.J. Öpik was educated at Tallinn High School and Moscow Imperial University. After four years at Moscow Observatory he became Director of the Astronomy Department, Tashkent. From 1921–1944 he was an Associate Professor at Tartu University, and from 1930–1934 visiting scientist at Harvard University. As a former volunteer to the White Russian army, he vehemently opposed the Bolshevik Revolution and, when Soviet occupation of Estonia was imminent, he moved, first to Hamburg, and lastly, in 1948, to Armagh Observatory, where he remained until 1981.

Öpik was one of the most outstanding astrophysicists of his generation, with wide-ranging interests in the physical sciences. Among his many pioneering discoveries were: (1) the first computation of the density of a degenerate body, namely the white dwarf 40 Eri B, in 1915; (2) the first accurate determination of the distance of an extra-galactic object (Andromeda Nebula) in 1922; (3) the prediction of the existence of a cloud of cometary bodies encircling the Solar System (1932), later known as the 'Oort Cloud'; (4) the first composite theoretical models of dwarf stars like the Sun which showed how they evolve into giants (1938); (5) a new theory of the origin of the Ice Ages (1952). Öpik made many contributions to our knowledge of the minor bodies of the Solar System, and founded the meteor research group at Harvard. His statistical studies of Earth-crossing comets and asteroids are fundamental to our understanding of the motions of these objects and how they impact on Earth. His predictions of cratering on Mars were dramatically confirmed fifteen years later by planetary probes. In recognition of his work, Minor *Planet Öpik* was named after him. Öpik was prolific in his output and often controversial in his opinions. Many of his later publications appear in the *Irish Astronomical Journal*, which he edited from 1950 until 1981.

Further reading:
Irish Astronomical Journal, **10**, Special Issue dedicated to Öpik (contains ten papers in remembrance of Öpik), 1972.
Irish Astronomical Journal, **17,** No 4, 1986.
Quarterly Journal of the Royal Astronomical Society, **27,** 508, 1986.

John Butler, Armagh Observatory, N. Ireland.

EDWARD CONWAY Physiologist, Biochemist and Mathematician

Born: Nenagh, Co. Tipperary, 3 July 1894.
Died: Dublin, 29 December 1968.

Family:
Married: Mabel Edith Hughes, 2 August 1934.
Children: Four daughters. His daughter Dorothy is also a biological scientist and was on the staff of the Department of Physiology, University College Dublin.

Addresses:
1922–1934 Alexander House, 26 Gilford Road, Sandymount
1934–1968 Woodbank, St George's Avenue, Killiney, Co. Dublin

Distinctions:
Member of Royal Irish Academy 1939; Fellow of the Royal Society 1947; Honorary DSc, Dublin University 1952; Honorary Fellow, Royal College of Physicians 1953; Fellow of Royal Institute of Chemistry 1957; Member of Academic Septentrionale, Paris 1958; Fellow of Royal Society of Arts 1958; Member of New York Academy of Sciences 1960; Member of Pontifical Academy of Sciences 1961; Boyle Medal of Royal Dublin Society 1968.

After matriculating in 1912, Edward Joseph Conway entered the medical school in University College Dublin where he also took a two-year course in physiology before graduating in 1921. Thereupon he joined Professor J.M. O'Connor's staff in the department of physiology. After taking his DSc in 1927 he spent a year in Professor Embden's laboratory in Frankfurt. In 1932 he was appointed to the newly-founded chair of biochemistry and pharmacology in University College Dublin, and he established an exceptionally successful research laboratory which attracted support from the Rockefeller Foundation, the US National Institute of Health and the US Air Force as well as the Medical Research Council of Ireland.

His research work began with an investigation of renal tubular function. Because of the necessity to make innumerable estimations on very small volumes of fluid, he had to develop his own methods. So successfully did he overcome this difficulty that his microburette and diffusion unit became a standard method of microanalysis for a generation.

The structural unit in the living organism is the cell. The contents of the cell are different from those of its environment. Molecules, e.g. proteins, not normally present in this environment, are present in the cell; or substances common to both media are present in different concentrations. Most cells have a high internal concentration of potassium salts. With chloride, sodium is the main ionic constituent of fluids surrounding the cells.

In 1941 Conway explained satisfactorily how these differences between fluids inside and outside the cells are establish and maintained. First the cell membrane is selectively permeable; it is a barrier restraining the movement of some but not all ions and water-soluble substances. Second, active transport (i.e. movement of an ion or molecule against an electrochemical gradient requiring the continuous expenditure of energy) places restraints on the migration of ions. As a result, a resting or membrane potential develops. Building on this work, Alan Hodgkin and Andrew Huxley explained the mechanism of conduction of the action potential in excitable tissues, an achievement which earned them the Nobel Prize in 1963.

The separation of positive and negative charges, in the form of protons (hydrogen ions) and electrons, is a fundamental biological process. That it might be related to the oxidation and reduction when cells respire was first surmised by Lund in 1928. In 1948 Conway suggested that the protons released by this reaction taking place in cells during metabolism of glucose in the presence of air might be the source of hydrochloric acid in gastric juice. Accumulating knowledge of the structure and chemistry of the mitochondrion (one of the constituents of cells) has allowed the present generation to take a broader view. Peter Mitchell's 'chemiosmotic' hypothesis, recognised by the award of the 1978 Nobel Prize, today suggests that the free energy of the electrons moving along the respiratory chain is used to transport hydrogen ions outward across the mitochondrial membrane setting up a concentration gradient. The potential energy stored in this gradient is used in the synthesis of the energy rich adenosine triphosphate.

The mathematical approach to biological problems was a feature of all Conway's work, from his early investigation of renal tubular function to his formidable theoretical analysis of the salinity of the ocean. With his abiding interest in the composition of body fluids, it was to be expected that he would examine critically the wild surmise that blood plasma was an oceanic remnant. Macallum's theory, published in 1926 in *Physiological Reviews*, suggested that when early vertebrate forms emerged from the ocean the composition of sea-water and blood plasma were similar. In 1941 Conway began a series of papers on the palaeochemistry of the ocean, and by combining a mathematical approach with a remarkably wide range of information, he showed that at the relevant geological periods sea-water had thrice the salinity of mammalian plasma and qualitatively it was very different; the romantic notion that Silurian seas still circulate in our blood stream is without scientific foundation.

Further reading:

M. Maizels: Edward Joseph Conway, 1894–1968, *Biographical Memoirs of Fellows of the Royal Society,* **15**, 69–82, 1969.

C.S. Breathnach: Edward Joseph Conway, 1894–1968, *Irish Journal of Medical Science,* **163**, 366–369, 1994.

Caoimhghín Breathnach, Department of Physiology, University College, Dublin.

135 JAMES DRUMM Chemist and industrial Technologist

Born: Dundrum, Co. Down, 25 January 1896.
Died: Dublin, 18 July 1974.

Family:
Married: Mary O'Reilly, 4 July 1928.
Children: Two sons, both of whom are graduates in Engineering from the National University of Ireland.

Addresses:

1905–1912	Derrygooney, Co. Monaghan
1912–1928	Dr Drumm resided at various addresses in Dublin, England and Belgium
1929–1974	70 Rathgar Road, Dublin 6

Distinctions:
DSc, NUI 1931; Member of Senate NUI 1934–59; Vice–President Federation of Irish Industries 1935; Member Board of Emergency Scientific Research Bureau 1941–45.

Most people if asked to associate a name with an electric cell would probably mention Leclanche or perhaps Daniell, but what many do not know is that there has also been a tradition of battery making in Ireland, with Nicholas Callan in Maynooth in the middle of the nineteenth century and, almost 70 years later in Dublin, J.J. Drumm.

The Drumm batteries were what would now be called alkaline cells, in which the positive plates were nickel oxide mixed with graphite, while the negative plates were very pure nickel. The electrolyte used was a composite of potassium and lithium hydroxide and zinc oxide. The advantage that these batteries had over others lay in the fact that they were capable of being charged at very high rates and as such were suitable for use for traction purposes. An interesting feature of later versions of the battery was that the plates were mounted horizontally rather than in the conventional vertical position.

In 1930 the Drumm battery was mounted in a demonstration rail coach built by the Great Southern Railways Company – a forerunner of the present Coras lompair Eireann – and a series of trials was carried out. As a result of these tests Dáil Eireann voted sums of money to finance the building of two of what later became known as 'Drumm Trains'. The first gave a demonstration run in November 1931, while the second entered service on the Dublin-Bray line in 1932. These trains contained a number of interesting developments including regenerative braking, which is a feature of the modern Dublin suburban electric trains.

In 1934 the batteries were mounted in breadvans for a number of Dublin bakeries, but the performance of the batteries did not reach the level expected, due to leakage of charge. They were also mounted with more success on a $2\frac{1}{2}$ ton lorry. This lorry was regularly able to cover 45 miles a day, with a one hour boost charge in the middle of the day while the driver had his lunch!

It was with the trains however that the successes were achieved. These trains, each of which could carry over 130 passengers, could travel up to 80 miles on a single charge with a maximum speed of 47 mph. By 1935 one train had completed over 95,000 miles and by the middle of 1945 the four trains (two more were built in 1938/39) had covered more than three-quarters of a million miles mainly on the Harcourt Street–Bray line, with the batteries being charged at both stations at the end of each run.

Great Southern Railways demonstration coach Train (photos courtesy of CIE)

By 1949, when the batteries had reached the end of their useful working life, the trains were withdrawn, as the cost of replacement, particularly the nickel, had become very expensive. The arrival on the scene of the diesel engine was another factor in the demise of the Drumm train.

One of the four 'Drumm Trains'
Note the battery mounted underneath

Dr Drumm was the first to import stainless steel into Ireland and he had to develop techniques, such as welding, for working with this material. His requirement of oxygen-free nickel for the batteries lead to the development of a method for producing it.

He was a great scientist and brilliant innovator who served Ireland well over a long period of time.

Further reading:

J.M. Fay and J.J. Drumm: Railway Electrification – Potentialities of the Drumm Battery, *Transport,* 8 July 1933.
New Drumm Battery-Driven Train on Trial, *Modern Transport,* 19 August 1933.
Coireall Mac An tSaoir: Traein Leictreachais i 1930, *Inniu,* 5 January 1973.

Tony Scott, Physics Department, University College, Dublin.

136 ROBERT (ROY) CHARLES GEARY Mathematical Statistician and Economist

Born: Dublin, 11 April 1896.
Died: Dublin, 8 February 1983.

Family:
The oldest child in the family of two sons and two daughters of Ned Geary, a Dublin civil servant, and Jennie O'Sullivan. Two of his cousins on his mother's side received honorary doctorates. Married: Mida Maura O'Brien 1927.
Children: Colm and Clodagh (Dooney).

Addresses:

1896–*ca*1923	25 Upper St. Brigid's Road, Drumcondra, Dublin
*ca*1923–*ca*1925	74 Drumcondra Road, Dublin
*ca*1925–*ca*1927	2 Iona Park, Glasnevin, Dublin
*ca*1927–1936	3 Parnell Square, Dublin
1936–1966	27 Leeson Park, Dublin
1966–1983	12 Court Flats, Wilton Place, Dublin

(He lived in Paris 1919–21, Cambridge 1946–47, and New York 1957–60.)

Distinctions:
President of the Statistical and Social Inquiry Society of Ireland 1946–50; Fellow of the Econometric Society 1951; Council Member 1962–4; President of the International Statistical Institute; Honorary Fellow of the Royal Statistical Society 1957; Vice-President of the Royal Irish Academy 1963–5; Honorary Fellow of the American Statistical Association; Chairman of Council, International Association for Research in Income and Wealth 1961–7; Honorary Doctorates National University of Ireland 1961, The Queen's University of Belfast 1968, Trinity College Dublin 1973; Boyle Medal of the Royal Dublin Society 1981.

The education of Ireland's greatest statistician, Roy Geary, began at the O'Connell Schools in Dublin and continued at University College Dublin, where he graduated with a first class BSc in 1916, and at the Sorbonne in Paris, 1919–21. He spent most of his working life in Dublin – from 1923 to 1949 in the Statistics Branch of the Department of Industry and Commerce, from 1949 to 1957 as director of the newly established Central Statistics Office, and from 1960 to 1966 as director of the Economic Research Institute. He remained there (it is now the Economic and Social Research Institute), as consultant, from 1966 until his death.

His international reputation rests mainly on his contributions to mathematical statistics, although he also contributed significantly to national accounting and headed the National Accounts Branch of the Statistical Office of the United Nations in New York from 1957 to 1960. His best known theoretical papers were published between 1925 and 1956, although he continued to publish theory much later into the 1960s and 1970s. While his work was almost always motivated by problems of practical importance, it contains many elegant and theoretically attractive results, e.g. that independence of mean and variance implies normality and that maximum likelihood estimators minimise the generalised variance for large samples.

A systematic analysis of his publications reveals three major identifiable though related strands. First, there was his continuing interest in the sampling theory of ratios and his 1930 derivation of the density of the ratio of two normal variates remains one of the few essential references in the field. The second major theme was testing for normality and enquiring into the robustness of inferential methods which depended formally on normality. The work is exemplified by his 1935 *Biometrika* paper where a suggested test is based on the ratio of the mean deviation to the standard deviation. The third major theme concerned the estimation of relationships where the variables are subject to errors of measurement. His influential 1949 paper has led to his being regarded, with Reiersol, as the leading pioneer of the now standard technique of instrumental variables, these being introduced by him in the context of errors in variables.

Roy Geary with daughter Clodagh Dooney and grand-daughter Jill Dooney at Trinity College Dublin in 1973, when he was conferred with an honorary doctorate

Later in life he turned more towards practical economics and the direct application of statistical methods. Perhaps this reflected his position as the first director in 1960 of the Economic Research Institute, founded to meet the need for more economic research in Ireland and to enlarge the knowledge of the economic and social conditions of Irish society. Among his first tasks as director, apart from administration, was a study of inter-industry relationships based on input-output data and a study of the Irish woollen industry. It is now generally accepted that he established a spirit of independent inquiry in the Institute from the beginning and that his own international reputation was of great significance in building its initial successes and reputation.

While his zest for work never diminished – he wrote more than half of his 112 published papers after the age of 65 – he had many other interests. Among his lifelong loves were music, the theatre, reading, humour, soccer and children. He greatly admired France, regarding the French as the most civilised of nations. He longed for more rationality in both Irish and international politics and in public affairs. He carried over his high standards of work to all areas of his life and was unsparing of his time with young research workers, many of whom were greatly encouraged by him.

Further reading:

D. Conniffe (ed.): *Roy Geary, 1896–1983: Irish Statistician*, Centenary lecture by John E. Spencer and Associated Papers, Oak Tree Press in association with the Economic and Social Research Institute, Dublin, 1997.

R.C. Geary: The Frequency Distribution of the Quotient of Two Normal Variates, *Journal of the Royal Statistical Society,* **43**, 442–6, 1930.

R.C. Geary: 'The Ratio of the Mean Deviation to the Standard Deviation as a Test of Normality'. *Biometrika,* **27**, 310–32, 1935.

R.C. Geary: Determination of Linear Relations between Systematic Parts of Variables with Errors of Observation the Variances of Which are Unknown. *Econometrica,* **17**, 30–58, 1949.

J. R. N. Stone: Robert Charles Geary (1896–1983), in J. Eatwell, M. Milgate and P. Newman (eds), *The New Palgrave: a Dictionary of Economics, Vol. 2*, 491, London, 1987.

John E. Spencer, Department of Economics, The Queen's University of Belfast.

137 JOHN LIGHTON SYNGE Mathematician

Born: Dublin, 23 March 1897.
Died: Dublin, 30 March 1995.

Family:
Synge's father Edward, brother to the dramatist John Millington Synge, was a land-agent, descended from a well-established Church of Ireland family, at least four of whom held bishoprics in the seventeenth and eighteenth centuries. One of these, Hugh Hamilton, before becoming Bishop of Ossory, was a Fellow of Trinity College Dublin, who wrote a book on the geometry of conic sections. Synge's mother, Ellen Price, was the daughter of James Price, a distinguished Irish engineer. Synge married Elizabeth Allen in 1918; they had three daughters, one of whom, Cathleen Morawetz, is a distinguished mathematician who has served as President of the American Mathematical Society and was awarded the US National Medal of Honor.

Addresses:
Synge lived in Toronto, Canada, from 1920 to 1925 and from 1930 to 1943, and in the US, in Columbus, Ohio, and in Pittsburgh, from 1943 to 1948, apart from a brief interlude during 1944 and 1945 in London and Paris as a ballistics mathematician with the US army air force. His Irish addresses were:

1897–1903 Rathe House, Kingscourt, Co. Cavan
1905–1913 3, Bayswater Terrace, Sandycove, Co. Dublin
1913–1916 Knockroe, Sydenham Road, Dundrum, Co. Dublin
1916–1918 TCD, House No. 26 and subsequently No. 16
1918–1920 118, Pembroke Road, Dublin
1925–1930 58, Marlborough Road, Donnybrook, Dublin
1948–1993 Torfan, Stillorgan Park, Blackrock, Co. Dublin.
1993–1995 Newtownpark Nursing Home, Blackrock, Co. Dublin

Distinctions:
Fellow of Trinity College Dublin and University Professor of Natural Philosophy 1925; Member of the Royal Irish Academy 1926; Professor of Applied Mathematics, University of Toronto, 1930; Professor of Mathematics, Carnegie Institute of Technology, 1946; Senior Professor in the School of Theoretical Physics of the Dublin Institute for Advanced Studies 1948; Fellow of the Royal Society 1943; President of the Royal Irish Academy 1961–1964; Tory Medal of the Royal Society of Canada 1943; Boyle Medal of the Royal Dublin Society 1972; Honorary Fellow of Trinity College Dublin 1954; Honorary degrees from the University of St Andrews (1966), The Queen's University of Belfast (1969) and the National University of Ireland (1970).

J.L. Synge was undoubtedly the most talented and distinguished Irish mathematician of his generation. He showed great originality and versatility in the application of mathematics to a wide variety of problems: his natural curiosity and his fascination with analysis and mathematical description and solution of problems remained with him and were still evident right up to his death at the age of 98. Geometry was central to his approach to mathematics – he liked to visualise things and had a unique capacity to see and express mathematical ideas in geometric form.

In 1920, the year after he graduated from Trinity, Synge moved to the University of Toronto as an Assistant Professor. The following year, the American Mathematical Society held a meeting in Toronto, an event of considerable importance in his mathematical development as it gave him the opportunity to meet many of the leading North American mathematicians. He had begun to take an interest in relativity: this was stimulated by meeting Veblen and Eisenhart from Princeton and discovering that they and their colleagues had also focused their attention on Einstein's theory.

Relativity remained a central interest for Synge. Although he always maintained a broad and open approach in his choice of research topics, and would have insisted that he was a mathematician, not a relativist in a specialist sense, it is his work in relativity which is most widely known and it was in relativity that his influence was most significant. His two books on relativity – *The Special Theory* and *The General Theory*, published in 1956 and in 1960, with their original and highly distinctive geometrical approach, profoundly influenced and shaped a whole generation of students of relativity and cosmology. Many of these came to work with him as Scholars or Visiting Professors at the Institute for Advanced Studies in Dublin.

Synge had little time for the narrowly abstract and formal approach to mathematics which he saw developing during his lifetime. He worried that 'a mathematics which feeds solely on itself will in time exhaust its interest' and preferred, like his eighteenth and nineteenth century predecessors, 'to roam abroad and seek mathematical sustenance in problems arising directly out of other fields of science'. This was an unfashionable view when he voiced it fifty years ago, but today it would find a much readier acceptance.

Further reading:

A.W. Conway and J.L. Synge (eds): *The Mathematical Papers of Sir W.R. Hamilton, Volume 1 – Geometric Optics*, Cambridge, 1931.

J.L. Synge and B.A. Griffith: *Principles of Mechanics*, McGraw-Hill, New York, 1942 (Second Edition 1949, Third Edition 1959).

J.L Synge and A. Schild: *Tensor Calculus*, University of Toronto Press, 1949 (Revised 1952, 1956, 1964 – Republished by Dover 1978).

J.L. Synge: *Science: Sense & Nonsense*, Cape, London, 1951.

J.L. Synge: *Relativity, the Special Theory*, North-Holland, Amsterdam, 1956 (Second Edition 1965).

J.L. Synge: *Kandelman's Krim* (A Realistic Fantasy), Jonathan Cape, London, 1957.

J.L. Synge: *Relativity, the General Theory*, North-Holland, Amsterdam, 1960

David Spearman, Trinity College, Dublin.

138 JOHN DESMOND BERNAL Physicist, Marxist, Science Populariser

Born: Near Nenagh, Co. Tipperary, 10 May 1901.
Died: London, 15 September 1971.

Family:
The Bernal family has Sephardic Jewish origins and was based in Limerick from the 1840s. His father, Samuel, went to Australia in 1884, worked a sheep-farm, and returned in 1898, buying the farm where Bernal was born. Bernal's mother, Elizabeth Miller, was American, with California and New Orleans background, and experience of Stanford and the Sorbonne. She converted to Catholicism to marry Samuel, the Bernal family having become Catholic in their Limerick period.
Married: Eileen Sprague in 1922.
Children: two sons – Michael (born 1926, physics Imperial College) and Egan (born 1930, farmer in Suffolk). Also Martin (born 1937 by Margaret Gardiner), and Jane (born 1953 by Margot Heinemann). Martin Bernal is the author of *Black Athena*, currently in Cornell. Jane is a consultant psychiatrist at St George's Hospital Medical School, London.

Addresses:
1901–1922 Family farm, Brookwatson, Portumna road, near Nenagh, Co. Tipperary
Subsequently Cambridge and London

Distinctions:
Fellow of the Royal Society 1937; Professor of Physics, Birkbeck College 1938; Scientific Adviser to Combined Operations under Lord Mountbatten 1943; President of the World Peace Council 1958–1963.

J.D. Bernal was initially educated locally, developing an early interest in science. He knew about the Birr telescope, and interacted with several local gentleman-amateur scientists, one of whom, Launcelot Bayly, introduced him to crystallography. His mother supported his scientific inclinations, researching the Irish educational opportunities, in the end sending him to boarding-school in England, whence in 1919 he went to Cambridge. He was, however, aware of what was going on in Ireland, and observed it sympathetically during his vacations, keeping a journal, which is archived in Cambridge.

In his Natural Science Tripos he took initially physics, chemistry, geology and mineralogy in 1922, then concentrating on crystallography, writing in his final-year papers on 'the vectorial geometry of space lattices', and 'the analytic theory of point systems'. This work drew him to the attention of Sir William Bragg, then setting up his research team on X-ray diffraction techniques. He worked with Bragg in the Royal Institution until 1927, designing the X-ray photogoniometer subsequently to be produced commercially as the standard domain tool.

Bernal went back to Cambridge in 1927 to a lectureship in structural crystallography, where for the next decade he worked on the structure of liquids, inventing the 'statistical geometric' approach to liquid modelling, and on solids of increasing complexity: pepsin, proteins, viruses, identifying the type of helical structures which subsequently led to the discovery of DNA, and to his appointment to the chair of physics at Birkbeck in 1938. He had little time to settle in, as the war started. He had been associated with the Cambridge Scientists Anti-war Group, criticising the government's civil defence policy, analysing the effects of bombs on cities, using scientific modelling and taking account of Spanish civil war experience. Thus began Bernal's involvement with the strategic application of scientific methods in the war, later to become known as 'Operational Research'. As adviser to Lord Mountbatten, he researched the Normandy beaches for landing troops and equipment, using historical evidence, geological knowledge, aerial photographs and hydrodynamics of wave motion. He also participated in the development of the Mulberry floating harbour. During his Cambridge period he had picked up experience of the interactions between science and government, with Marxist insight into the historical background, leading to his seminal *Social Function of Science* published in 1939. This was celebrated in 1964 in *The Science of Science* (Edited by Goldsmith and McKay) as the founding text for the scientific study of science itself in a social context.

After the war, back in Birkbeck, he built up the crystallography group, occupying the Chair from 1964 up to his retirement in 1968, though increasingly incapacitated from 1965 by a stroke. He had earlier supervised work by Rosalind Franklin, who subsequently contributed to the DNA work for which Watson and Crick received the Nobel Prize. After the war politically Bernal became isolated by the 'cold war' environment. He supported the World Peace Council and the nuclear disarmament movement, and was a prime mover in initiating the Pugwash conference, a channel between the USA and the USSR during the worst period of the cold war.

In his declining years, despite increasing communication difficulties, he continued to publish scientific papers up to 1969, his last being a letter to *Nature* (with Barnes, Cherry and Finney) on 'anomalous' water. Bernal had an integrated, egalitarian approach to science and to politics: for him the work of the technician and craftsman was as important as that of the scientist. He regarded this teamwork process as the model for the socialist society of the future, rather than the flawed state-centralist model in the east.

Further reading:

Dorothy Hodgkin: *Biographical Memoirs of Fellows of the Royal Society*, **26**, Dec 1980 (see also her paper in *Proceedings of the Royal Irish Academy,* **81B** (3), 1981).
Helena Sheehan: *Marxism and the Philosophy of Science, a Critical History,* Humanities Press International, 1985 and 1993.
Brenda Swann and Francis Aprahamian (eds): *J.D. Bernal: a Life in Science and Politics*, Verso, 1999.

Bernal's own publications include:

The World, the Flesh and the Devil, Cape, 1929.
The Social Function of Science, Routledge Kegan Paul (RKP), 1939.
The Freedom of Necessity, RKP, 1949.
Marx and Science, Lawrence and Wishart, 1952.
Science and Industry in the Nineteenth Century, RKP, 1953, Indiana University Press, 1970.
World Without War, RKP, 1958.
Science in History, Watts 1954, 1957 and 1965.
The Origin of Life, Wiedenfelt & Nicholson (W&N), 1967.
(Posthumously) The Extension of Man: Physics Before 1900, W&N, 1973.

Roy Johnston, P.O. Box 1881, Rathmines, Dublin 6.

139 PHYLLIS CLINCH Botanist and Plant Virologist 1901–1984

Born: Dublin, 12 September 1901.
Died: Dublin, 19 October 1984.

Family:
Father was James Clinch, Mother was Mary Powell, the daughter of Major Powell who is believed to be the character on whom Joyce modelled Major Tweedy, the father of the famous Molly Bloom.
She never married.

Addresses:
Grew up at Mariae on Leinster Road, Dublin
Lived at Lissarda, Granville Road, Foxrock, Co. Dublin

Distinctions:
DSc National University of Ireland for her published works on the potato viruses; Boyle Medal of the Royal Dublin Society for her work on Potato virus diseases 1961; Vice-President Royal Irish Academy 1975–1977; Vice-President Royal Dublin Society 1977.

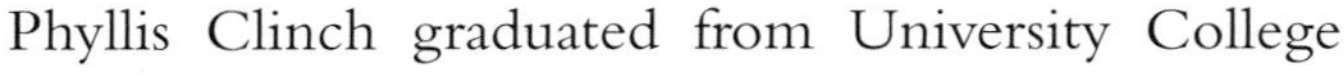

Phyllis Clinch graduated from University College Dublin with First Class Honours in Chemistry and Botany. She studied Plant Pathology under Professor V.H. Blackman at Imperial College London, was awarded her PhD in 1928, and that same year was appointed Assistant to the Professor of Biology in University College Galway. A year later she was appointed as a Research Assistant in the Department of Plant Pathology at the Albert College, University College Dublin, where she devoted her energies to the study of potato virus diseases. During the early thirties she spent six months visiting the eminent cytologist Alexandre Guillermond at the Sorbonne in Paris to advance her studies on cytological effects of potato viruses. In 1950 she was appointed by Professor Joseph Doyle as a statutory lecturer in Botany, a position she retained till she succeeded him to the Chair of Botany in 1961.

Professor Clinch played a key role in the transfer of the Botany Department to the new buildings at Belfield in 1964. As well as being a distinguished scientist, she had a great vision and understanding of the importance of cellular biology and she actively encouraged its development. Her years in plant pathology gave her a good understanding of microbiology and she encouraged the expansion of mycology in her department. Equally, when the opportunity arrived in the form of J.J. Moore SJ (later Professor), she again encouraged the development and expansion of the study of phytosociology.

She was, apart from her science, a keen golfer and bridge player. A reserved and conventional woman, she was extremely avuncular and kind-hearted: she was referred to by her students and many others as 'Auntie P'. She left an indelible mark on botanical studies in Ireland.

Matthew Harmey, Department of Botany, University College Dublin.

140 KATHLEEN LONSDALE X-ray Crystallographer

Born: Newbridge, Co. Kildare, 28 January 1903.
Died: London, 1 April 1971.

Family:
Tenth child of Harry Yardley, the postmaster at Newbridge, Co. Kildare, and his wife, Jessie Cameron. Kathleen's was a poor family, and four of her six brothers died in infancy. The girls were tougher, though Kathleen herself suffered from rickets as a child, and never grew tall. Kathleen's mother was a 'considerable character', a Baptist, with an independent mind, and the marriage was not too happy. The family moved to Essex in 1908.
Married: Thomas Lonsdale; three children.

Kathleen Lonsdale lectures at the Dublin Institute for Advanced Studies in 1943 (courtesy of the DIAS)

Addresses:
Newbridge, Co. Kildare; Ilford, Essex; London; Leeds; London

Distinctions:
In 1945, Kathleen broke new ground for women, by being the first of her sex to be admitted Fellow of the 285-year-old Royal Society. (Also admitted to Fellowship on 17 May was biochemist Marjorie Stephenson – they were both elected 22 March 1945.) She was created a Dame Commander of the Order of the British Empire in 1956. She was the first woman President of the British Association for the Advancement of Science in 1968, and she received many other honours.

After her family moved to Essex, Kathleen attended classes in physics, chemistry and higher mathematics at the County High School for Boys in Ilford, the only girl from the girls' school to do so. Anxious to progress, she entered Bedford College for Women in London at the age of sixteen, and came first in the honours BSc exam in 1922. Indeed she obtained the highest mark for ten years in her final examination in London University. She made good use of her small size by coxing the Bedford College eight, and she was also secretary of the Music Society.

On graduation, she was immediately offered a place in the research team of William Bragg, one of her examiners, then at University College London, where she began her work on X-ray diffraction of crystals of organic molecules. By an odd coincidence, the first compound she began to measure was succinic acid, the di-basic acid which had first been synthesised by Maxwell Simpson. In 1923 she transferred, with Bragg, to the Royal Institution, but her marriage to Thomas Lonsdale resulted in a move to Leeds, where he was working. She considered giving up science to become a good wife and mother, but Thomas would have none of it. He had not married, he said, to get a free housekeeper. Indeed, he served the additional useful function of cutting her unruly hair! So, while they had their three children, she progressed in her career. It was at Leeds that she made her most memorable contribution to chemical history, for she showed, by her analysis of hexamethyl benzene, in 1929, that the benzene ring was flat.

Her return to London interrupted her experimental work but, in 1931, Bragg offered her £200 to get help at home so that she could work with him. She calculated that it would cost £277 to replace her at home – so he came up with £300! Using a large electro-magnet she studied diamagnetic anisotrophy, and she was able to provide experimental verification of the postulated delocalisation of electrons and the existence of molecular orbitals – results of considerable importance to theoreticians.

E. Schrödinger, M. Born, K. Lonsdale, D. Hyde (President of Ireland), P. Ewald, and E. de Valera (Taoiseach) at a reception at Arus an Uachtaráin in 1943 (courtesy of the DIAS)

After the war, in 1946, she was appointed reader in crystallography at University College, London (UCL), and was Professor of Chemistry there from 1949 to 1968, developing her own research school, which dealt with a wide variety of topics, including solid state reactions, pharmacological compounds, and the constitution of bladder and kidney stones. While at UCL, she edited three volumes of *International Tables for X-ray Crystallography* (1952, 1959, 1962), the standard work in the field.

Meanwhile, she had become a Quaker and thus a pacifist, and this led to her imprisonment in 1943 for refusing to register for civil defence duties, and then refusing to pay the £2 fine – 'Do the police come for one or do I just have to go to prison myself?' Following her internment, she wrote to the Governor suggesting how conditions for the prisoners could be improved, and she later became a prison visitor.

She got out of prison in time to attend a 1943 scientific meeting at The Dublin Institute for Advanced Studies (DIAS). This was chaired by Nobel Prizewinner Erwin Schrödinger, and was attended by Eamon de Valera. While in Ireland she visited her birthplace at Newbridge.

Further reading:

Dictionary of National Biography.

Dorothy Hodgkin: Kathleen Lonsdale, *Biographical Memoirs of Fellows of the Royal Society*, **21**, 447–484, 1975.

Charles Mollan, Blackrock, Co. Dublin.

141 ERNEST THOMAS SINTON WALTON Physicist

Born: Dungarvan, Co. Waterford, 6 October 1903.
Died: Belfast, 25 June 1995.

Family:
Married: Winifred Isabel Wilson (Freda), 1934.
Children: Alan (physicist, Cambridge University), Marian Woods (retired Vice-Principal and Head of Science, Methodist College, Belfast), Philip (Professor of Applied Physics, NUI Galway) and Jean Clarke (formerly science teacher).

Addresses:

1903–1922	Manses in Dungarvan, Rathkeale, Castleblayney, Banbridge, Cookstown, Tandragee and Lurgan
1922–1926	Trinity College, Dublin
1926–1934	Trinity College, Cambridge
1934–1940	36 Merlyn Park, Ballsbridge, Dublin
1940–1992	26 St Kevin's Park, Dartry, Dublin 6
1992–1995	69 Osborne Park, Malone Road, Belfast BT9 6JT

Distinctions:
Fellow of Trinity College Dublin 1934; Member of the Royal Irish Academy 1935 (Secretary 1952–1957); Hughes Medal of the Royal Society 1938; Nobel Prize for Physics 1951; Honorary Fellow of the Institute of Physics 1987; Honorary DSc: The Queen's University of Belfast, University of Ulster, Gustavus Adolphus College (Minnesota) and Dublin City University.

E.T.S. Walton's parents were from Ulster but, as his father was a Methodist minister, the family frequently moved around Ireland from one manse to the next. After attending Methodist College, Belfast, Walton won an entrance scholarship to Trinity College, Dublin, gaining there a double first moderatorship (honours degree) in mathematics and experimental science, and an MSc for research into hydrodynamics. In 1927 John Joly recommended him to the great pioneer of nuclear physics, Ernest Rutherford. Under his supervision in the Cavendish Laboratory, Cambridge, Walton received a PhD in 1931, remaining there until 1934.

There he joined the legendary band of nuclear physicists who made 1932 an *annus mirabilis*. With John Cockcroft he grabbed the world's attention when he 'split the atom' (more properly its nucleus) by artificial means. His achievement realised the alchemist's age-old dream of transmutation in a dramatically new way. Rutherford had already in 1919 induced a nuclear reaction which changed nitrogen into oxygen using alpha particles (helium nuclei) from a radioactive source. Encouraged by Rutherford, Walton and Cockcroft built an accelerator in the Cavendish which provided far more particles than had been available previously. Furthermore, the particles, being accelerated protons, could induce nuclear reactions much more easily than alpha particles.

The accelerator was supplied with several hundred kilovolts from a voltage multiplier circuit designed and built by themselves (the design is still used today). Old equipment and even recycled pieces of wood and nails were used, as was standard practice there, and its successful operation stemmed directly from Walton's great manual ability. Part of the apparatus is now in the Science Museum in South Kensington.

Walton seated at his accelerator in Cambridge in 1932

On 14 April 1932, Walton, sitting in a small box, turned the proton beam onto a lithium target, and observed on a screen scintillations characteristic of alpha particles. These arose from lithium nuclei splitting apart when struck by protons. Walton loved to recount how, after Cockcroft came to look, Rutherford hurried over, as he always did when something important was happening, and how the two young physicists had to manoeuvre his imposing bulk into their little box. 'These look mighty like alpha particles', boomed Rutherford. 'I should know – I was at their birth'. 'Indeed', Walton would say, 'he should have added that he was at their christening also, for it was he who named them'. Walton himself, when inside the box on that day, was present at another birth – that of the era of accelerator-based experimental nuclear physics. Other achievements were a major verification of Einstein's mass-energy equivalence in relativity, $E = mc^2$, and the demonstration that, as predicted from Schrödinger's new theory of wave mechanics, the accelerator voltage needed was much lower than expected from classical considerations. In 1951 Walton and Cockcroft were awarded the Nobel Prize in Physics for this work.

Afterwards Walton could easily have gone to laboratories in the USA or elsewhere – during the second world war he was invited to join the Manhattan project for developing a nuclear bomb. Instead, he returned to Dublin in 1934 as a Fellow of Trinity College for the rest of his career. In 1947 he also became head of Physics and the Erasmus Smith's Professor. Funding was extremely limited and his duties were onerous. Nevertheless, his many lectures, often accompanied by practical demonstrations, were meticulously prepared and presented, and were an inspiration to many generations of students. Walton was a pacifist, becoming president of the Irish Pugwash Group, concerned with reducing the dangers of nuclear war. A modest, devout man, he never craved the great fame which he acquired. He possessed a keen mind, love of physics, and exceptional dexterity, qualities which remained with him throughout his long life.

Further reading:

Brian Cathcart: Ernest Walton, Atomic Scientist, in G. O'Brien and P. Roebuck, *Nine Ulster Lives*, Ulster Historical Foundation, 1992.

E. Finch and D. Weaire: *Physics World*, **8**, No. 11, 62–63, 1995.

Eric C. Finch, Department of Physics, Trinity College, Dublin 2.

WALTER HEITLER Physicist 1904–1981

Born: Karlsruhe, 2 January 1904
Died: Zürich, 15 November 1981

Walter Heitler, senior professor at the Dublin Institute for Advanced Studies 1945–49

Family:
Son of Adolf and Ottille (née Rudolf) Heitler.
Married: Kathleen Nicholson 1942.
Children: One son, Eric.

Addresses:

1941–1949	21 Seapark Road, Clontarf, Dublin
1949–1958	Drusbergstrasse 59, Zürich
1958–1981	Am Guggenberg 5, Zürich

Distinctions:
Membership of the following academies: Royal Irish 1943, Royal Society 1948, Leopoldina in Halle 1968, Mainz 1970, Norwegian 1974. Honorary doctorates of National University of Ireland 1954, University of Göttingen, University of Uppsala. Medals: Max Planck 1968, Marcel Benoist Prize 1970, Literaturpreis der Stiftung für Abendländische Besinnung 1977, Gold Medal of Humboldt Gesellschaft 1979.

Walter Heitler was born in Karlsruhe, Baden, Germany, of a Bohemian-Jewish family, nearly all of whom perished in the Nazi holocaust. His early education was classical but, at about the age of eleven, he began to develop a personal interest in the natural sciences. He studied at universities in Karlsruhe, Berlin and Munich, where he took his PhD degree under the supervision of Herzfeld. After a brief stay in Copenhagen he arrived in Zürich just a few months before Schrödinger left for Berlin in 1927. Having mastered Schrödinger's papers on quantum mechanics he set about applying them to calculate the Van der Waals interaction between two atoms. He collaborated with another research worker Fritz London, and the result was the Heitler-London theory of chemical bond in which the two electrons forming the bond are indistinguishable from each other.

In 1927 Heitler went to Göttingen as assistant to Max Born. When Hitler came to power in 1933, Heitler left Germany for Bristol where he remained until he transferred to Dublin in 1941. In the meantime he had been recognised as the world's leading authority on the quantum theory of radiation. In Dublin he devoted his energies to the theory of the newly discovered particle whose mass was about two hundred times that of electron – now called the muon. He gathered about himself an active group that included J. Hamilton, N. Hu, S.T. Ma, H.W Peng, S.C. Power and P. Walsh. Though Schrödinger and Heitler were together in Zürich and Dublin, their research interests at any time did not coincide. Heitler became an Irish citizen in 1946 and retained Irish citizenship when he left for Zürich in 1949.

One of Heitler's regrets in his latter years was that he had chosen to specialise in physics rather than in biology or philosophy. Of his seven books, four deal with philosophy and religion, and he died a member of the Swiss Reformed Church. The number of his scientific papers exceeds eighty. It was a matter of surprise to his contemporaries that Heitler was not awarded the Nobel Prize.

Reverend Monsigneur James McConnell, 1912–1999.

143 WILLIAM HUNTER McCREA Theoretical Astronomer 1904–1999

Courtesy of the Royal Astronomical Society

Born: Dublin, 13 December 1904.
Died: Lewes, 25 April 1999.

Family:
Eldest child of Robert Hunter McCrea and Margaret (née Hutton).
Married: Marian Nicol Core Webster in 1933.
Children: Two daughters and a son.

Addresses:

1904–1907	Boyne Villa, Oakley Road, Ranelagh, Dublin
1966–1999	87 Houndean Rise, Lewes, Sussex

Distinctions:
Member of the Royal Irish Academy 1938; Fellow of the Royal Society 1952; Royal Astronomical Society (RAS) Gold Medallist 1985; Knight Bachelor 1985; Freeman of the City of London 1988; Honorary doctorates from the National University of Ireland, The Queen's University of Belfast, Cordoba, Dublin & Sussex.

Bill McCrea was born in Dublin and, when he was two, his father, a teacher, moved to England. Educated at Chesterfield Grammar School, he won an Entrance Scholarship to Trinity College Cambridge where he earned high distinctions in mathematics. Graduating in 1926, McCrea studied for a year in Göttingen before returning to Cambridge to receive his PhD in 1929. After lecturing in Edinburgh and Imperial College London, he was appointed to the chair of mathematics at Queen's University, Belfast, in 1936. Although he held this post formally until 1944, he served in the Operations Research group of the British Admiralty from 1943 to 1945. After the War, he became Professor of Mathematics at the Royal Holloway College of the University of London, a post he held until 1966 when he was appointed Research Professor of Theoretical Astronomy at the University of Sussex. He retired in 1972.

McCrea made many fundamental contributions to mathematics, quantum mechanics, stellar astronomy and cosmology. His interest in astronomy started in Cambridge under R.H. Fowler. In Edinburgh he collaborated with G.C. McVittie in applying relativity to cosmology. With E.A. Milne, he explored the use of Newtonian methods in cosmology to provide simple derivations of relativistic results. Other investigations included the evolution of galaxies, mechanisms for forming planets and satellites, the formation of molecules in the interstellar medium, and the emission of neutrinos from the Sun.

Further reading:
D. McNally: Sir William McCrea at 90, *Irish Astronomical Journal*, **24**, 49–54, 1997.

Ian Elliott, Dunsink Observatory, Dublin 15.

144 E. ESTYN EVANS Geographer 1905–1989

Born: Shrewsbury, England, 29 May 1905.
Died: Belfast, 12 August 1989.

Courtesy of Lilliput Press

Family:
Father: Rev. George Owen Evans (1865–1919), Methodist Minister; Mother: Elizabeth Jones (1866–1945).
Married: Gwyneth Lyon Jones on 26 August 1931.
Children: David Wyn, Colin Ivor, Edwin Owen, and Alun Estyn.

Addresses:
Childhood years in Shrewsbury
1922–1925 Student years in Aberystwyth, Wales
1926–1928 Wiltshire
1928–1989 The Queen's University of Belfast

Distinctions:
Honorary Degrees: DLitt, Trinity College Dublin 1940; DLaws, The Queen's University of Belfast 1943; SCD, Bowdoin College, Maine, USA, 1949; DLitt, University of Wales 1974; DLitt, University College Dublin 1975; DSc, New University of Ulster 1980.
Other Distinctions: Fellow, Society of Antiquaries (London) 1931; Member, Royal Irish Academy 1934; Honorary Member, Royal Town Planning Institute 1950; Commander of the Order of the British Empire 1970; Victoria Medal, Royal Geographical Society 1973; Merit Award, American Geographical Society 1979; Honorary Member, Royal Dublin Society 1981; six Honorary Doctorates.

Estyn Evans graduated from the University College of Wales, Aberystwyth, in 1925 and assumed a lectureship at Queen's University Belfast in 1928. There he founded the first Department of Geography in Ireland and held the Chair from 1948 to 1968. His publications, such as *France* (1937), *Irish Heritage* (1942), *Mourne County* (1951, 1967), *Irish Folk Ways* (1957), *Prehistoric and Early Christian Ireland* (1967), and *The Personality of Ireland* (1973) won international acclaim. A keen promoter of archaeology and anthropology, he featured in Radio Éireann and BBC NI television programmes. As Visiting Professor at Bowdoin College, Maine (1948–1949), he studied the North American experiences of eighteenth century Scots-Irish emigrants. His long-standing concern for planning is evident in several papers he wrote on Belfast's development, as well as in *The Ulster Countryside*, published in 1947 by the government of Northern Ireland, of which he was the principal author. In 1955 he participated in the famous conference at Princeton, sponsored by Wenner Gren Foundation, on *Man's Role in Changing the Face of the Earth*. He was a key mover in establishing the Ulster Folk and Transport Museum which opened at Cultra in 1963. In 1970, he became the first Director of the Institute of Irish Studies at Queen's. E. Estyn Evans was an innovator, 'The Prof.' of Irish Geography, pioneer voice of environmental sensitivity, landscape heritage, and cultural history.

Further reading:
E.E. Evans: *Ireland and the Atlantic Heritage. Selected Writings,* Lilliput Press, Dublin, 1996.

Anne Buttimer, Department of Geography, University College, Dublin 4.

145 ERIC MERVYN LINDSAY Astronomer 1907–1974

Born: Portadown, Co Armagh, 26 January 1907.
Died: Armagh, 27 July 1974.

Family:
Seventh son of Richard and Susan Lindsay.
Married: Sylvia Mussells, Cape Town, 20 May 1935.
Children: Derek Michael, born 1944.

Address:
1937–1974 Armagh Observatory

Distinctions:
Member of the Royal Irish Academy 1939; Order of the British Empire 1963.

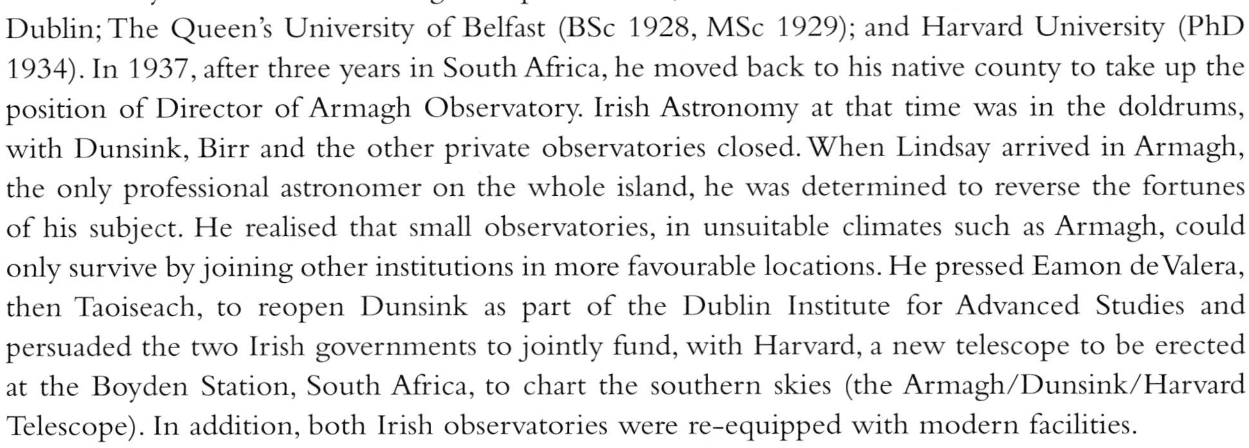

Eric Lindsay was educated at King's Hospital School, Dublin; The Queen's University of Belfast (BSc 1928, MSc 1929); and Harvard University (PhD 1934). In 1937, after three years in South Africa, he moved back to his native county to take up the position of Director of Armagh Observatory. Irish Astronomy at that time was in the doldrums, with Dunsink, Birr and the other private observatories closed. When Lindsay arrived in Armagh, the only professional astronomer on the whole island, he was determined to reverse the fortunes of his subject. He realised that small observatories, in unsuitable climates such as Armagh, could only survive by joining other institutions in more favourable locations. He pressed Eamon de Valera, then Taoiseach, to reopen Dunsink as part of the Dublin Institute for Advanced Studies and persuaded the two Irish governments to jointly fund, with Harvard, a new telescope to be erected at the Boyden Station, South Africa, to chart the southern skies (the Armagh/Dunsink/Harvard Telescope). In addition, both Irish observatories were re-equipped with modern facilities.

In 1954, after Harvard threatened to withdraw from Boyden, Lindsay, together with Herman Brück, persuaded Sweden, Belgium, Germany and the USA to join Ireland in the first international observatory at Boyden – a forerunner of the European Southern Observatory. In 1948, Lindsay brought to Armagh from a refugee camp in Europe a former colleague at Harvard, the renowned Estonian astrophysicist, Ernst Julius Öpik. Öpik's fundamental work in astronomy brought world-wide recognition to the small institution which Lindsay had, through his diplomatic skill, rescued from oblivion. Another achievement was the foundation of the Planetarium in Armagh, the first in Ireland and one of only two in the UK at that time.

Though Lindsay made no great astronomical discoveries, his influence on the politicians of his day and their adoption of his imaginative schemes were crucial to the renaissance of Irish Astronomy in the second half of the Twentieth Century.

Further reading:
J.A. Bennett: *Church, State and Astronomy in Ireland*, Armagh Observatory, 1990.
Eric Mervyn Lindsay Memorial Issue, *Irish Astronomical Journal*, **12**, Nos 3/4, 1975.

John Butler, Armagh Observatory, N. Ireland.

146 VINCENT BARRY Chemist and Medical Researcher 1908–1975

Born: Cork, 17 May 1908.
Died: Dublin, 4 September 1975.

Family:
Barry's father worked in the Post office in Cork city. Vincent was the youngest of a family of seven brothers and four sisters. He was educated at the North Monastery Christian Brothers' School and University College, Dublin.
Married: Angela O'Connor, June 1931.
Children: Two sons and four daughters.

Vincent Barry with Sister Rose Hourahan, Gambo Leprosy Control Centre, Ethiopia

Addresses:

1908-1928 3 Lower Janemount, Sunday's Well, Cork
1929-1943 10 Presentation Road, Galway
1944-1955 23 Belgrave Road, Dublin
1956-1975 67 Garville Avenue, Rathgar, Dublin

Distinctions:
Boyle Medal of the Royal Dublin Society 1969; President of the Royal Irish Academy 1970–1973; ScD (hc) University of Dublin 1972.

Vincent Christopher Barry graduated in chemistry in University College Dublin, in 1928 and the year following became assistant to the Professor of Chemistry at University College Galway. Barry was a pure scientist with a keen sense of the relevance of his fundamental work. His work on sugars at Galway quickly led to the establishment of an industry based on seaweed. On the advice of the Medical Research Council of Ireland (MRCI), established in 1936, Dr Barry was appointed to a Fellowship in Organic Chemistry to carry out investigations into the chemotherapy of tuberculosis. In 1950 he was appointed Director of the new laboratories for medical research established by the Council. Over the next two decades Barry and his team of researchers had some notable successes. They synthesised and tested hundreds of new compounds against *mycobacteria*. Several of them were found to be effective against experimental tuberculosis. Another proved most effective against leprosy but was costly to make. However, it is now established as one of the three first-line drugs in the treatment of the disease. Dr Barry's interests also extended to the chemotherapy of cancer: work in this area as well as in leprosy continues to be carried out by the MRCI research team which, in 1984, moved into new laboratories in the Chemistry Department, Trinity College.

Further reading:

V.C. Barry: *Chemotherapy of Tuberculosis*, Lectures, Monographs and Reports, No. 2, Royal Institute of Chemistry, 1952.
V.C. Barry: Boyle Medal Lecture: Synthetic Phenazine Derivatives and Mycobacterial Disease: A Twenty Year Investigation, *Scientific Proceedings of The Royal Dublin Society*, **A3** (16), 153-170, 1969.
Award of the Boyle Medal, *Scientific Proceedings of the Royal Dublin Society*, **A3** (15), 149-151, 1969.

William J. Davis, University Chemical Laboratory, Trinity College Dublin.

147 DENIS PARSONS BURKITT Surgeon and Medical Geographer 1911–1993

Born: Enniskillen, 28 February 1911.
Died: Bisley, Gloucestershire, 23 March 1993.

Family:
His father, James, was a county surveyor and naturalist who invented the ringing of birds to plot their territories; his mother was Gwendoline Hill. Married: Olive Mary Rogers 1943; three daughters.

Addresses:
1947–1966 Uganda
1966–1986 London
1986–1993 Bisley, Gloucestershire

Distinctions:
Fellow of the Royal Society 1972; Commander of the Order of St Michael and St George 1974; Honorary Fellow of the Royal College of Surgeons in Ireland 1973; Fellow of the Royal College of Physicians of Ireland 1976; Honorary Fellow of Trinity College Dublin 1979; DSc (London) 1984; Member *Académie des Sciences* 1990.

Denis Burkitt was educated at Portora Royal School, Enniskillen, where he lost an eye in a playground accident, and Dean Close School, Cheltenham. He entered TCD in 1929 to study engineering but, after his Christian conversion, he changed to medicine. He served in Royal Army Medical Corps from 1941 to 1946 and after demobilisation went to Africa; there 'God, in his mercy, enabled me with my one eye to see things which my predecessors had missed with two'.

In Uganda in 1957 he began his work on cancer when Hugh Trowell showed him a five-year-old boy with extensive facial swellings. This led to Burkitt's discovery of the 'lymphoma belt' across equatorial Africa, which pointed to an infectious agent. His 1961 lecture in London on the geographical distribution of the tumour caught the attention of virologist Anthony Epstein, who later confirmed that the Epstein-Barr virus was the cause of that particular cancer (Yvonne Barr from Carlow, BA TCD 1953). Moreover Burkitt showed that the lymphoma was curable with chemotherapy. But this first demonstration of a link between a virus and a human cancer was revolutionary.

During the 1970s Burkitt collaborated further with Trowell and with Peter Cleave in propounding the 'fibre hypothesis' that suggested a link between the low fibre (indigestible remnants of plant foods) content of Western diets and the prevalence of the 'diseases of civilisation', more especially bowel cancer, gall stones and coronary heart disease. They published *Refined Carbohydrate Foods and Disease* (1975), *Western Diseases* (1981) and *Dietary Fibre, Fibre-Depleted Foods and Disease* (1985). The original hypothesis has had to be greatly modified, but Burkitt put nutrition and eating habits on the map, more by his witty lectures than his books.

Further reading:
Obituaries in *British Medical Journal,* **306**, 996, 1993; *Lancet,* **341**, 951, 1993.
Clive Lee: *Journal Irish Colleges of Physicians and Surgeons,* **25**, 126–30, 1996.

Caoimhghín Breathnach, Department of Physiology, University College, Dublin 2.

CORMAC O CEALLAIGH Physicist 1912–1996

Born: Dublin, July 1912.
Died: Dublin, 10 October 1996.

Family:
The family originated in South Co. Derry. He had one sister. His father, Seamus, was both a consultant obstetrician and a distinguished early Irish historian, with strong Nationalist connections, being doctor to St Enda's, Patrick Pearse's school. He was also a friend of Eoin McNeill who, on the eve of the Rising of 1916, issued orders countermanding the rebellion from the O Ceallaigh home at Rathgar Road. Cormac was brought up an Irish speaker and had a remarkable gift for languages.
Married: Millie Carr, a teacher whom he met in Cambridge. They had three daughters.

Addresses:

1920–1934	Upper Fitzwilliam Street, Dublin
1934–1935	Paris
1935–1938	Trinity College Cambridge
1940s	Lower Montenotte Cork
1949–1951	Nailsea and Wroxall, near Bristol
1953–1996	Killiney Road, Killiney, Co. Dublin

Distinctions:
Senior Professor Dublin Institute for Advanced Studies; Hon. DSc The Queen's University of Belfast 1975; Royal Dublin Society Boyle Medal 1979; Member of Euratom Scientific and Technical Committee 1974; Visiting Professor Tata Institute Bombay 1975–6.

Cormac O Ceallaigh had an ebullient temperament and told jokes in five languages. It might be expected therefore that his Physics would show the same flamboyance. This was not so. Rather, it was meticulous in the extreme. He had a particular interest in statistics, and applied this in sophisticated ways to extract the maximum information from sparse data. His most important work was done in Bristol as a member of C.F. Powell's group (1949–51). In the new and exciting field of elementary particles, Bristol was then the best department in the world. They used the nuclear emulsion technique exposed to cosmic rays. The patience of O Ceallaigh's work was illustrated by his observation of a new meson, the kappa. He measured seven hundred meson decays to find two kappas.

Later, with his colleagues in Dublin, he continued the cosmic ray work, then turned to particles from accelerators. Later again he took up a new technique which had some similarities with nuclear emulsions, but was sensitive to very heavy cosmic rays.

Further Reading:
Various papers in *The Philosophical Magazine*, 1949–1954.

Neil Porter, Dundrum, Dublin 16.

149 DAVID ALLARDICE WEBB Botanist

Born: Dublin, 12 August 1912.
Died: Near Oxford, 26 September 1994.

Family:
Son of George R. Webb (died 1929), Fellow of Trinity College Dublin, and Dr Ella Webb (died 1946), an eminent Dublin medical doctor and founder of the Children's Sunshine Home.

Addresses:

1912–1933	20 Hatch Street, Dublin
1933–1948	39 St Kevin's Park, Dartry, Dublin
1948–1994	No. 23, Trinity College Dublin

Distinctions:
Member Royal Irish Academy 1945; Fellow of Trinity College Dublin 1945; Boyle Medal of the Royal Dublin Society 1982; Honorary degree from Stirling University (DSc, 1976); Foreign Member Linnean Society of London; President Botanical Society of the British Isles (1989–1991).

David Webb spent most of his life in Dublin and devoted his career to research and teaching at Trinity College, first as a zoologist and then converting to botany to become one of Ireland's most internationally renowned botanists. He was educated at Castle Park School in Dalkey and at Charterhouse in Surrey, UK. He subsequently graduated with a first in Natural Sciences at Trinity College Dublin in 1935. At Trinity, and then at the Department of Zoology, Cambridge, and the Stazione Zoologica in Naples, Italy, his research focused on biochemistry and the physiology of marine invertebrates. He gained PhDs from Trinity College (1937) and from the University of Cambridge (1939).

In 1939, while recovering from diphtheria in Dublin, he was invited by H.H. Dixon, Professor of Botany at Trinity College, to become Assistant Lecturer in Botany. One of his first contributions was in the introduction of botanical field excursions into the curriculum for Trinity botany students. This led to his early botanical work in the preparation of a field guide to the Irish flora, *An Irish Flora*, first published in 1943 and then updated regularly throughout his life, which appeared as the original and six revised editions and a seventh posthumous edition in 1996. He was an excellent teacher, lecturing clearly on many subjects.

In 1950 he became Professor of Plant Biology at Trinity College and in 1954 succeeded Dixon as University Professor of Botany, both of which chairs he held until 1965. In 1966 he became Professor of Systematic Botany until his retirement in 1979, after which he held an honorary chair of the same name until his death. In the mid 1940s he began studies of Irish Saxifrages which led to a lifetime interest in the taxonomy and ecology of the group that culminated in the publication (with Richard J. Gornall of Leicester) of a major book on European *Saxifraga* in 1989. Always committed to understanding Ireland's flora in its European context, Webb led the 1949

International Phytogeographical Excursion (with Frank Mitchell) which stimulated his interest in plant geography. In the 1950s and 1960s he led in the recording of the distribution of the Irish flora, the results of which were published as an atlas of the flora of Britain and Ireland in 1962.

Professor David Webb in conversation with Professor Gordon Herries Davies

In 1954 he became involved in a project to prepare a complete Flora of Europe, of which he became one of the driving forces. Published as five volumes between 1964 and 1980, *Flora Europaea* was the first comprehensive continental Flora ever completed. His taxonomic work in the *Flora* included general editorship, editorship of 43 families and author of 1,022 species accounts. Webb was also given the task of checking the geographical distribution of each plant included in the *Flora*. He also made a major contribution to the development of the Trinity College taxonomic library and herbarium, which under his curatorship grew to contain some 200,000 specimens. Many of the plants he collected in Ireland and throughout mainland Europe were also added to the living collections of the Trinity College Botanic Garden.

With *Flora Europaea* completed, his attention returned to Ireland to complete the classic book on the flora of the west of Ireland, *Flora of Connemara and the Burren* (with M.J.P. Scannell), published in 1983, and *Trinity College Dublin, 1592–1952, an academic history* (with R.B. McDowell), 1982.

One of his greatest contributions to Irish botany was the way in which he promoted botanical field studies and encouraged many young botanists over several generations to follow in his footsteps in the academic study of the Irish and European flora.

Further Reading:

J.R. Akeroyd & P.S. Wyse Jackson: Obituary, *Watsonia*, **21**, 3–6, 1996.
M.H.P. Webb: Obituary, *Glasra* (new series), **3**, 173–174, 1998.
M.B. Wyse Jackson: Bibliography (of David Allardice Webb), *Watsonia*, **21**, 7–13.
M.B. Wyse Jackson & P.N. Wyse Jackson: David Allardice Webb (1912–1994): Bibliography of Published Writings – Additions, *Watsonia*, **22**, 206–207,1998.
W.A. Watts: Obituary, *Irish Naturalists' Journal*, **25**(4), 121–122.

Peter Wyse Jackson, Botanic Gardens Conservation International, Descanso House, 199 Kew Road, Richmond, Surrey, UK.

150 GEORGE FRANCIS (FRANK) MITCHELL Environmental Historian

Courtesy of Town House and Country House

Born: Dublin, 15 October 1912.
Died: Drogheda, Co. Meath, 25 November 1997.

Family:
Son of David William Mitchell and Francis Elizabeth Kirby. Married: 1940 Lucy Margaret [Pic] (née Gwynn) (1911–1987), daughter of E.J. Gwynn, Provost Trinity College, Dublin.
Children: two daughters Lucy and Rosamund.
His eldest brother David was President of the Royal College of Physicians and his younger sister Lillias is a noted weaver.

Addresses:

1912–1914	15 Casimir Road, Harold's Cross, Dublin
1915–1927	2 Garville Avenue, Rathgar, Dublin
1928–1936	Templeville, Orwell Road, Rathgar, Dublin
1937–1947	Milverton, Temple Road, Dartry, Dublin
1948–1968	63 Merrion Square, Dublin
1968–1997	Townley Hall, Drogheda, Co. Meath

Distinctions:
Member Royal Irish Academy 1939; President Dublin Naturalists' Field Club 1945–1946; Fellow Trinity College Dublin (TCD) 1945; President Royal Society of Antiquaries of Ireland 1957–1960; President Royal Zoological Society of Ireland 1958–1961; President International Union for Quaternary Research 1969–1973; Fellow Royal Society 1973; President Royal Irish Academy 1976–1979; President An Taisce 1991–1993; Boyle Medal Royal Dublin Society 1978; Pro-Chancellor TCD 1985–1987; Personal Chair in Quaternary Studies TCD 1965–1979; Honorary Member Royal Hibernian Academy 1981; Honorary Life Member Royal Dublin Society 1981; Honorary Member Prehistoric Society 1983; Honorary Member Quaternary Research Association 1983; Honorary Member International Association for Quaternary Research 1985; Honorary Fellow Royal Society Edinburgh 1984; Honorary degrees from Queen's University, Belfast (DSc 1976), National Museum of Ireland (DSc 1977), and University of Uppsala (filD 1977); Cunningham Medal Royal Irish Academy 1989.

Born into a Dublin merchant family, Frank Mitchell was to become Ireland's foremost natural scientist of the latter half of the twentieth century. From an early age he was interested in ornithology and struck up a friendship with T.H. Mason and later with the entomologist A.W. Stelfox who encouraged him further in his natural history studies.

Educated at the High School and Trinity College, Dublin, he was destined to join his father's business upon graduation, and so read languages. However, he switched to natural sciences and followed courses in zoology and geology. His undergraduate career culminated with the award of Scholarship in 1933 and a Gold Medal the following year. On graduation in 1934 he was appointed as Assistant to the Professor of Geology, and remained at Trinity for the rest of his life. He was appointed Lecturer in Geology in 1940; Reader in Irish Archaeology in 1959 – a unique position; culminating in a personal chair in Quaternary Studies in 1965. He served the College administration successively as Registrar, Senior Lecturer and Tutor and, after retirement, was briefly a Pro-Chancellor of the University. Frank Mitchell's research interests included geology, ornithology, archaeology and early farming, but his most important legacy are his studies on the evolution of Ireland during the last two million years, particularly during the Quaternary since the retreat of the glacial ice over Ireland, and the effect man has had on this landscape. In *Who's Who* (1985) he styled himself an 'environmental historian' rather than naturalist, geologist or palynologist, and it is a label that cogently reflects his wide interests.

His interest in Quaternary studies began when he was appointed as an assistant to the Danish palaeobotanist Knud Jessen in 1934. Jessen studied the distribution of ancient pollen grains preserved in post-glacial sediments and bogs and used them to reconstruct vegetational history. These palynological methods were similarly used in 1965 by Mitchell, who clearly demonstrated the vegetational changes that occurred over the last 12,000 years at Littleton Bog in County Tipperary. Pine declined 7500 years ago and was replaced by deciduous woodland. Man arrived around 3000 BC and began clearing woodland for agriculture. The present interglacial is now known as the Littletonian Warm Stage.

From 1940 in the *Proceedings of the Royal Irish Academy* he published a series of papers under the general title *Studies in Irish Quaternary Deposits* which examined diverse topics such as the distribution of Irish Giant Deer and Reindeer remains in Ireland, lake deposits in Co. Meath, interglacial deposits in south-east Ireland, the palynology of Irish raised bogs, and cave deposits. In that serial publication and elsewhere he published on the remains of fossil pingos (mounds of earth or gravel) in Co.Wexford, on bog flows, on fossil shells, and on the deposits of the older Pleistocene Period in Ireland. He did not confine his studies to Ireland but also carried out work in Glasgow, Cornwall and the Isle of Man. He was heavily involved in the mammoth task of organising the 1957 British Association for the Advancement of Science meeting in Dublin, and was chairman of the Geological Society of London sub-commission on British and Irish Quaternary stratigraphy that produced a comprehensive report in 1973.

Frank Mitchell was a frequent broadcaster on radio, a prolific author of academic papers and, in later life, several acclaimed books. These included *The Irish Landscape* (1976) which went through two further editions as *Reading the Irish Landscape* (1986, 1997), and a volume of semi-autobiography *The Way That I Followed* (1990).

Although the recipient of many honours including Fellowship of the Royal Society, Frank Mitchell was modest about his achievements. He had an infectious enthusiasm for many branches of natural history and archaeology, and stimulated, encouraged, and facilitated the studies of students and scholars alike.

Further reading:

G.F. Mitchell: *The Way That I Followed*, Dublin, 1990.

Patrick N. Wyse Jackson, Department of Geology, Trinity College, Dublin.

151 WILLIAM HAYES Physician, Microbiologist and Geneticist 1913–1994

Born: Dublin, 18 January 1913.
Died: Sydney, 7 January 1994.

Courtesy of the Royal Society

Family:
Only child of William Hayes, founder of the pharmaceutical company Hayes, Conyngham and Robinson (1897) and his second wife, Miriam Harris.
Married: Honora (Nora) Lee.
Children: One son, Michael, a doctor who practised in Sydney.

Addresses:
1913–1950 Edmondston House, Rathfarnham, Co. Dublin
Between 1941 and 1985 he held positions in India, England, Scotland, Australia, and USA.

Distinctions:
Elected to the Royal College of Physicians of Ireland (1943), the Royal Society of London (1964), the Australian Academy of Science (1976); Honorary degrees from Leicester, Trinity College Dublin, Kent, National University of Ireland; Royal Society Leeuwenhoek Lecturer (1965); Genetical Society Mendel Lecturer (1965); Burnet Medal of the Australian Academy of Science (1977); President, Genetical Society (1971–73).

Hayes, a founder of the field of bacterial genetics, had a major influence on 20th century science: molecular biology and genetic engineering, which derive from bacterial genetics, owe much to him. He was educated at Castlepark School and St Columba's College in Dublin. Aged about 14, his radio-building nearly caused a serious fire at Columba's. At Trinity he took degrees in natural science (1935) and medicine (1937), and was an Assistant in Bacteriology at Trinity under J.W. Bigger. He became fascinated by the genetic variations in bacterial antigens.

In the Royal Army Medical Corps from wartime India, Hayes published 11 scientific papers, as well as valuable reports on the use of the revolutionary antibiotic penicillin. He returned to Trinity in 1946, the year in which Tatum and Lederberg discovered bacterial conjugation, a process by which genes move from one bacterial cell to another. At Hammersmith Hospital in London (1952) Hayes made the remarkable and crucial discovery that during conjugation one strain is a gene donor and the other a recipient – gene transfer is unidirectional, variable and partial. From 1957, working in London and Edinburgh, Hayes founded and built up the hugely influential British school of microbial genetics. In 1964, he published his masterpiece *The Genetics of Bacteria and their Viruses*. He retired in 1978. In later years he bore the stresses of Alzheimer's Disease with courage.

Further reading:
Biographical Memoirs of Fellows of the Royal Society, **42**, 173–189, 1996.
Obituaries in: *The Times*, 27 January 1994; *The Irish Times*, 8 February 1994; *The Guardian*, 13 January 1994.

David McConnell, Department of Genetics, Trinity College, Dublin.

JAMES ROBERT McCONNELL Theoretical Physicist 1915–1999

Born: 3 Duke Row, Dublin, 25 February 1915.
Died: Dublin, 13 February 1999.

James McConnell receives the Boyle Medal of the Royal Dublin Society in 1986 (courtesy of the RDS)

Family:
Second child of Robert McConnell, pharmacist, and Frances (née Lennon).

Addresses:
1915–1939 Dublin City
1939–1959 Harbour Road, Dalkey, Co. Dublin
1959–1994 Avondale Park, Killiney, Co. Dublin

Distinctions:
Secretary of the Royal Irish Academy 1967–1972; Membro Corrispondiente Instituto de Estudios Avanzados Cordoba Argentina 1968; Boyle Medal of the Royal Dublin Society 1986; Pontifical Academician 1990; Papal Prelate of Honour 1991.

Educated at O'Connell School, Dublin, James McConnell entered University College, Dublin, in 1932 to study mathematics and was strongly influenced by A.W. Conway who aroused his interest in theoretical physics. After obtaining an MA in 1936 he decided to seek ordination and entered Holy Cross College, Clonliffe. A year later he was sent to the Lateran University of Rome: he was ordained in 1939 in Rome, completing his studies at the Lateran in 1940. Whilst there he enrolled at the Royal University of Rome and in 1941 was awarded the degree of Doctor of Mathematical Sciences. He returned to Dublin in 1942 and began his researches at the School of Theoretical Physics of the Dublin Institute for Advanced Studies (DIAS) where he studied under Erwin Schrödinger and Walter Heitler.

From 1945 to 1968 he was Professor of Mathematical Physics at St Patrick's College, Maynooth, and from 1968 to 1989 a Senior Professor in the School of Theoretical Physics DIAS. McConnell wrote extensively on the possible existence of the negative proton (discovered in 1955) and on the theory of fundamental particles. On his appointment to DIAS, his interest turned to the theory of the electric and magnetic properties of materials. In 1992, following a very active and productive research career, he was appointed to the Pontifical Academy of Sciences by Pope John Paul II, who also made him a Prelate of Honour.

Further reading:
Who's Who, What's What and Where in Ireland, Geoffrey Chapman Publishers, London and Dublin, 203, 1973.
J. McConnell: *Quantum Particle Dynamics*, second edition, North-Holland, Amsterdam, 1960.
J. McConnell: *Rotational Brownian Motion and Dielectric Theory*, Academic Press, London, 1980.
J. McConnell: *The Theory of Nuclear Magnetic Relaxation in Liquids*, University Press, Cambridge, 1987.
J. McConnell: From nonlinear optics to nuclear magnetics, Boyle Medal Lecture 1986, *Occasional Papers in Irish Science and Technology* **3**, Royal Dublin Society, 1987.

Brendan Scaife, Engineering School, Trinity College, Dublin.

153 PATRICK ARTHUR WAYMAN Astronomer 1927–1998

Born: Bromley, Kent, 8 October 1927.
Died: Dublin, 21 December 1998.

Family:
One of three sons of Lewis Wayman and Mary (née Palmer).
Married: Mavis Gibson (1954).
Children: A son and two daughters.

Addresses:
1964–1992 Dunsink Observatory, Dublin 15
1992–1998 Wicklow Town

Distinctions:
Member Royal Irish Academy 1966; Associate Royal Astronomical Society 1982; Honorary Andrews' Professor Trinity College Dublin 1984; Honorary DSc National University of Ireland 1993.

Photo by W. Dumpleton

Educated at the City of London School and Emmanuel College Cambridge, Wayman studied the geometrical optics of Schmidt telescopes for his PhD (1952). He held brief appointments at two US observatories before joining the Royal Greenwich Observatory at Herstmonceux. After a tour of duty in South Africa (1957–1960), he returned to RGO and later became Head of the Meridian Department.

In 1964 he was appointed Senior Professor in the Dublin Institute for Advanced Studies and resident Director of Dunsink Observatory. During his tenure he transformed the standing of astronomy and astrophysics in Ireland and he made significant contributions internationally. From the start he pursued a policy of using overseas observing facilities, first at Boyden Observatory in South Africa and later at La Palma in the Canary Islands. The buildings at Dunsink were adapted and restored to cater for the increased level of activity and to give support in electronics and computing. He fostered links with astronomical research groups in the universities and it was on his initiative that the Astronomical Science Group of Ireland was founded in 1974. Wayman made important contributions to the work of the International Astronomical Union. As General Secretary (1979–1982), he supervised the establishment of its permanent secretariat in Paris and he solved the long-standing problem of the adherence of the two parts of China to the Union.

He played an active part in the affairs of the Royal Irish Academy as Chairman of its National Committees for Astronomy and for the History and Philosophy of Science. His keen interest in the history of astronomy led to the publication of his bicentennial history of Dunsink. In retirement he was working on a biographical volume on Thomas and Howard Grubb.

Further reading:

P.A. Wayman: *Dunsink Observatory, 1785–1985 – A Bicentennial History*, Dublin Institute for Advanced Studies/Royal Dublin Society, 1987.

Ian Elliott, Dunsink Observatory, Dublin 15.

154 JOHN STEWART BELL Physicist and Philosopher of the Quantum Theory

Born: Belfast, 28 July 1928.
Died: Geneva, 1 October 1990.

Family:
Son of a working class father, John Bell, and mother Annie Bell (née Stewart), both of whom were of Scottish extraction, John grew up with his elder sister Ruby, and two brothers, David and Robert. David was also to show scientific leanings, and became a Professor of Engineering in Canada. As a young researcher, in 1954, John married Mary Ross, another mathematician and physicist, working in the same accelerator design team at Malvern.

Distinctions:
Fellow of the Royal Society 1972; Reality Foundation Prize 1982; Honorary Foreign Member of the American Academy of Arts and Sciences 1987; Dirac Medal of the Institute of Physics 1988; Doctor of Science, *hc*, (Queen's University of Belfast and Dublin University) 1988; Dannie Heinemann Prize for Mathematical Physics, American Physical Society 1989; Hughes Medal of the Royal Society 1989.

Addresses:
Tate's Avenue, Belfast, Northern Ireland
Malvern, England
Harwell, England
Geneva, Switzerland

A bright schoolboy at his local primary schools of Ulsterville School and Fane Street, even at eleven John showed all the intellectual prowess which was to be evident in his scientific work (at home he was known as 'The Prof'). However, his family could not find the money to send him to a good secondary school. Instead he was enrolled at 'The Tech' (the Belfast Technical High School). Four years later, at age 16, he was qualified for entrance to Queen's University. Again, because his family could not afford it, he became a laboratory technician in the Physics Department of the University, rather than an undergraduate. His talents were quickly recognised by the then head of the Physics Department, K.G. Emeleus, and he was encouraged to attend the first year lectures while still a laboratory technician. The following year (1945) he became an undergraduate, and obtained a first class honours degree in experimental physics in 1948, and another first in mathematical physics the following year.

Even at this early stage as a student, in an effort to gain a better understanding of the quantum theory, he had started to query how its ideas and concepts were being presented. This was a fundamental problem, for the theory appears contrary to our 'instinct' as manifested in our everyday experience. It was an area Bell was to make very much his own in his later years, and in which he was a brilliant communicator. It was one of much contentious debate, in which Bell constantly sought rational answers from other physicists and philosophers, albeit with diffidence and a dry North-Irish humour, as exemplified in his book *Speakable and Unspeakable in Quantum Mechanics*. He regarded work in this sphere as almost a hobby, while at the same time making important contributions to the more bread-and-butter work of synchroton design, which was to lead to his eventual employment by CERN (The European Organisation for Nuclear Research) in 1960.

His first employment was at the Atomic Energy Establishment at Harwell, where he joined the section headed by Klaus Fuchs. A skilled mathematician, with a superb capacity for numerical computation, during this period in England he spent two years at Birmingham University researching for a PhD degree, where he made an important discovery concerning the anomalous symmetry behaviour of high-energy particles. This so-called CPT theorem was simultaneously discovered by two other physicists of note, Gerhard Lüders and Wolfgang Pauli. Bell's work throughout the 1950s embraced both experiment and theory, but the latter was gradually becoming the more important feature of his interests.

In 1960, John and Mary Bell moved to Geneva to work at CERN (for which he was already a consultant). He was now probing the theoretical foundations of physics, but ever mindful of their practical applications; he became interested in practical experiments on such esoteric particles as the neutrino and other high-energy particles. Constantly he concentrated on the way in which the 'idiosyncratic' behaviour of particles or states such as the spin of a particle (what Einstein called 'the spooky action at a distance' and which he also found difficult to accept), could be measured by experiment. One of his pieces of work disposed of an earlier, erroneous 'proof' by von Neumann of the impossibility of a hidden-variable interpretation of Quantum Mechanics. The second, Bell's Theorem (or Bell's 'inequalities'), showed that it should be possible to detect an instantaneous communication between sub-atomic particles which are far apart. This theorem inspired many other physicists, such as Alain Aspect in Paris in 1980, to attempt to set up such demonstrations. He showed that, as predicted by Bell, when an atom ejects two photons in opposite directions, the measurement of the spin of one immediately fixes the (opposite) spin of the other. Put simply, even when they are apart, these photons are entangled with each other.

John Bell died at the early age of 62. The ferment he created about how we should approach the notoriously difficult task of describing the microscopic world of atomic particles lives on. At a recent (August 1999) international conference on the philosophy of science, his work was the subject of three speakers, from Italy, from Poland, and from Hungary.

Further Reading:

J.S. Bell: *Speakable and Unspeakable in Quantum Mechanics*: collected papers, Cambridge University Press, 1987.

J.T. Cushing & E. McMullin (eds.): *Philosophical Consequences of Quantum Theory: Reflections on Bell's Theorem*, University of Notre Dame Press, 1989.

P.G. Burke & I.C. Percival: *John Stewart Bell*, in *Biographical Memoirs of Fellows of the Royal Society, London*, **45**, 1–17, 1999.

A. Whitaker: John Bell and the most profound discovery of science, *Physics World*, **11** (no. 12), 29–34, 1998.

William J. Davis, University Chemical Laboratory, Trinity College, Dublin.